Eyewitness to Space

1. *Paul Calle.* THE MOMENT. Pen and ink on paper, 30×22″

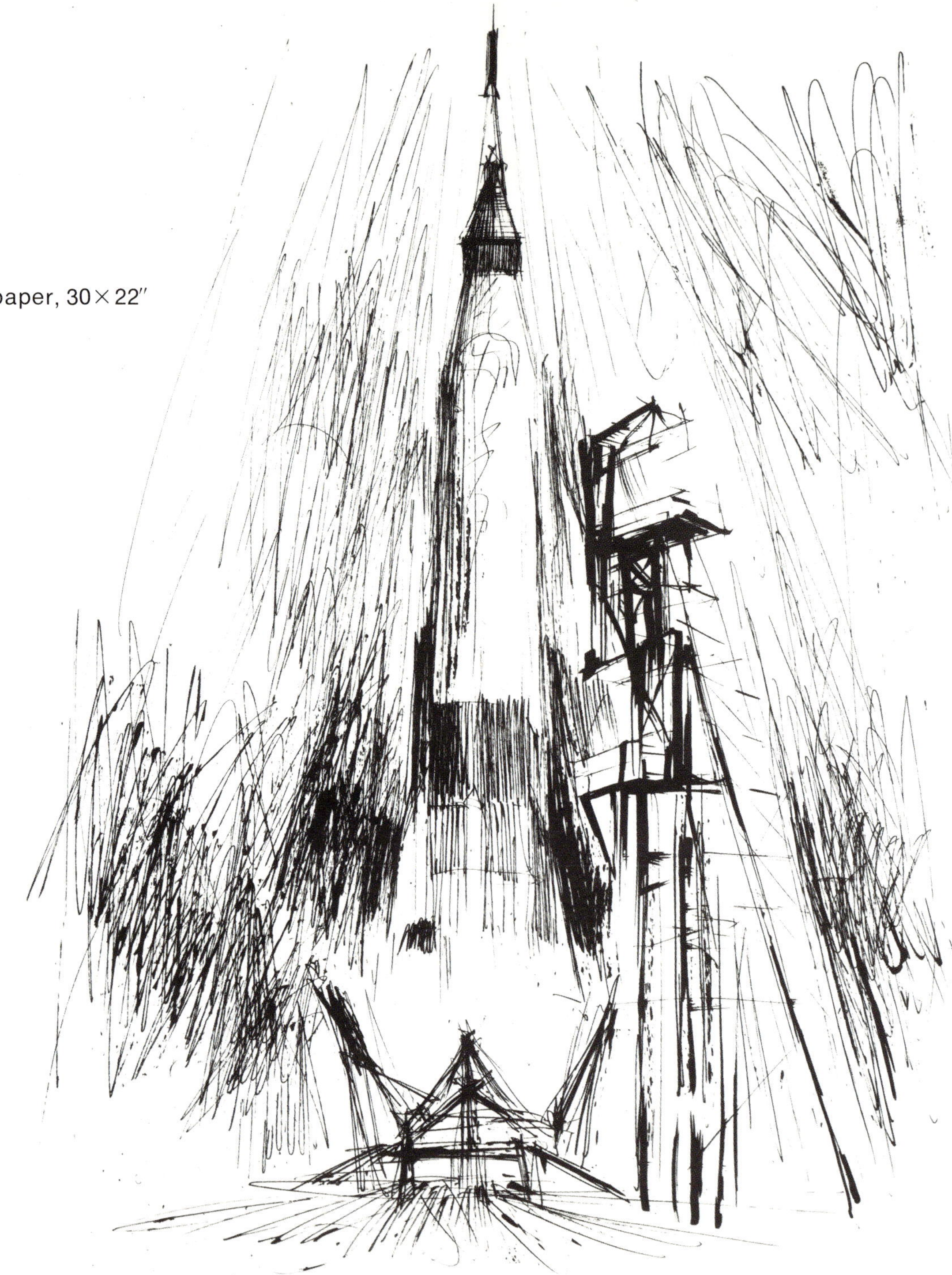

PAINTINGS AND DRAWINGS RELATED TO THE APOLLO MISSION TO THE MOON, SELECTED, WITH A FEW EXCEPTIONS,

FOREWORD BY: *J. Carter Brown*

Director, National Gallery of Art, Washington, D.C.

PREFACE BY: *Thomas O. Paine*

Former Administrator, National Aeronautics and Space Administration

Eyewitness to Space

FROM THE ART PROGRAM OF THE NATIONAL AERONAUTICS AND SPACE ADMINISTRATION (1963 TO 1970).

WRITTEN BY: *Hereward Lester Cooke*

Curator of Painting, National Gallery of Art, Washington, D.C.

WITH THE COLLABORATION OF: *James D. Dean*

Deputy Director, Media Development Division, National Aeronautics and Space Administration

HARRY N. ABRAMS, INC., PUBLISHERS, NEW YORK

DEDICATED TO THE SPIRIT OF COLUMBUS

MILTON S. FOX, Editor-in-Chief

Book design by NAI Y. CHANG

Standard Book Number: 8109–0112–9

Library of Congress Catalogue Number: 76–141761

HARRY N. ABRAMS, Incorporated, New York

Printed and bound in Japan by Toppan Printing Co., Ltd.

CONTENTS

PREFACE

Those of us involved in the day-to-day fulfillment of missions are, by the nature of our tasks, concerned with the immediate future and have had too little time to pause and consider what the distant future may think of our efforts. And yet we have a solemn obligation to leave for future generations a truthful record at all levels of perception, and for this the imagination and insight of the artist are invaluable assets. We at NASA feel that we made a wise decision when we called upon the artists for help. Machines can duplicate and preserve the cold facts, but the emotional impact and hidden meaning remain within the scope of the artist and the poet. The best scientific talent in the Western world has helped us reach our goals, and we are proud that some of America's most talented artists have in their way also been our allies and assistants.

Thomas O. Paine

Eyewitness to Space

2. George Weymouth *On the Beach*
3. Robert Rauschenberg *Trust Zone*
4. Robert Rauschenberg *Horn*
5. John W. McCoy II *North*
6. James Wyeth *Apollo 11—One, Two, Three*
7. Paul Calle *Suiting Up*
8. Paul Calle *Command Destruct Center*
9. Nicholas Solovioff *Crawler*
10. Franklin McMahon *VAB Exterior*
11. William Thon *VAB Interior*
12. Paul Calle *Road Block 17*
13. Paul Calle *Countdown*
14. Nicholas Solovioff *Back to Earth*
15. Franklin McMahon *Mission Control at Houston*
16. Robert Rauschenberg *Earth Tie*
17. Robert Rauschenberg *White Walk*
18. Paul Calle *Flotation Gear*
19. Robert McCall *Eye Check*

FOREWORD

The National Gallery of Art is proud to have been associated with the National Aeronautics and Space Administration in one small aspect of what surely must rank among the great achievements of all time. The artist, it has often been said, is the weathervane and prophet of history in the making. He records what he sees, adds what he knows, remembers, and feels, and thus immeasurably expands our knowledge.

There are many precedents in American history for NASA's decision to ask artists to help document and interpret the space program. Explorers from the seventeenth to the nineteenth centuries, Civil War magazine editors, and official historians during both world wars have all called on the artist. Looking back on these momentous events we realize how much our knowledge and appreciation depends on the sensitivity and perception of the artist-on-the-spot. His insight is unique, and the awesome events of our day are a worthy challenge to his abilities.

When Apollo 11 left the earth in July, 1969, to achieve man's first landing on the moon, the world watched through the medium of television. Among those present at Cape Kennedy were a group of artists who interpreted the historic event in their own way. Artists were at Houston in the control room during the landing on the moon and on the carrier in the Pacific for the splashdown. Thus we have a record for the future of some of the greatest moments of the present through a medium that is as old as our recorded past—the intimate, human medium of the artist's eye and hand.

J. Carter Brown

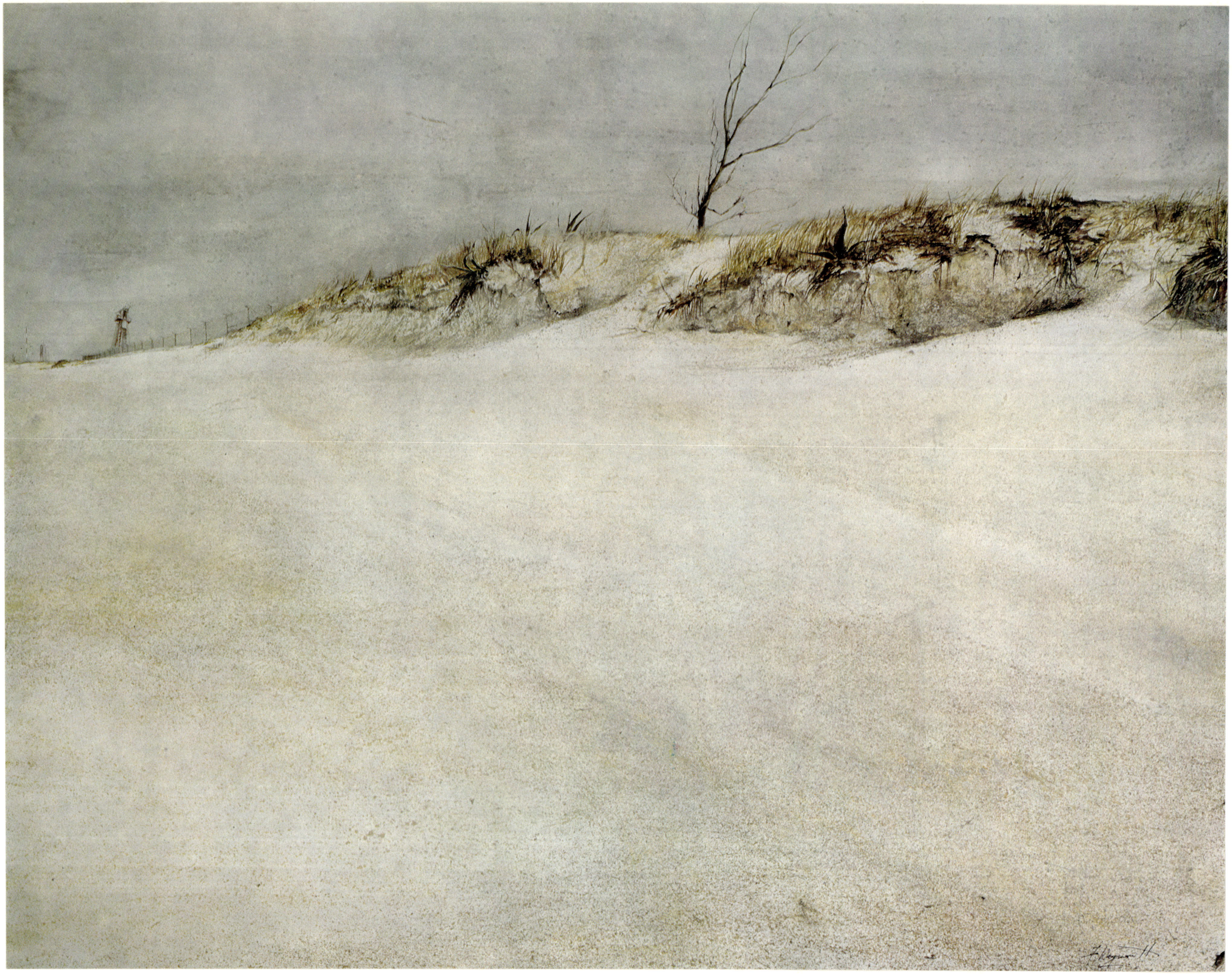

EYEWITNESS TO SPACE

2.
George Weymouth
ON THE BEACH
Watercolor on paper, 26×35″

Before the Space Age, the sand beach stretching the length of Cape Kennedy was the domain of honeymooners, surf casters, and bird watchers. Today the beach is unchanged, except that over the nodding grasses and the sand dunes occasionally a red gantry is silhouetted against the sky.

IT SEEMS CERTAIN THAT THE AGE we are living in will go down as a red-letter epoch because the first steps in space exploration are going to mark a hinge in the fate of mankind. Not since the lung-fish slithered out of the Paleozoic slime have living creatures sought to change their environment, and not since Columbus have men dared to enter more dangerous and mysterious regions. Perhaps in space we will find greener pastures; perhaps we will find nothing but infinite hostility and will return to realize that life on this planet is more precious than we ever imagined. Whatever the results, space exploration is changing history.

The effort we are making in this country represents the frontiers not only of technical achievements, but also of the imagination. The space effort in every way is a worthy theme for the artist. Space travel started in the imagination of the artist, and it is reasonable that artists should continue to be the witnesses and recorders of our efforts in this field.

NASA decided to ask artists to supplement the record after reviewing the documentation of the first few years of the space age. It was realized that important aspects of the story were missing. When a major launch takes place at Cape Kennedy, more than two hundred cameras record every split second of the activity. Every nut, bolt, and miniaturized electronic device is photographed from every angle. The artist can add very little to all this in the way of factual record. But, as Daumier pointed out about a century ago, the camera sees everything and understands nothing. It is the emotional impact, interpretation, and hidden significance of these events which lie within the scope of the artist's vision. An artist may depict exactly what he thinks he sees, but the image has still gone through the catalyst of his imagination and has been transformed in the process. So style in this assignment is unimportant. If you see events in non-objective terms, represent them this way. NASA is commissioning your imagination, and we want records of fleeting impressions and poetic by-products of thought as much as precise documents of optical experience. What is important is that the artist gives us his personal and sincere interpretation. His work will not be juried, judged, or evaluated when it arrives at NASA, at least not by this generation, but it will become a permanent record of an individual's talent and perception, and in this way is a sliver of immortality.

There are no restrictions regarding medium, style, size, subject, shape, except that the materials used must be of proven permanence. I do not want faded and flaking pictures in a government archive.

The artist, like the scientist, must reorient some of his thinking in

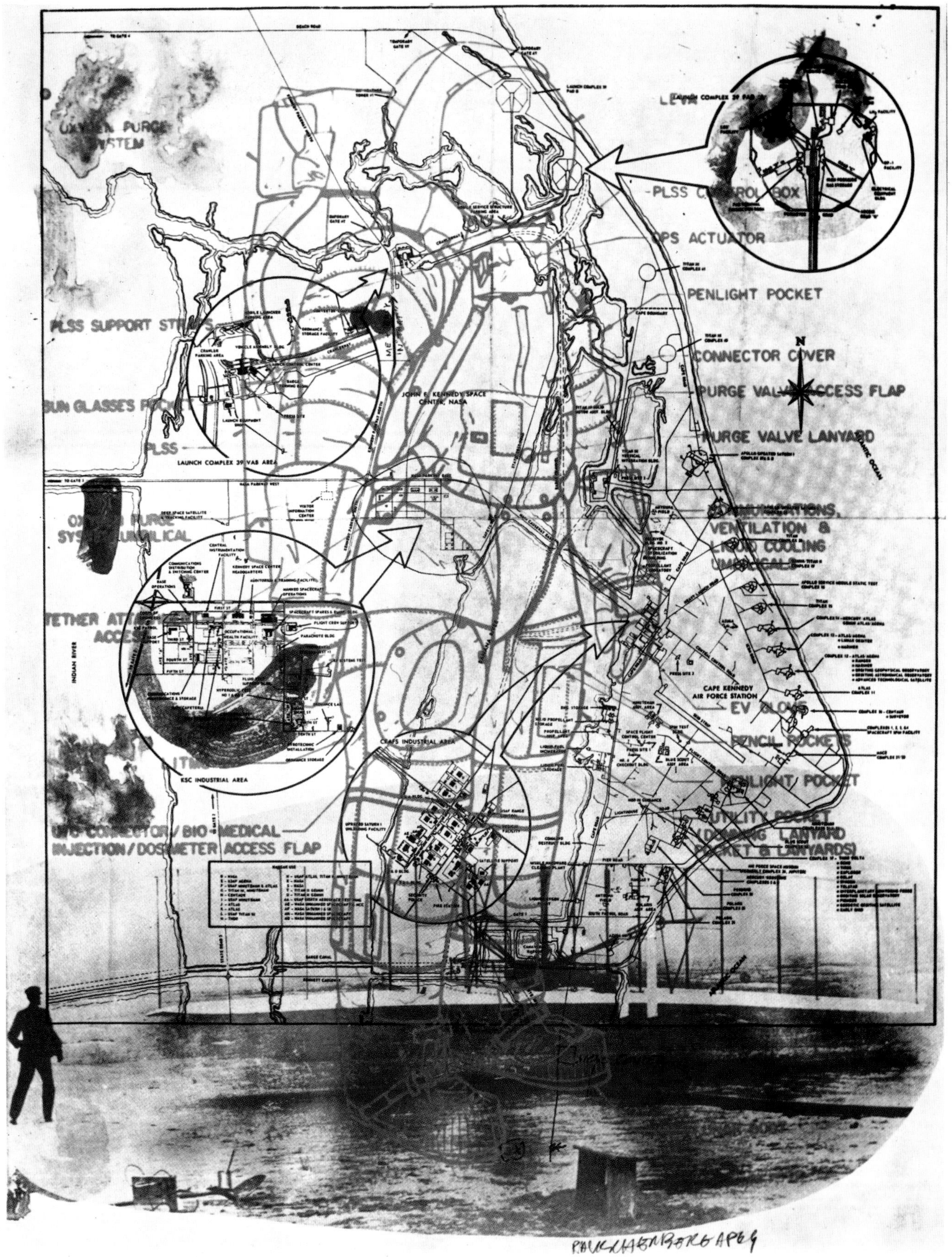

3. *Robert Rauschenberg.* TRUST ZONE. Color lithograph on paper, 40×33″
From the Stoned Moon series reproduced courtesy of Gemini G.E.L., Los Angeles, Calif.

the space environment. Have you ever thought what the human body really looks like in a state of weightlessness? Or what it means to have no up or down? That a glass of water is just as right upside down as in any other position? Or that Leonardo's optical perspective does not apply to the moon landscape and a distant mountain is just as sharp and colorful as a pebble six inches away? Or that space is a whirlpool of unseen waves through which the earth swims, leaving a wake like a ship? Or that the electrical impulses which send pictures from the moon are about as strong as the static generated by a butterfly's wings? Or that the earth in relation to a solar flare-up is about the same as a pinhead in the flame of a match? All of our concepts of size, speed, number, mass, and even life are quaint and obsolete in this age. And have you thought of the role of man, who not only controls but is also caught in the web of the monstrous machines he has made? The metaphysical aspects of space are as fascinating as the physical and are equally within the scope of the artist.

There are several more direct reasons for commissioning drawings and paintings. First, NASA is anxious to document the space effort in every way possible. From my own point of view, I am convinced that artists should be key witnesses to history in the making, and that in the long run the truth seen by an artist is more meaningful than any other type of record. Second, I want to build up a collection of drawings and paintings which will convey to future generations some of the excitement and wonder that we feel as we cross the space threshold. I also want these paintings to stand on their own merit as works of art, quite apart from the subject matter. I hope future generations will realize that we have not only the scientists and engineers capable of shaping the destiny of our age but artists worthy to keep them company.

For best results, I am sure that it is good policy to leave the artists alone. Therefore, I may indicate subjects which seem adaptable, but in general you will be completely free to draw, paint, and record whatever strikes you as significant and unique. We are commissioning not only the artist's skill, but also his perception and his imagination, and for best results neither should be dictated or forced; therefore you are free to use whatever style you wish.

One thing I insist on. Every drawing made on site, regardless of how slight, must be saved and eventually added to the permanent archive. The reason for this is that on-the-spot sketches often have an impact and immediacy which finished works of art lack. I became aware of this when working on an exhibit of artwork which had been commissioned during the Civil War. Often it was the crumpled scrap of paper with the sketch of a figure or incidental group which, after a hundred years, had the most meaning in conveying the real character of that war. Therefore, please save every drawing, and, after you have finished, send it along. It will be carefully preserved and eventually may fill a gap in the future's estimate of us and our endeavors.

MOST OF THE ARTISTS whose work is reproduced in this book received invitations similar to the foregoing, and most elected to visit Cape Kennedy first for their artistic introduction to the strange new world of space. Although the network of specialists whose work makes a moon voyage possible is nationwide, even worldwide, the focal point of their efforts, at least from an artist's point of view, tends to be Cape Kennedy.

The question most often raised is, "What can an artist reveal that has not already been covered by photography?" The answer is simple. Any moment in history is made up of myriad sensations: visual, auditory, tactile, sensual, emotional. The camera records one moment on the visual scale with incomparable accuracy and skill, but the artist has a far wider range to choose from. Moreover, he is not limited in time. His memory supplies a backlog of vital information, and he is spared the photographer's anguish of realizing that he did not press the button at the right time. He can also prophesy and invent. Whereas the glass-eyed Mata Hari of photography mixes the trivial with the important, the artist has the option of selecting what he wishes from what he sees around him. When the camera is invented that not only can see through the eyes of an artist but is also equipped with the imagination, understanding, emotions, and memory of an artist and, moreover, can edit, interpret, emphasize, omit, and invent, then the artist faces competition.

No guidelines were laid down for the space-faring artist except a wish that the work be intelligible both now and in the future. Although artists had recorded America at war since the days of Valley Forge, and there had been in peacetime active and successful art programs in several of the armed services, no civilian agency of the government had ever invited artists to document its activities. Perhaps it is an indication of a change in the traditional American philosophy of pragmatism that NASA's art program was both welcomed and understood. The scientists in particular realized that the artists had a point of view which was

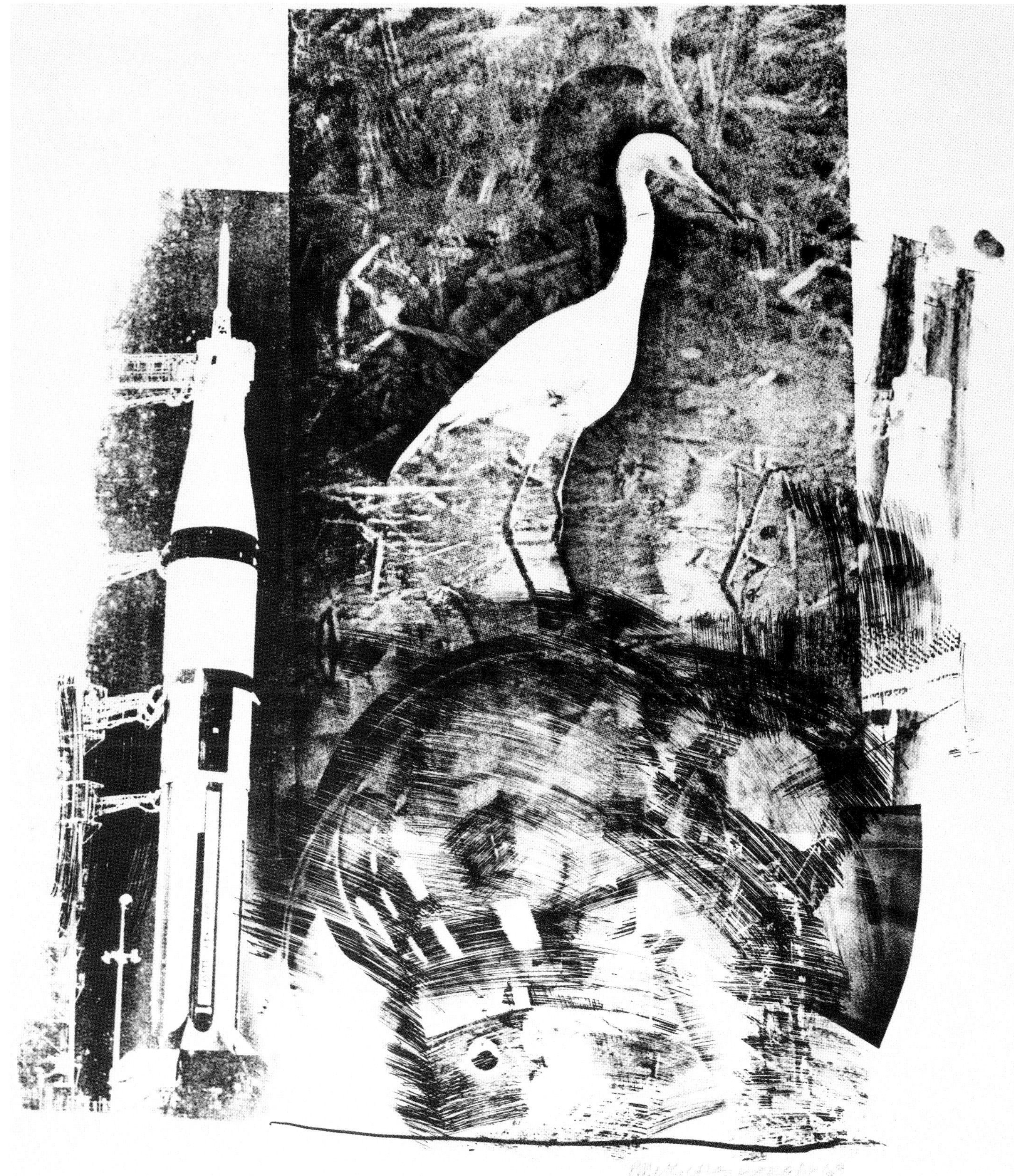

4. *Robert Rauschenberg.* HORN. Color lithograph on paper, 41¼ × 34″
From the Stoned Moon series reproduced courtesy of Gemini G.E.L., Los Angeles, Calif.

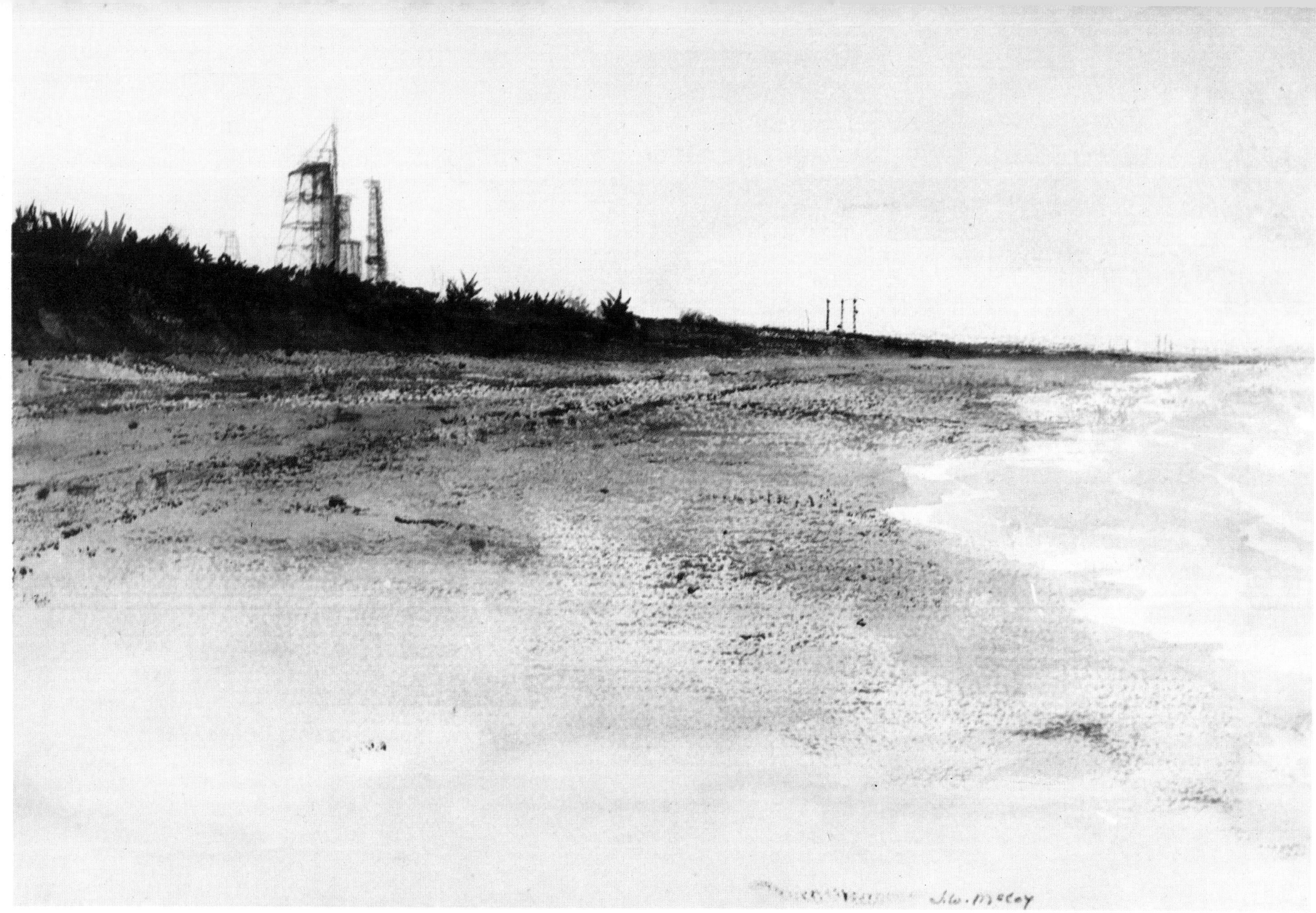

5.
John W. McCoy II
NORTH
Watercolor on paper, 18×26″

not unlike their own. Also, since NASA has a mandate to keep the public informed about its progress in space exploration, it was realized that the artists, through their work, could provide new insights into various space-program activities.

Responsibility for choosing the artists for the NASA program is shared by both the National Gallery of Art and NASA. In fact, the system has not been formal. An artist who has participated, a curator who makes the rounds in contemporary art, or a sympathetic art historian suggests an artist whose work reflects an interest in some aspect of space. Occasionally the artist himself makes the first move. Photographs of works are sent in. There are telephone calls. If there seems to be a possibility for a meeting of the minds, there is a round of paper work and the artist is invited to participate. Some of the artists chosen have been realists of both the Classical and Neo-Realist schools; others have been completely nonobjective in their approach. Many different media have been used, ranging from lead pencil to molten aluminum.

Cape Kennedy is a flat spit of sandy marshland jutting into the Atlantic about halfway down the east coast of Florida. Its beaches are not unlike countless other stretches of shore between Sandy Hook and Key West. Beautifully broken shells, Portuguese men-of-war, and flights of terns mix with the flotsam and jetsam that a warm-hearted Gulf Stream spews at the high-water mark. Inland, gentle sand dunes topped by nodding grasses ring the shore, and farther on, among the marshes, great blue heron and giant turtles make their nests. Just south of the steel-plated lighthouse on the Cape, a channel opens to Port Canaveral, where a fleet of shrimp boats has made Port Canaveral shrimp famous as a gourmet's delight.

Conservationists and the Audubon Society friends shook their heads in sorrow and their fists in anger when it was decided that

Cape Kennedy would be the base for America's entry into the Space Age. As it has turned out, their misgivings were unjustified. Wildlife has adjusted to rocketry more easily than to sunbathers, real-estate promoters, and speeding cars. Today, safe behind the guardrails surrounding Cape Kennedy, great colonies of rare and beautiful birds have come to regard a launch as a new kind of thunderstorm that represents no peril and the rusting gantries as ideal nesting grounds.

Reasons for choosing Cape Kennedy were compelling indeed. A rocket is launched in the same direction as the turning of the earth in order to take advantage of the velocity imparted to the launch vehicle by the earth's rotation. Therefore, the launch is to the east. Cape Kennedy, then Cape Canaveral, was an inevitable choice because, first, it is the most easterly point on the Florida coast and, second, downrange along the path which a rocket must follow there is open water for six thousand miles. This is important: rockets, in order to lighten their loads, shed used parts, ranging from first-stage components weighing tons to explosive bolt heads that, until the vehicle is above its atmosphere, fall back to earth. To have this potentially lethal hardware rain from the sky on populated areas in Europe or Africa would obviously have been bad foreign policy, and therefore a launching site with hundreds of miles of open water downrange was a must. As Jules Verne had predicted a hundred years earlier, a central Florida site was the natural choice.

The Atlantic testing range was established shortly after World War II, when policymakers came to the full realization that, with refinements and apart from its obvious military role, the rocket had vast potential in the fields of communication and transportation. The German research staff of Peenemünde, with their years of valuable experience, was asked to continue their work in Huntsville, Alabama, and Cape Canaveral was used as a testing ground assigned to their project.

Old-timers at the Cape enjoy telling how it was back in the early days of the space program, during the big rush of the fifties. Rockets exploded on their pads; rockets exploded in the air; rockets turned upside down; rockets turned sideways and, belching fire like vengeful dragons, roared a few hundred feet over their petrified launch crews. In 1958, stung by Sputnik, the American effort accelerated and the pressure and the funds increased. It became increasingly obvious that America could not afford to lose the space race, and the time had come to initiate a great national

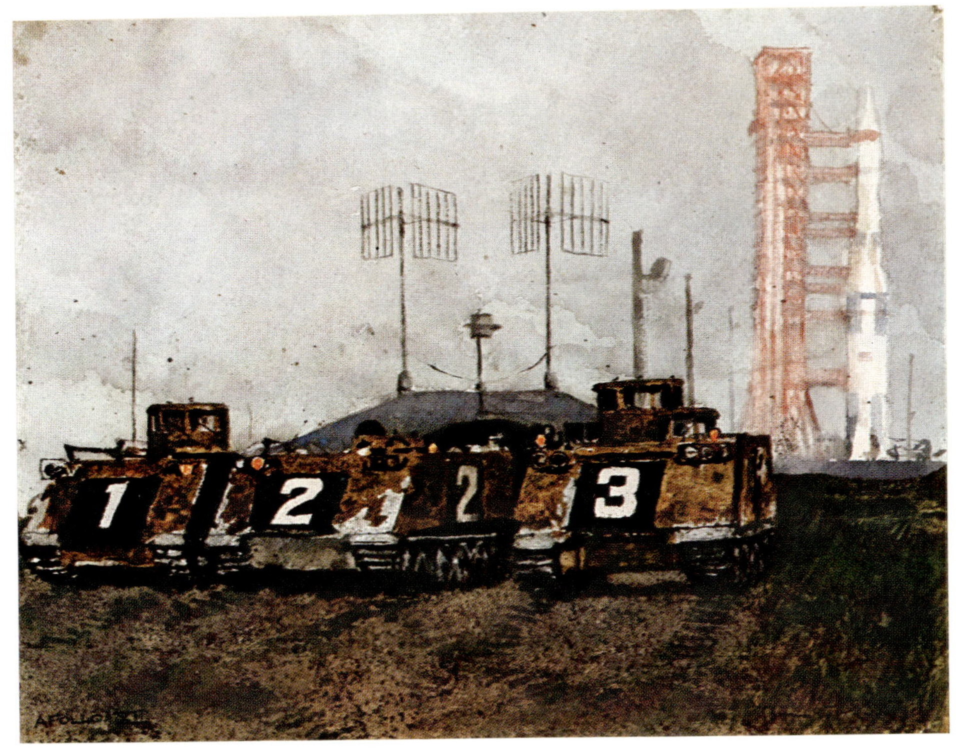

6.
James Wyeth
APOLLO 11—ONE, TWO, THREE
Watercolor on paper, 17¼ × 20″

effort to win and hold the crown of space. Wernher von Braun was authorized to try to orbit a satellite with the army's Redstone rocket. He succeeded from Cape Canaveral in 1958, with Explorer I. The first American satellites were not very large. Premier Khrushchev, speaking to a delighted communist world and confident of a commanding lead, compared them to grapefruit. The Russian lead was undisputed: Yuri Gagarin in 1961 became the first man in space.

Following the successful suborbital flight of Alan Shepard, America's first man in space, in 1961, the American effort got into high gear. A galloping technology made equipment obsolete almost as soon as it was built. Launch towers, each more strange and wonderful than the one before, sprouted along the Florida coast.

Outside the gates of the government compound, the town of Cocoa Beach, a sleepy haunt of honeymooners, fishermen, and bird watchers, mushroomed overnight into a frontier city like those of the gold-rush days. Speakeasies, motels, drive-ins, and souvenir clip joints rose from the sand overnight, each with a space-oriented clientele, nomenclature, and décor. The "Spaceburger" and the "Liftoff Fizz" were featured by the Moon Shack Drive-In or the Astrocraft Motel.

It was at this point in space history, toward the end of 1962, that the NASA art program was born. The idea of having artists help document the record of America's space effort was that of James E. Webb, Administrator of the National Aeronautics and

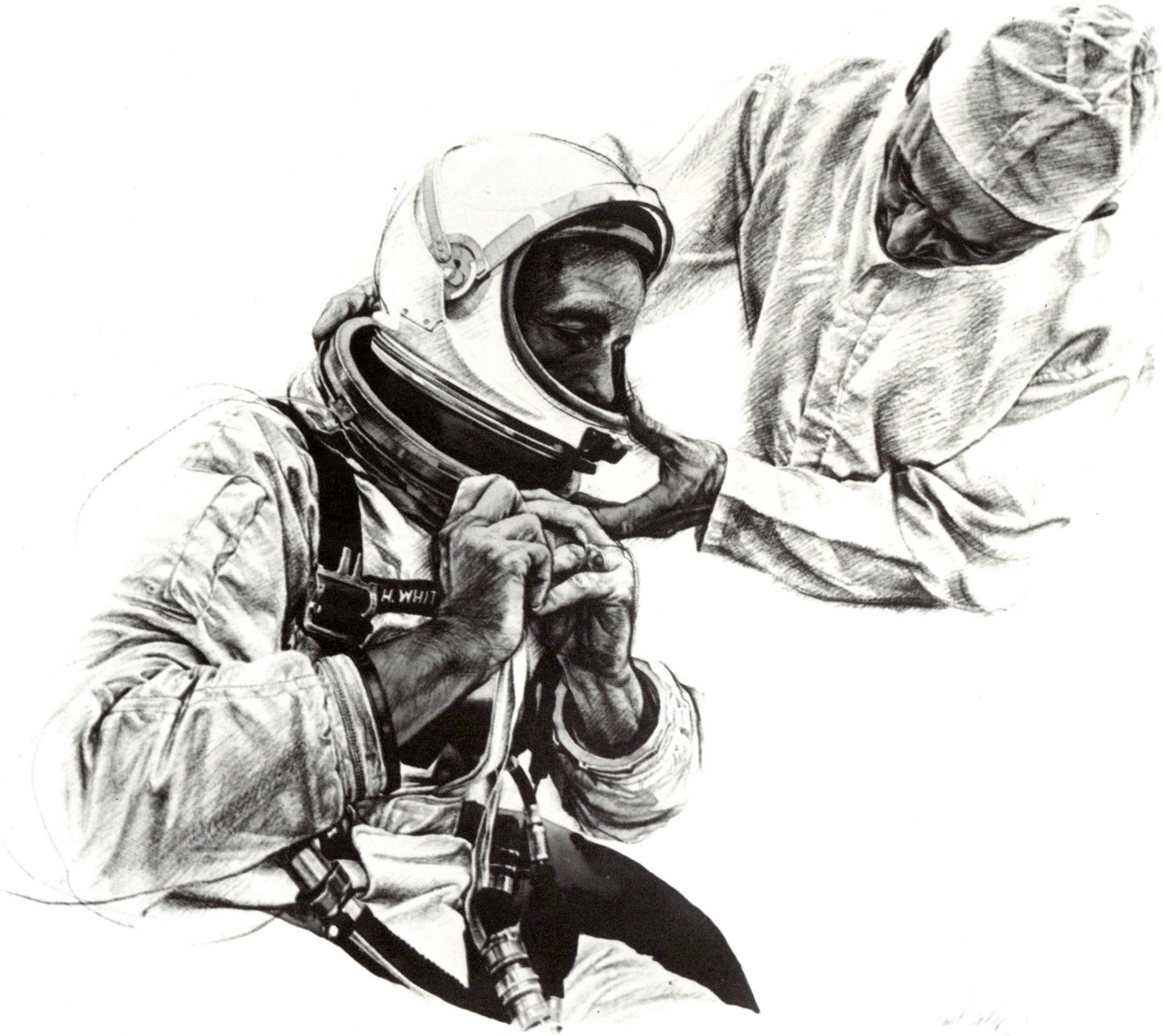

7. *Paul Calle.* SUITING UP
Pencil and wash on paper, 30 × 22″

Space Administration from 1960 to 1968. He consulted the Commission of Fine Arts in Washington, under David Finley, Director, and Linton Wilson, Secretary, who recommended me to act as consultant to NASA in this matter. My counterpart on the NASA side has been James Dean, Deputy Director of the Media Development Division.

The first actual testing of the NASA art program occurred in May of 1963, when the last one-man space flight—Gordon Cooper's earth orbital mission in Mercury-Atlas 9—was to take place. The group of seven artists, covering a wide range in age and style and headed by Peter Hurd and Lamar Dodd, assembled beside the swimming pool at their motel in Cocoa Beach, Florida, a few days before the launch and wondered what was expected of them. Neither the representatives of the National Gallery nor those of NASA, or any of the artists had a clear idea, except that space was a new challenge. The next day, armed with ciphered badges bearing the pioneer title "NASA Artist," the group set out in a convoy of cars, probably with somewhat the same feelings that Lewis and Clark had heading into the Rockies. Official guides and photographers were assigned to the group, and drivers remained politely mystified as the artists, hypnotized by the stranger-than-fiction creations of space engineers, peeled off from the group to make sketches.

Guides and security guards, accustomed to docile flocks of dignitaries, were frantic as the ranks were broken. By noon it was obvious that artists prefer societies of one, and for the next seven years no one tried to take the painters around as a group again. That evening, at a strategy meeting, it was decided that each artist should be turned loose and be left to find out for himself just where the security barriers fell. A map and a telephone number for help were the only survival-kit equipment issued. This system was successful. An unescorted artist was soon classified as a harmless curiosity by security guards. Tired of the barbed queries of scientists, engineers, and pressmen, they regarded the simple questions of the artist as a chance to expound in the realms of space fiction. The artists were welcomed as waifs from another world and admitted into areas far beyond the reach of officially guided VIPs, since, upon close scrutiny, they could prove they were working on a NASA assignment.

The result was that an artist was present while Gordon Cooper was being suited up, and an artist was also in the White Room on top of the Pad 14 gantry as the astronaut was buttoned into his tiny Mercury spacecraft. As the work platforms were folded back from the gantry, again an artist was perched on the uppermost level. As Cooper returned to earth after twenty-two orbits, there was an artist waiting on the carrier deck in the Pacific to record every action in the drama of recovery. An artist accompanied Gordon Cooper on his triumphal welcome in Hawaii and on the plane back to Cape Kennedy. The result was more than one hundred on-the-spot sketches. Mementoes of those events seen through the eyes and imagination of artists have little in common with the results turned in by photographers and newsmen. In a hundred years it is likely that Gordon Cooper's flight and perhaps much of the Mercury program will be seen mostly through the medium of these on-the-spot drawings. There is an impact, immediacy, and truth in these sketches that reveal, and will continue to reveal, the importance of the elements which constitute the historic event. Work done on later assignments has reinforced this conclusion. Many painters, particularly those of the postwar generation, prefer to work from photographs, but by so doing they tend to miss the essential emotional overtones of an on-the-spot painting. Work done back in the studio may be technically superior and even more accurate, but it is rarely so moving or meaningful and is less likely to transport the viewer's imagination to Cape Kennedy and carry him along in the excitement of a countdown.

The great majority of the artists chosen for the NASA program have found more material at Cape Kennedy than they could possibly hope to record during their stay. Some chose the landscape itself as a theme. The Cape consists of miles of flat, sandy wasteland, covered with palmettos and scrub growth. Many roads are unpaved. Installations are widely dispersed to minimize the risks in the event of explosions and fires. Very few of the structures bear any resemblance to anything built by man before, because the problems have not been similar to anything man has faced before. An actual launch pad will consist of a concrete base, often shaped vaguely like a Maya temple, on which the rocket stands slightly above the surrounding terrain. Under the rocket there is a flame deflector, a brilliant yellow structure reminiscent of a mammoth caterpillar hunched to divide and divert the fury of the rocket's blast. Next to the rocket there is the service tower, the vital link with the rocket and spacecraft during the countdown—a crutch on which the rocket is supported before it springs to life. This long, narrow, gangling tower and a companion gantry carry

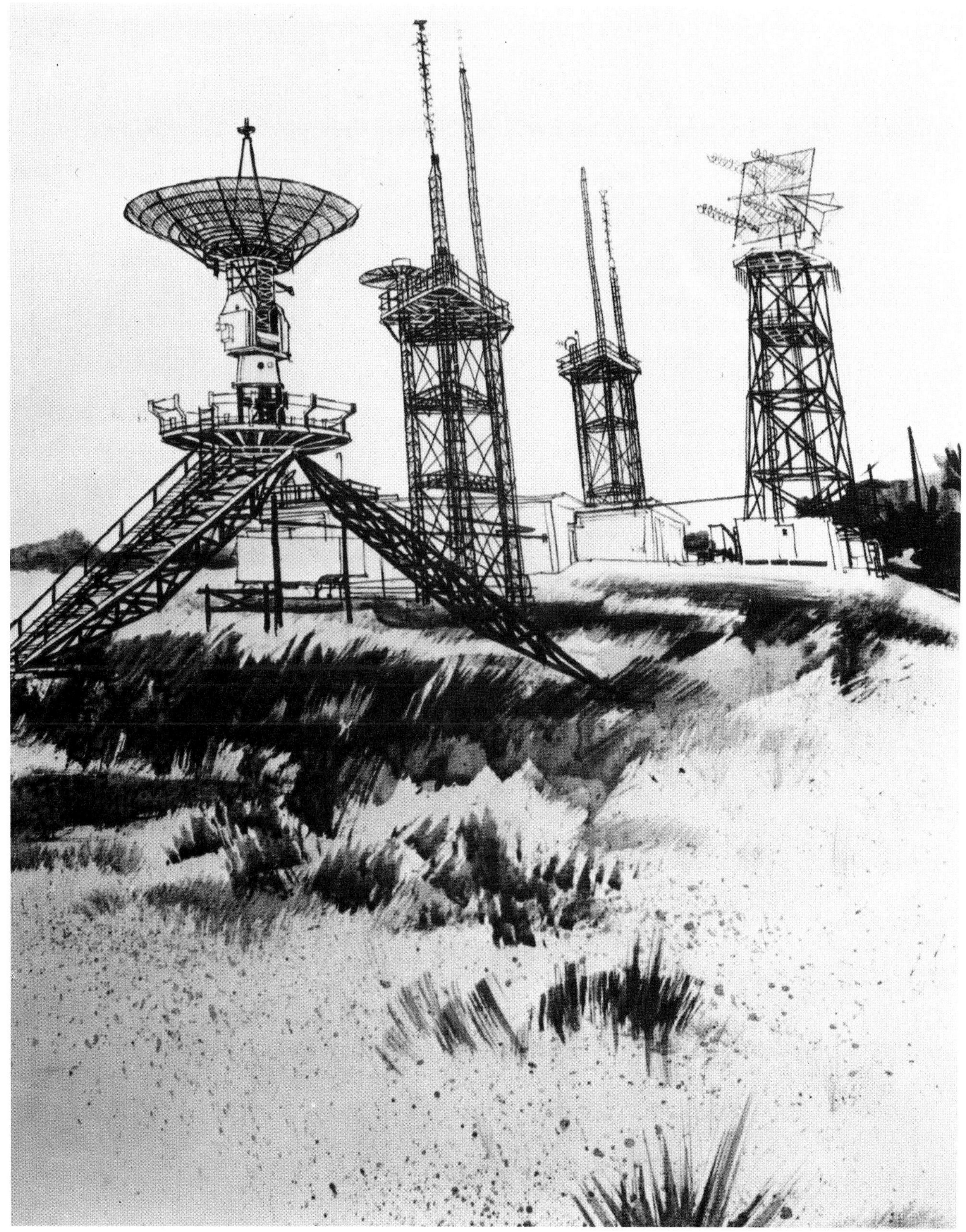

8. *Paul Calle.* COMMAND DESTRUCT CENTER. Pen, ink, and watercolor on paper, 35×28″

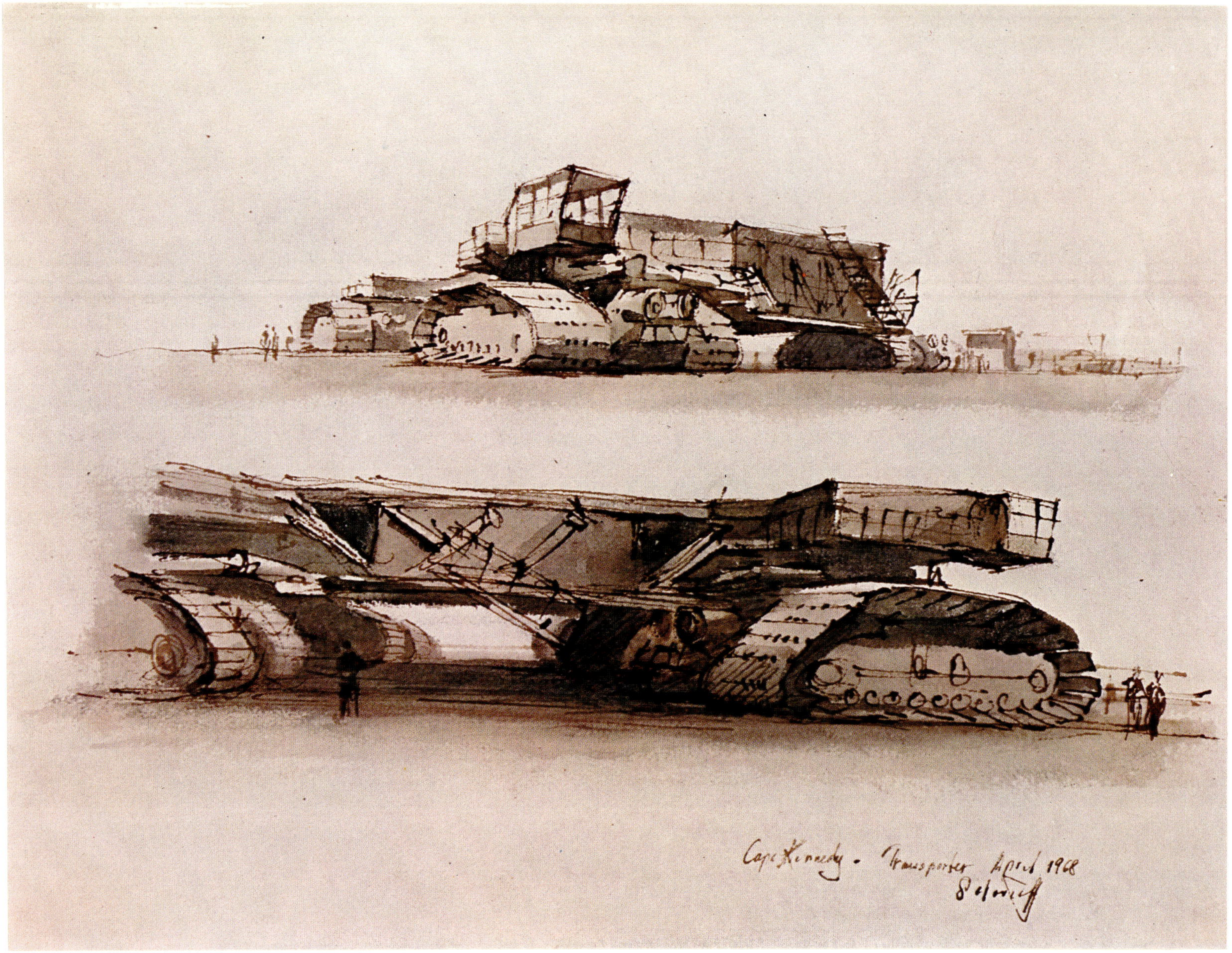

9. *Nicholas Solovioff.* CRAWLER. Sepia ink on paper, 12×18″

up to various levels fuels, control lines, electrical wires, technicians, and finally the astronauts themselves. Unlike the svelte beauty of the rocket, the service tower and gantry look like medieval siege machines—red, clumsy, and uncouth. Fuel is stored in huge round tanks located at a discreet distance from the pyrotechnics. The launch area itself is interlaced with a water-cooling apparatus, designed to douse with millions of gallons of water every section of the launch pad to prevent its being incinerated by the rocket's fire. In earlier launch complexes, the launch crews huddled in a beehive-shaped fortress known as the firing room, located a few hundred feet from the launch site. Here they watched their monster through periscopes. Other actors in the Cape Kennedy drama sit in asbestos-covered tanks, ready to dash into the flames and rescue the astronauts if there is a fire on the pad. Other unsung heroes wait in amphibious craft to pluck them from the Atlantic if their spacecraft should come down in the water.

From the artists' point of view, among the most fascinating creations of the Space Age are the antennae that dot the Cape area. Reigning monarch among these is the sixty-foot dish antenna that swings noiselessly on its triangular turret, grasping firmly in electronic tentacles some invisible object speeding through a cold black yonder. The most admired of the antennae are known as the "Four Horsemen of Destruction." These are four structures of strange and wonderful design, which sit on metal lattice towers not far from the Indian burial mounds on the Cape. The function of these antennae is to send impulses that will cause a rocket to blow up in midflight if it is seriously off course or has other malfunctions which might be a hazard to people on earth. Dishes, spikes, forks, coils, grids, and whips are combined in fantastic shapes that survey and stab at the sky. It is hard to believe that these antennae were not designed by abstract sculptors rather than engineers. Arranged in a solemn semicircle, they seem like totem poles of a new religion and are an irresistible theme to artists.

For those who prefer their antennae in less skeletal form, there are the tracking ships anchored alongside the shrimp and party fishing boats in Port Canaveral. Tracking ships are converted merchant vessels of about ten thousand tons that have been plastered with Byzantine onion domes, not housing images of saints but protecting delicate antennae. Bigger-dish antennae are exposed to the elements in all their lacy finery as they yaw, pitch, and roll with the ship to keep track of their distant charges. Perhaps the strangest feature of these ships is that they are named after Air Force generals and operated by Air Force personnel.

For the artist gathering material in the Port Canaveral area there is often a bonus. Out of the inky depths of the Atlantic rise at intervals nuclear submarines, which slink into Port Canaveral looking like the quintessence of evil in their oily black grace. Moored alongside the jettys, they open their hatches and out of each ocean-going silo there is hoisted a Polaris or Poseidon missile to have its electronic guidance system and its awesome load checked, tested, and realigned.

At the Kennedy Space Center, a less sinister monster is the crawler transporter, which, compared to electronic devices, seems almost a creation of the Stone Age. In order to move the giant Saturn V rocket and its service tower three miles from the Vehicle Assembly Building to the launch site, it was necessary to build the biggest moving machine ever designed by man. Four Brobdingnagian tractors support a slab platform on which, in great dignity, the rocket and its service tower rest. The driver of this giant, his course directed by radar, perches in a glass-sided control booth as he roars down the crawlerway at about one mile an hour. As with most other machines at Cape Kennedy, aesthetics did not play a part in the design. Mesozoic in scale, titanic in power, reptilian in movement, and terrible beyond belief in the roaring, grinding, and squealing of its sound effects, the crawler has received more than its share of the artists' attention.

Another creation stranger than any the wildest imagination could conjure up is known in space vernacular as the "Train Wreck." Inside a building devoted to the astronauts' training is a jumble of squared shapes standing perhaps forty feet high. A ladder leads to a small doorway in the center. Alongside, there is what appears to be a mammoth four-poster bed upside down supporting a crumpled bedspread modeled to represent the surface of the moon. Astronauts entering through the doorway stand in an exact replica of the Lunar Module, the spacecraft that takes astronauts to the moon's surface. Looking out of the small triangular windows, they see the surface of the moon as it will appear to them in moon orbital flight. As they move the controls, the moonscape below appears to move under them, drawing nearer or receding as they fire their imaginary rockets. The Lunar Module Mission Simulator is a true marvel of the Space Age, in which the senses are duped in every way, except that it cannot

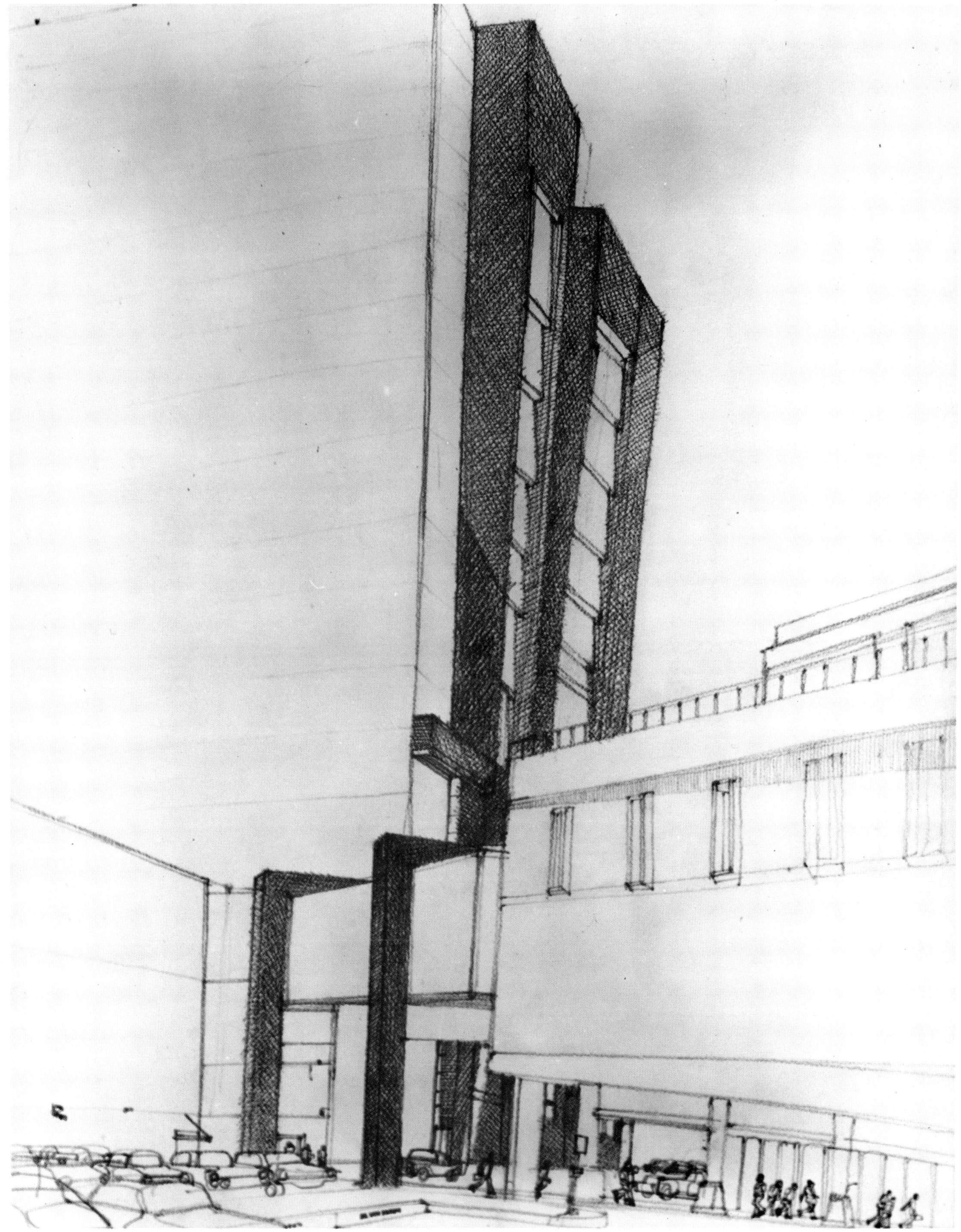

10. *Franklin McMahon.* VAB EXTERIOR. Pencil on paper, 30×22″

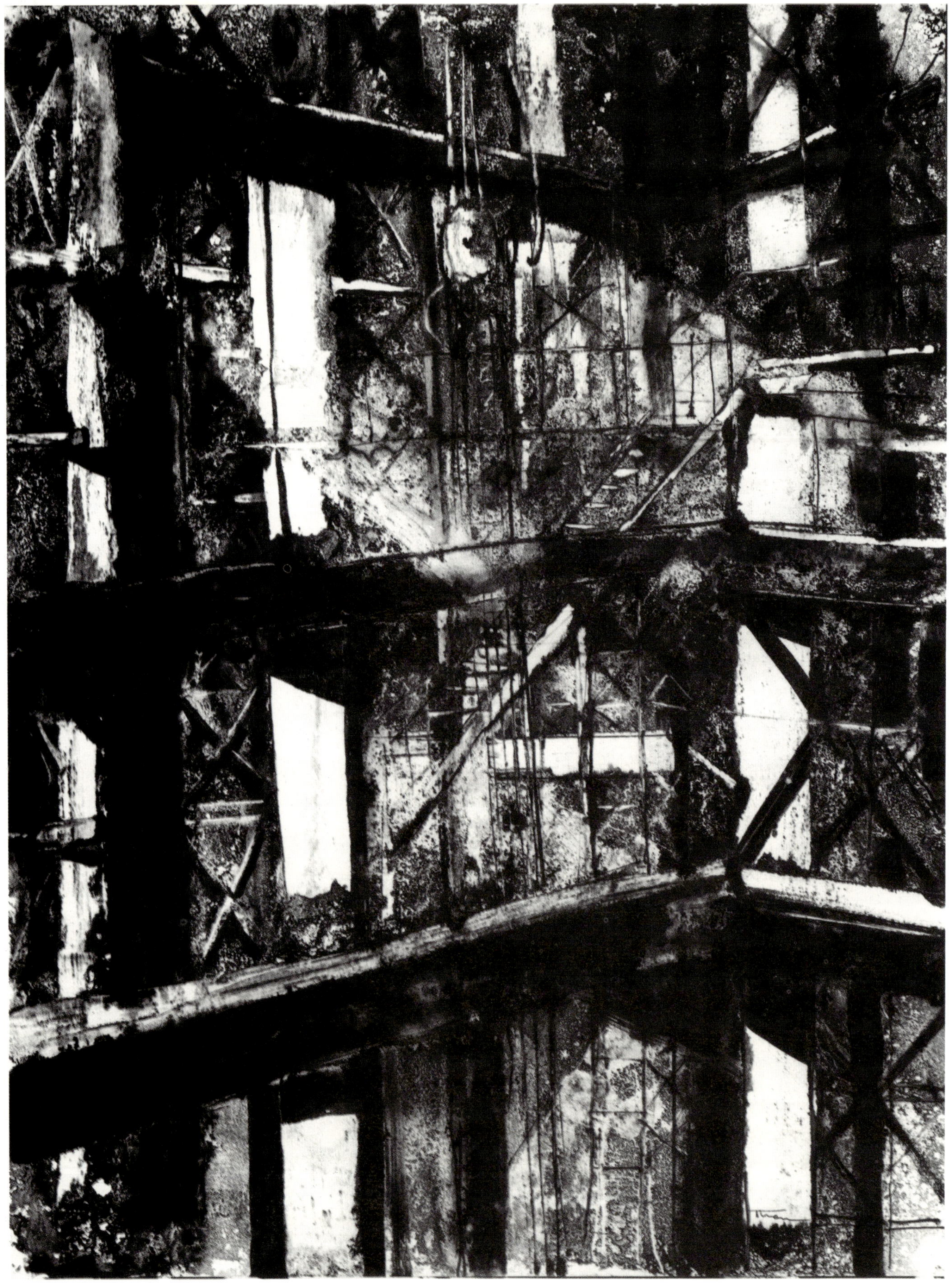

11. *William Thon.* VAB INTERIOR. Ink on paper, 14×11″

supply the sensation of weightlessness or of reduced gravity that one would experience in space flight or on the moon's surface. The technicians in charge of this facility, perhaps because they also are involved in make-believe, established a close bond with the artists and allowed most of them to try at least one landing on the moon or one blastoff in the Apollo Spacecraft Simulator mounted in the same room.

Of the static installations on the Cape, none can compare with the Vehicle Assembly Building (VAB), located near the center of Merritt Island, about four miles from the sea. From a distance, it looks like an isolated highrise apartment building. As one gets nearer, it begins to look like a huge windowless box. What is not apparent is that it is the biggest building ever to be constructed in the world. Three Empire State Buildings could be cut up and fitted inside it. The Great Pyramid of Cheops would leave room to spare. The Secretariat building of the United Nations could be wheeled through any one of the four doorways. The reason for its monstrous size is that within its walls must be assembled vertically all the elements that make up the Saturn-Apollo moon rockets. Before the Apollo program, lesser rockets were assembled on their launch pads, the components being hoisted and stacked by cranes. However, this left the delicate mechanisms unprotected from the voracious salt air and the technicians working on them exposed to the merciless Florida sun. Therefore, since assembly of the Saturn-Apollo took much longer and was infinitely more complex, it was decided to assemble the parts, flown in and barged in from all over the country, under cover inside a giant building. The VAB is a testament and a monument to the principle that form follows function. Although architects were involved, their role was probably that of frustrated onlookers. Engineers and budget-watchers reigned supreme. While the building was under construction, it was a Bruegelesque fantasy, with cobwebs of steel girders silhouetted against the sky. Nets to catch men and their tools were swathed from one pinnacle to another. Workmen crawled through the maze like ants. When the outer sheathing was bolted over the skeleton, fantasy was replaced by dull, straight, flat-planed reality. There is no scale, no means of measuring height or volume. The trees and cars near the base give some idea, but they no longer help as the eye proceeds upward. Where there are windows, the panes are elephantine in size—meaning that an elephant could pass comfortably through any one. There are no rails or stairs, nothing to relate the construction to the human scale. Some vertical elements, like the metal sheathing beside the huge doorways, lean outward, counteracting the effects of perspective. Tour guides with awe-inspiring facts and photographers with cloud shadows and trick angles have done their best to convey an idea of the behemoth's size, but without much success. The VAB remains a gigantic box in the middle of a scrubby marshland. Its grandeur is so well camouflaged behind its flat triteness that no one is very impressed.

The inside is a different story. Like a medieval cathedral, much of the interior is open, so that the eye can travel the full length and height of echoing vastness. It might be compared to a Piranesi prison scene: immense chains hang from monstrous hooks; catwalks crisscross dizzy heights; rows of arc lamps wink and glow in the misty twilight; birds occasionally fly through shafts of sunlight; loudspeakers deliver messages with the voice of doom. In the midst of a ganglion of steel girders and fittings, parts of the moon rocket can be glimpsed—white, pristine, and seemingly disassociated from the surrounding confusion. The visitor enters a glass-sided elevator and soars upward through space, watching the earth fall away below. This is no place for a person prone to vertigo. Each of the thirty-four levels is a world apart, dedicated to servicing a segment of the Saturn-Apollo moon vehicle. On the uppermost level is the escape rocket, designed to pull the astronauts inside the Apollo spacecraft free in an emergency during launch. Below that, the Apollo White Room, where the actual spacecraft, like a precious form of life in a snow-white womb, is ministered to by its own special cadre of midwives. Below that is the Lunar Module. Below that is the third-stage engine. Below that is the second-stage booster. Finally, some 360 feet below the space capsule, are the main booster engines capable of generating more than 7.5 million pounds of thrust. One enters among them feeling like an archaeologist entering a prehistoric temple. Strange and beautiful bell shapes, covered with friezes of wires, tubes, and hoses, and bedecked with red tags and encased in plastic bags, are placed like sculptural fantasies at irregular intervals. Outside, great webbing nets are festooned in swooping arcs like opera curtains. Truly, the interior of the VAB was made to order for the artist.

Although the peripheral material is fascinating, of course the climax of any trip to Cape Kennedy is the blastoff. As the countdown proceeds for important launches, the feeling becomes almost religious and the tension builds dramatically.

Representatives of the press from every quarter of the globe aim an extraordinary array of TV and photo lenses at the rocket smoking on its pad, three miles distant across the marsh. The liquid oxygen fuel chills the outer casing, so the rocket is sheathed in ice and gleams like a neon tube. The noise level of the crowd falls to a hushed silence as the tension builds. During the last minutes, only the loudspeaker broadcasting every detail of the countdown can be heard. During the Mercury program, seemingly endless holds were almost standard, and the waiting crowd fidgeted and sweated under the subtropic sun. The Apollo-Saturn V moon rocket, however, has a different level of reliability. Months ahead of the launch, the exact second of ignition is programmed and published, and with the inevitability of an eclipse the event has taken place to within a split second of schedule. The actual launch always comes as a surprise to visitors, in spite of the fact that they think they know what to expect from having watched it on television. The first surprise is the blinding brightness of the

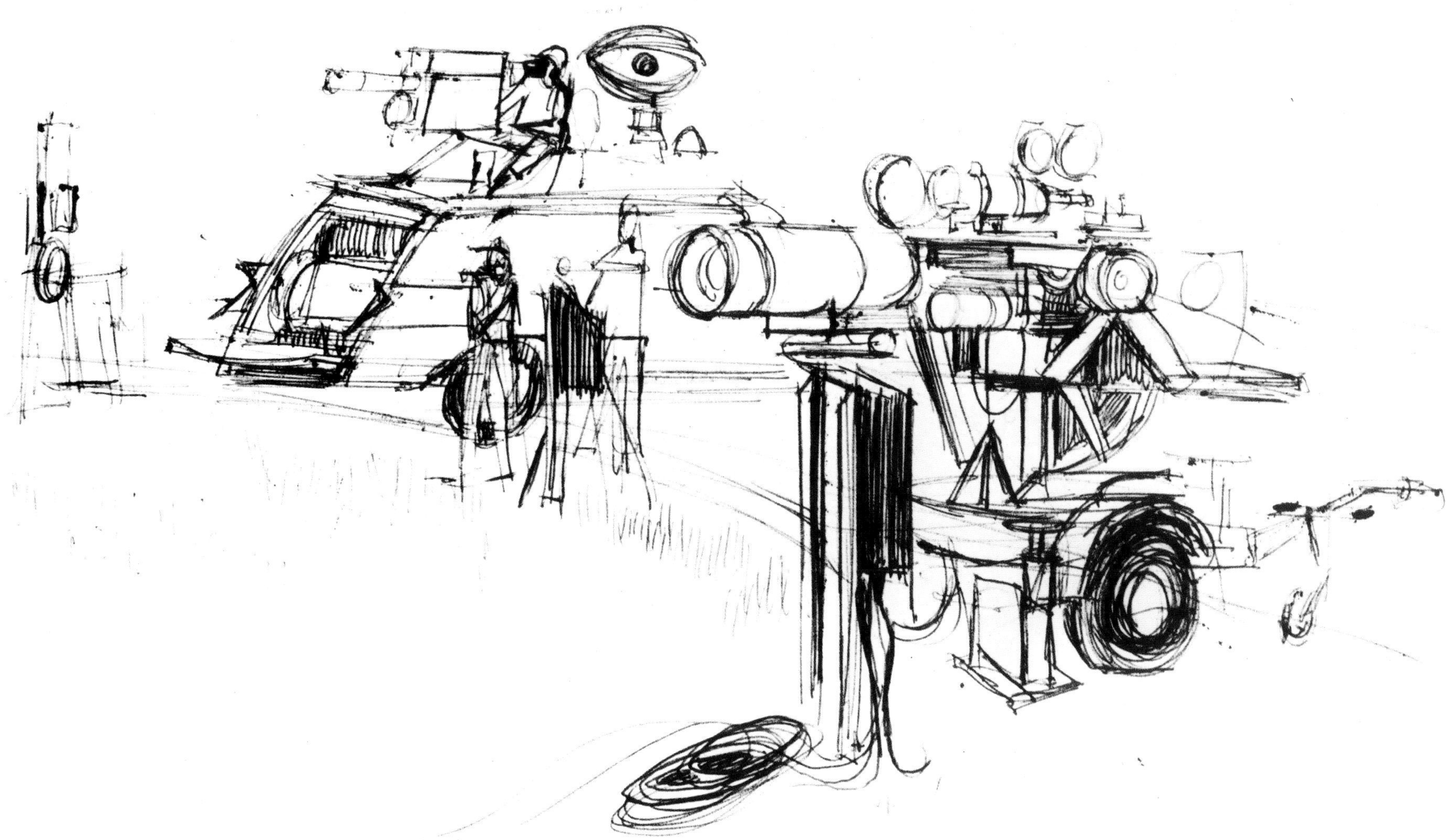

12.
Paul Calle
ROAD BLOCK 17
Pen and ink on paper, 9×14″

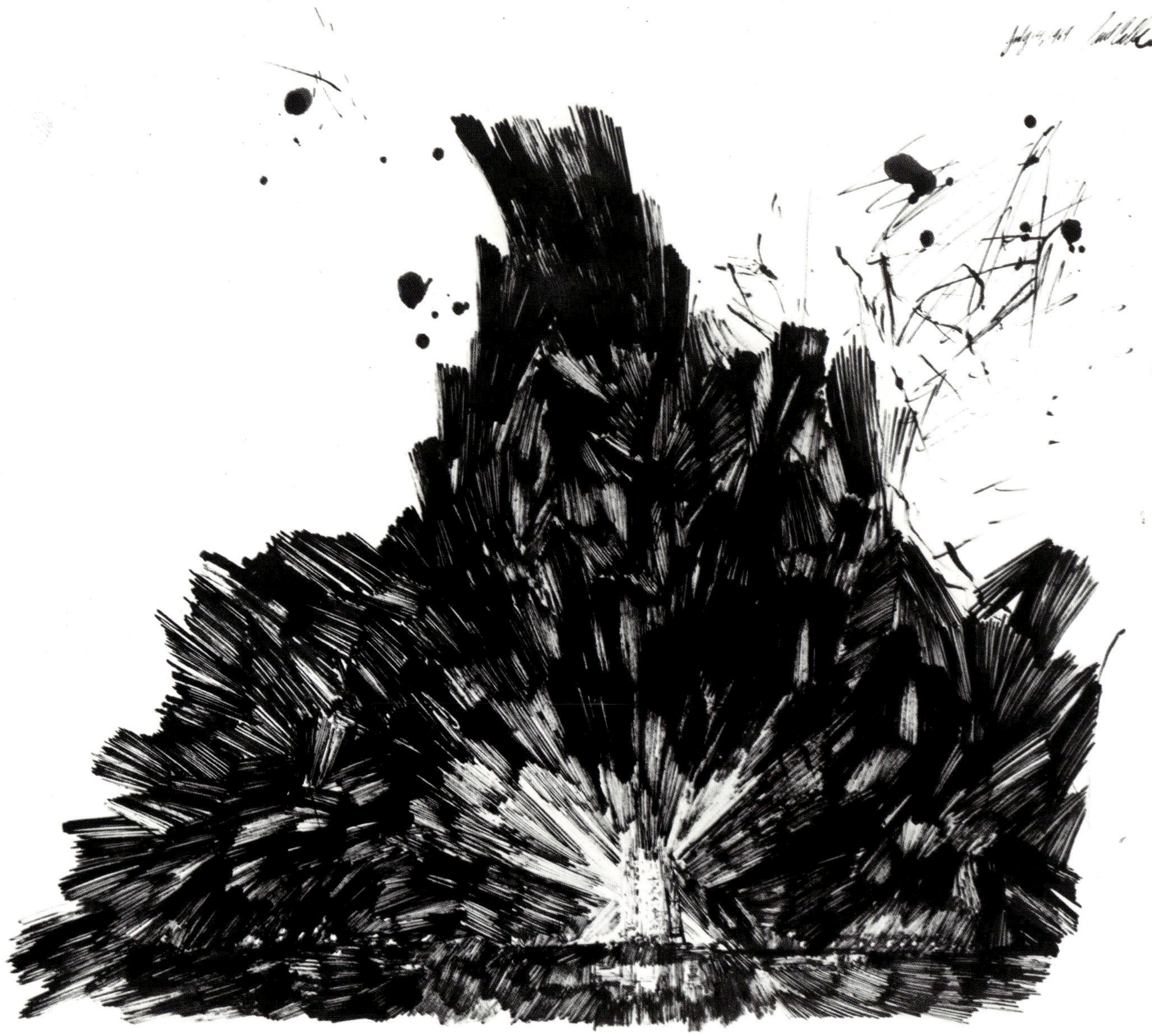

13.
Paul Calle. COUNTDOWN
Ink on paper, 17×22″

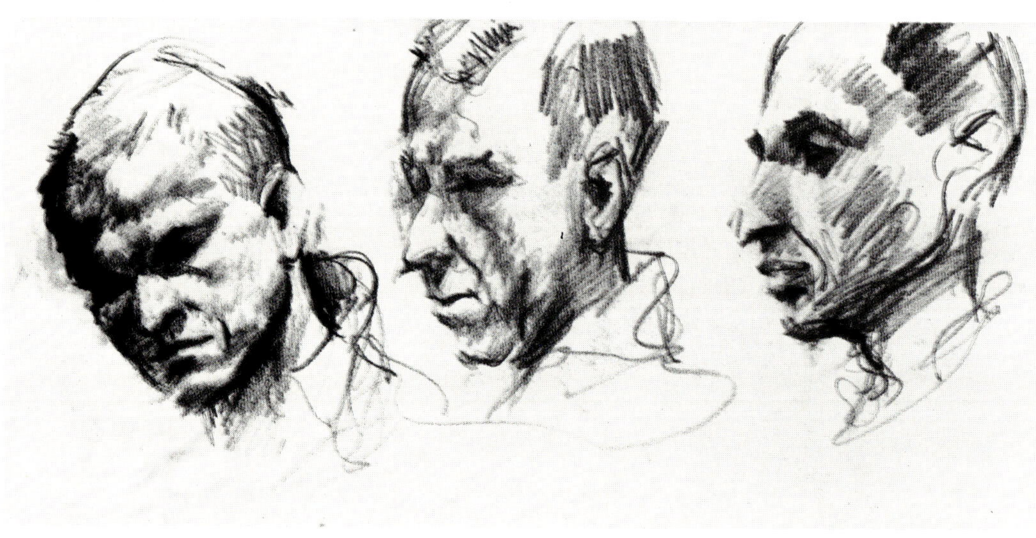

14.
Nicholas Solovioff
BACK TO EARTH
Pencil on paper, 12×18″

rocket's fireball. It is as brilliant as the sun and looking at it hurts the eyes. The second surprise is the slowness of the launch. The thirteen seconds for Saturn V to reach the top of the service tower seem much longer. Last and not least is the roar that cascades across the marsh in waves, making the water ripple and the crowd sway before its buffeting. As the rocket rises in a majestic arc, burning its way through any clouds, the crowd cheers, stamps, and applauds. Seconds later, there is nothing left to see or hear.

The Manned Spacecraft Center, Houston, consists of about twenty flat-topped buildings on the scrubby pasturelands a short distance inland from the sultry waters of the Gulf of Mexico. The nerve center of the complex is a four-story windowless white building entered through a sculptured portico. Here, thirteen seconds after the Saturn moon rocket leaves its pad at Cape Kennedy a thousand miles to the east, control of the mission is taken over. The inner sanctum of the Mission Control Center is a kind of theater with translucent screens forming one end of the auditorium. On these fourteen screens are projected typewritten data, TV images, diagrams, and rows of magical ciphers which, to the initiated, give an exact profile of how the mission is progressing. The flight controllers sit at rows of consoles, linked by phone, closed-circuit television, and radio with every part of the space organization throughout the world. And here the high priests of space, working in four-hour shifts, preside in solemn conclave during a flight. Those responsible for every part of the mission watch over their machines and the astronauts in them, and with calm recommendations rule by remote control. The flight surgeons watch golden pinpoints dancing over a dark-red screen, recording every quiver of the hearts of their patients over two hundred thousand miles away. Most of the critical information relevant to the mission is projected onto the twenty-foot screens at the end. Ghostly ten-foot hands appear and paste up the cold facts of history on typewritten pages. Another screen carries the TV pictures taken by the astronauts. When a critical maneuver like landing on the moon is performed, the projected pathway of the spacecraft is drawn in a golden arc on a black screen. The actual flight path then is superimposed over this in miniature blue and red crosses. The effect is that of an invisible Sindbad scattering a trail of sapphires and rubies over a jet-black beach. Actual voice communication with the astronauts is restricted to one man, himself an astronaut, who stands near the center, with every strand of control at his fingertips. He is the astronauts' chosen vicar on earth, and for a four-hour period he alone holds the keys to their cold, black kingdom. He is surrounded by the college of space cardinals, mostly young men like himself, each with a closed-circuit TV screen on which the second-by-second performance of the particular part of the mission assigned to his responsibility is recorded. If a question comes from outer space the astronaut in charge relays it to the scientist responsible for the equipment being questioned, to the engineer who designed it, or in some cases even to the technician who built it. These people may be hundreds of miles away in laboratories and factories, but they are standing by, linked to this same closed circuit. There is a nationwide conference, and their answer is given to the astronaut in Mission Control, who relays it to his colleagues two hundred thousand miles away. Never has so much information been available so fast to so few people. From the

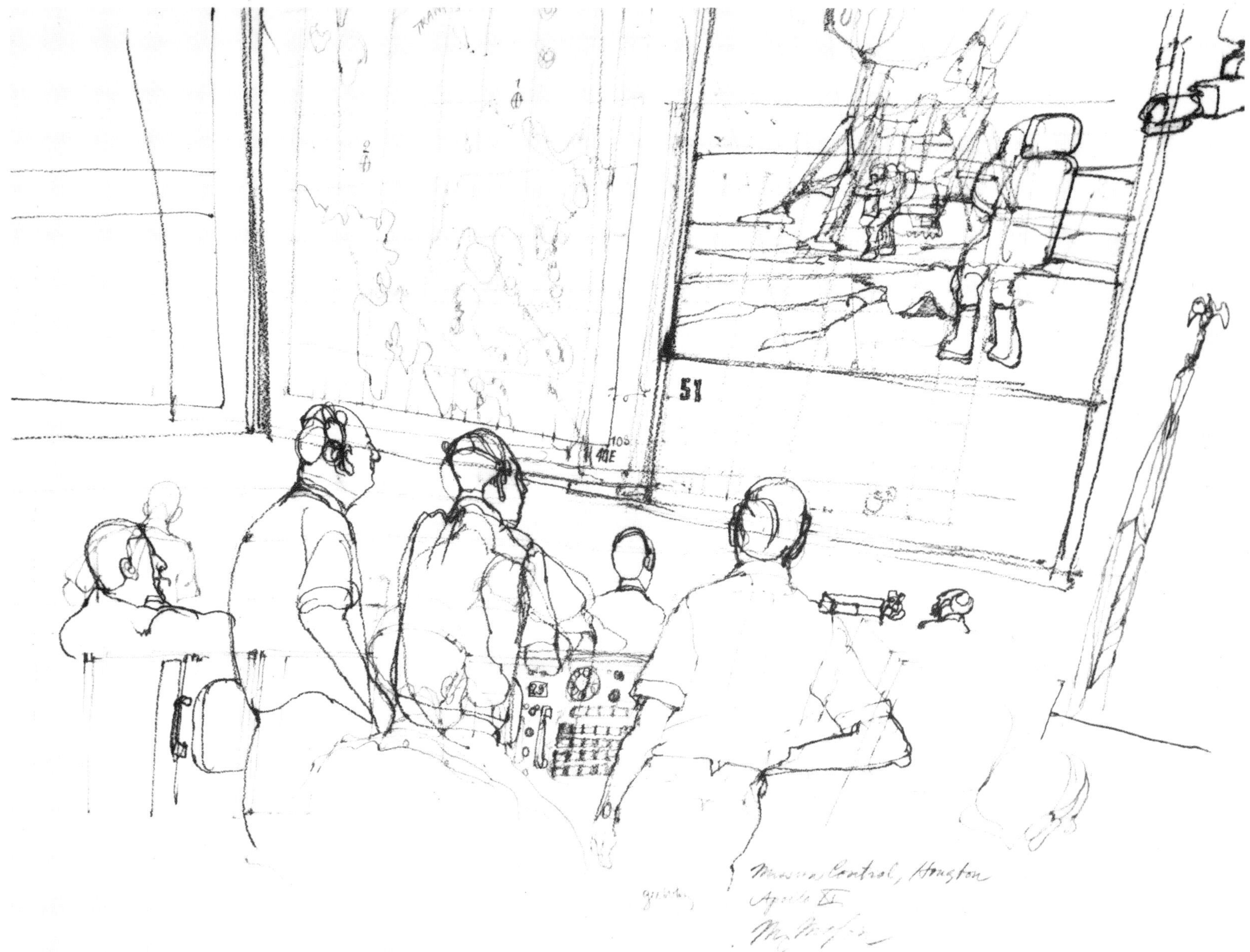

15. *Franklin McMahon.* MISSION CONTROL AT HOUSTON. Pencil on paper, 14×17″

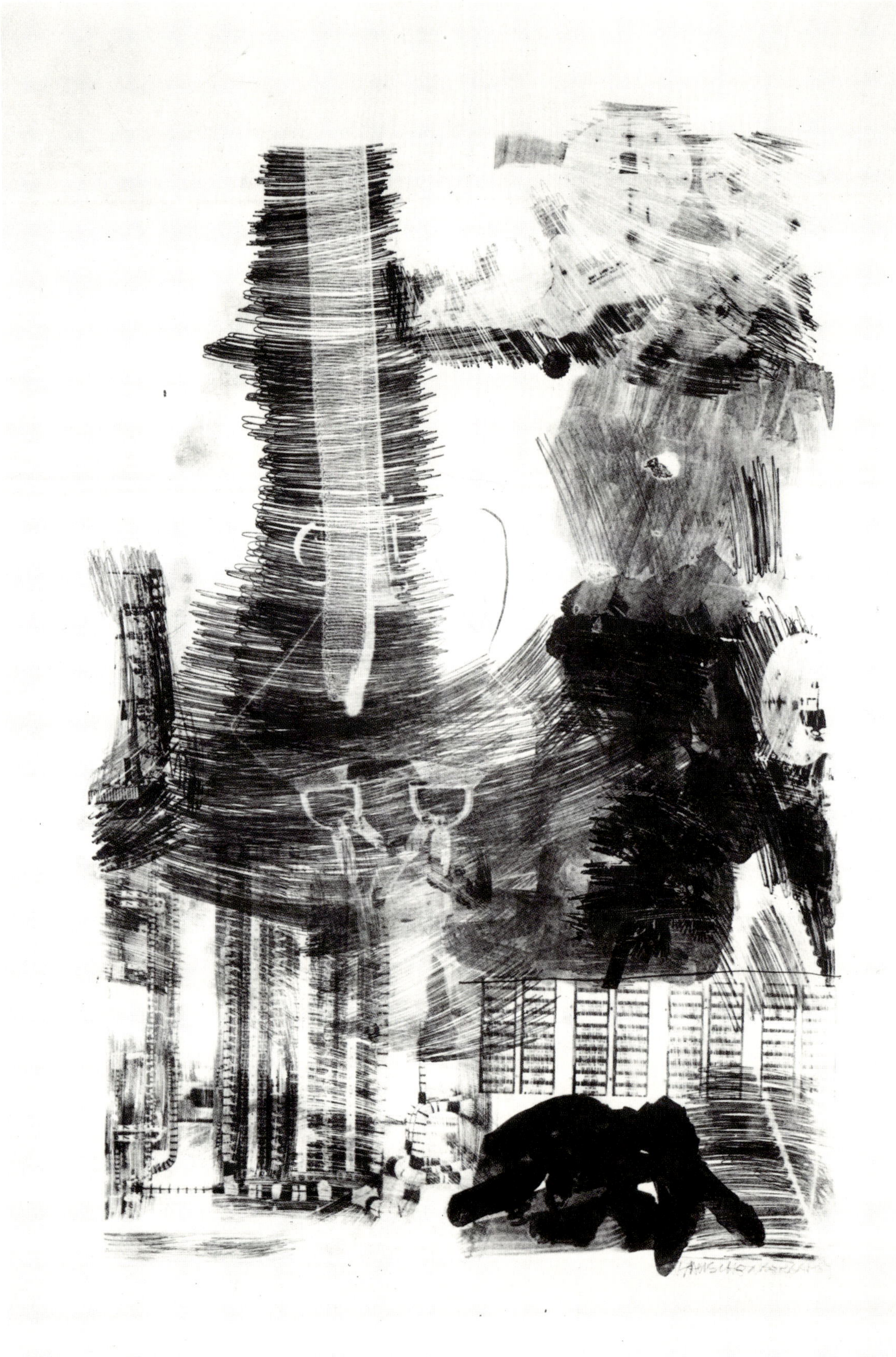

16.
Robert Rauschenberg
EARTH TIE
Color lithograph on paper, 48×34″
From the Stoned Moon series
reproduced courtesy of Gemini G.E.L.,
Los Angeles, Calif.

artists' point of view, as for the glassed-in few who are privileged to watch from the visitors' gallery, this is a ringside seat to history in the making. During important flights the room becomes a Who's Who in Space. Semilegendary figures like Wernher von Braun, Robert Gilruth, Kurt Debus, Chris Kraft, and astronauts past and present come in for brief periods to watch and wonder. For the artist in particular, the Mission Control Center presents a challenge. The physical facts—the yaws, rolls, pitches, pressures, circuits, and heartbeats of the lonely voyagers—here are converted into impulses, sounds, lights, figures, and graphs. They take the form of nonobjective art. Whether the artist can unscramble or reconstruct the data into a form intelligible to the nonspecialist is a test of the imagination still to be attempted.

From a portraitist's point of view the Mission Control Center has great possibilities. During critical times, like the Apollo 11 lunar landing flight, the key figures of space gather there, reminding one of medieval monks come to say their orisons. The specters of Goddard and Tsiolkovsky seem to walk the aisles, spellbound like their successors by the flashing lights and ciphers.

The Houston complex has other shows to offer, but none on such a cosmic scale. There is the long, pitch-black nave, where, high above the spectators' viewing booth, a Lunar Module mounted on rails curtsies and bows before the ghostly white bosom of the Command Module. Here astronauts are taught the science of maneuvering in space and the ticklish technique of docking two spacecraft. Several astronauts have told me that the sweetest sound in space is the click when two spacecraft have successfully completed their docking maneuver. So far, no artist has tried to portray the minuet-like advances of these two delicate yet awkward monsters.

Another segment of the Houston complex, which in time may develop into the most exciting of all, is known now as the Lunar Receiving Laboratory. Here, out-of-this-world treasures that have been captured in their native habitat and brought back to earth are examined in vacuum chambers. It is the Ellis Island for the secrets of the universe. Who knows what discoveries lie ahead for the scientists working in their lead-lined gloves.

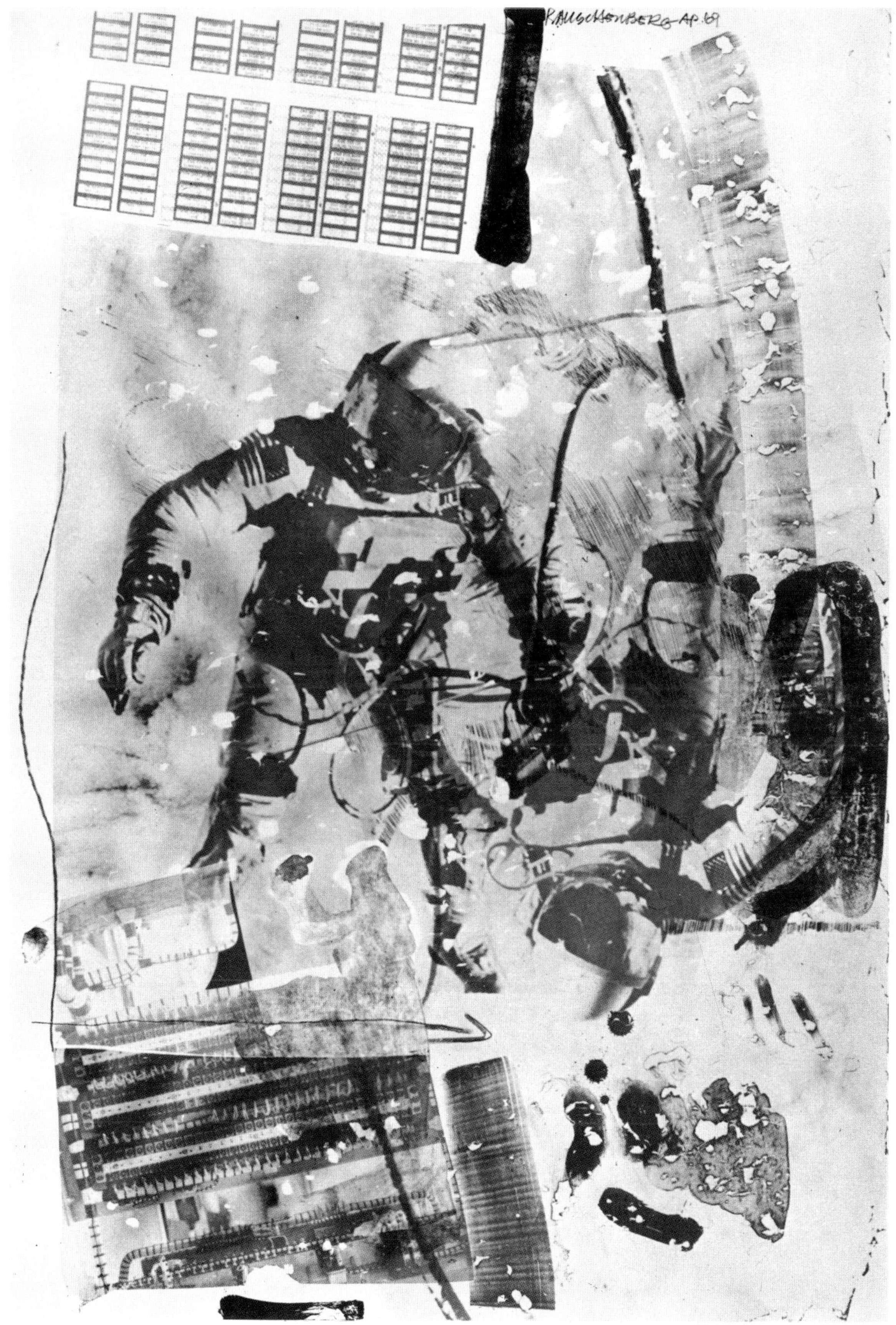

17.
Robert Rauschenberg
WHITE WALK
Color lithograph on paper, 42¼ × 29½″
From the Stoned Moon series
reproduced courtesy of Gemini G.E.L.,
Los Angeles, Calif.

The final stages of an Apollo lunar mission, compared with the exquisite refinement of the earlier phases, is almost a handcrafted operation. Earthward-bound in response to the pull of gravity, and certainly in keeping with the wishes of the astronauts, the spacecraft increases its speed from a leisurely six thousand miles per hour to a breakneck twenty-five thousand miles per hour. Probably the astronauts have little or no interest in the fact that they are breaking every speed record held by man since the dawn of time. Their interest, understandably, is focused on an imaginary slice of space into which their frail craft must plunge with unerring accuracy. The penalties for missing this keyhole in the atmosphere are truly Dantesque. To undershoot is to be burned alive in the friction heat of the atmosphere; to overshoot is to orbit for eternity around the earth. For reasons that are beyond the comprehension of ordinary mortals, the scientists in Houston, with their phalanxes of computers, can predict almost to within a fraction of a mile where the hurtling spacecraft will reenter the atmosphere. Shortly before reentering the earth's atmosphere, they cast off like the wings of a new Icarus the tried-and-true Service Module, and from this point on they are at the mercy of Newton's first law. The spacecraft has its ablative heat shield aimed toward the atmosphere. As it hurtles toward its landing site, minirockets with bantam nozzles can affect the yaw, roll, and pitch, but not the main direction of flight. However, by minute changes in attitude, the spacecraft can be made to skip over or plunge into the mantle which cloaks this planet and thus change by hundreds of miles the exact spot where it will finally settle on the ocean. To those waiting in the cold light of predawn on the carrier deck, the arrival of their guest from the black yonder is announced by two heavenly signs worthy of classical mythology. A streak of fire arcs across the night sky as the Service Module is burned up in the atmosphere and is followed by twin thunderclaps of sonic booms, the first from the spacecraft and the second from the echo off the water. Finally, the orange-and-white chutes, like the petals of a flower, float through the clouds and, home from heaven, man finally returns to his own.

The final act in the drama, in spite of the skill involved, lacks dignity. After the splashdown, sausage-like flotation gears are attached by Navy frogmen, and the heroes of space are winched up into the helicopters like old boots snagged on a fisherman's line. Back on the carrier deck a red carpet, a band, an honor guard, a quarantine trailer, and welcome from the ship's captain are the first of a series of homecoming ceremonies in store for the astronauts. Below deck follow physical exams, isolation, and a well-earned rest. The newest electronic media capture the pageantry of smiling triumph, but artists record the utter exhaustion and contemplative mood. Men who have seen sights too wonderful to describe in words or paint and have gone through experiences too marvelous to relate become a race apart.

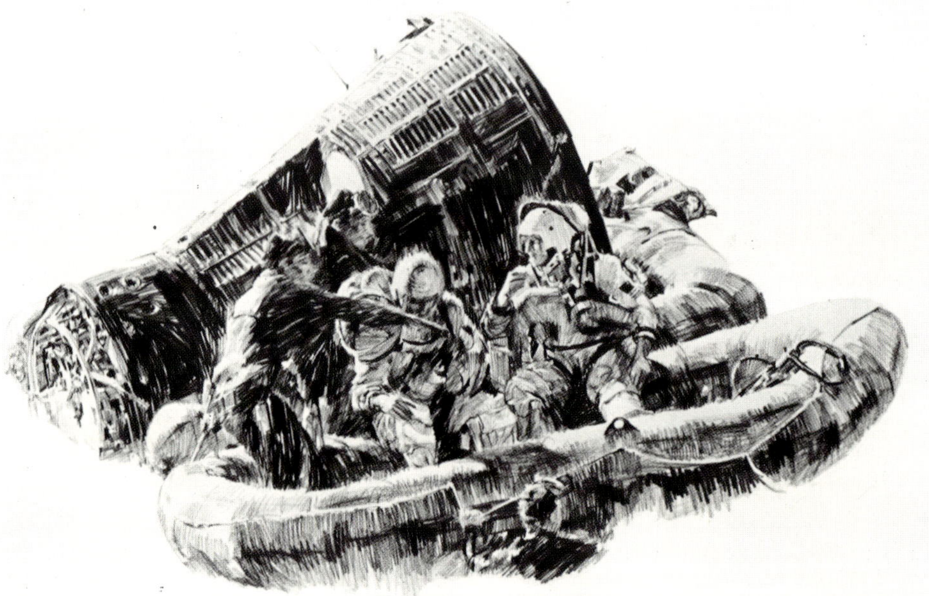

18. *Paul Calle*. FLOTATION GEAR. Pencil on paper, 22×30″

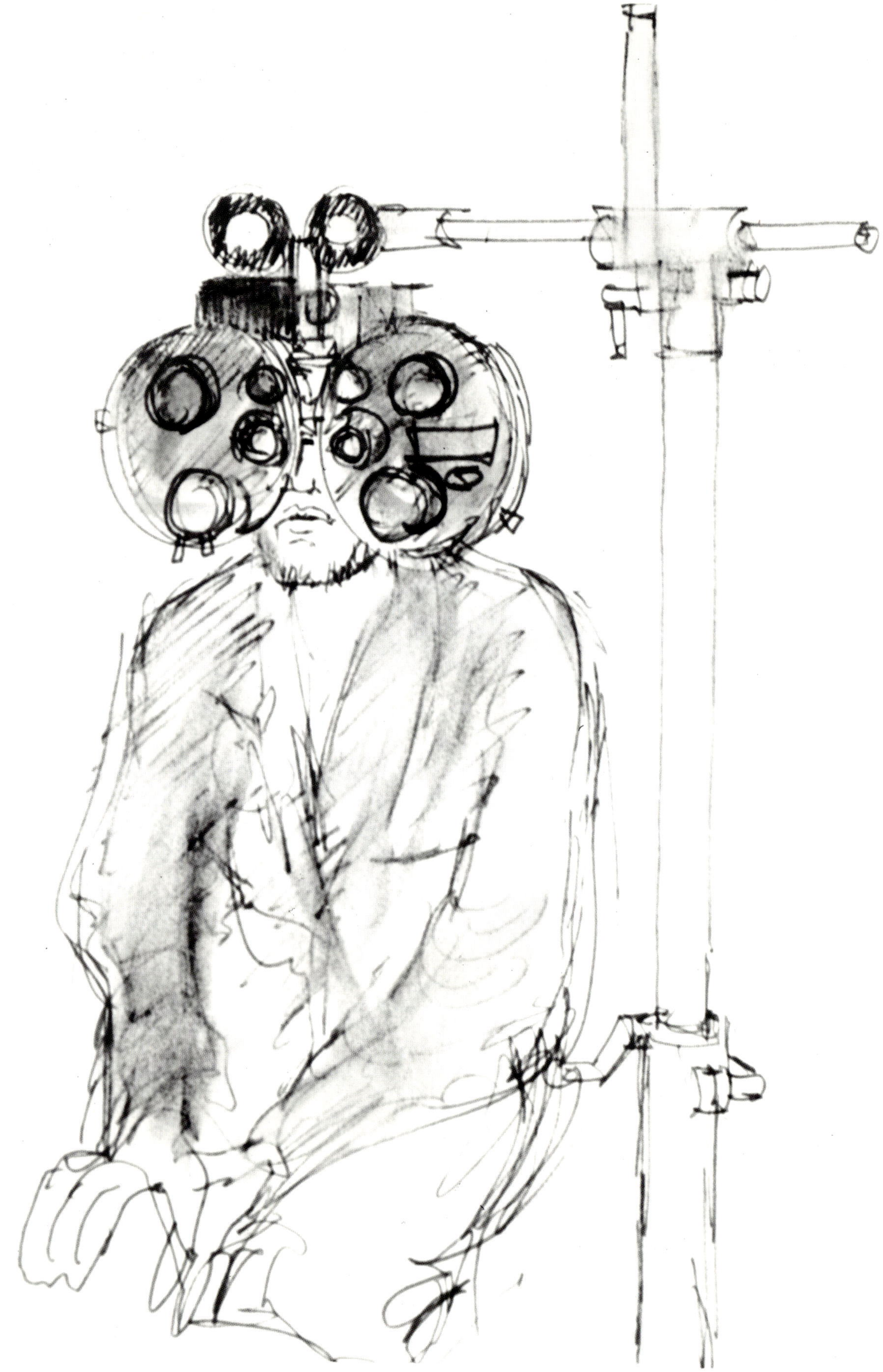

19. *Robert McCall*. EYE CHECK. Pen and ink on paper, 14×11″

Eyewitness to Space

I

20. Hugh Laidman *Along the Beach after Hurricane*
21. Alfred McAdams *South Along the Shore*
22. John W. McCoy II *First Light*
23. James Wyeth *Blockhouse 34*
24. George Weymouth *The Start*
25. Robert McCall *Securing the Hatch*
26. Robert Shore *Atlas Held*
27. Robert McCall *Gantry White Room*
28. Robert Shore *Gordon Cooper*
29. Robert McCall *MA 9—Suit-Up*
30. Paul Calle *Power*
31. Hugh Laidman *Tracking Antenna*
32. Vincent Cavallaro *T Plus 30 Seconds*
33. Paul Calle *Blastoff*
34. Franklin McMahon *Press Site*
35. James Wyeth *Gemini Launch Pad*
36. Mitchell Jamieson *Inside Blockhouse*
37. Fred Freeman *Saturn Blockhouse*
38. Lamar Dodd *Max Q*
39. Robert Shore *Back-up Spacecraft*
40. Robert McCall *Mercury Suit Test*
41. Paul Sample *Neighbors*
42. Fred Freeman *Saturn IB*
43. Franklin McMahon *Orbit Number 1*
44. Franklin McMahon *Canary Island Tracking Station*
45. Lamar Dodd *Listening*
46. Alfred McAdams *Range Safety*
47. Robert Shore *From the Inside*
48. Lamar Dodd *Launch Pad*
49. Alfred McAdams *Fire Damage*
50. Franklin McMahon *Motel Sign*
51. Alfred McAdams *Rollback*
52. Alfred McAdams *Cape Kennedy*
53. Robert McCall *Rollback*
54. Lamar Dodd *White Room*
55. Fred Freeman *Roadside Viewing Stand*
56. Hugh Laidman *T Minus 80*
57. Lamar Dodd *Watching*
58. Robert McCall *Waiting for GT-5*
59. Mitchell Jamieson *Helicopter Crewmen*
60. Mitchell Jamieson *First Sighting*
61. Mitchell Jamieson *Pacific Recovery*
62. Mitchell Jamieson *Flight Deck*
63. Leonard Dermott *Waiting for Pickup*
64. Mitchell Jamieson *Afterthoughts*
65. Mitchell Jamieson *First Steps*
66. Hugh Laidman *Mila Under Construction*
67. Theodore Hancock *Engines at Huntsville*
68. John W. McCoy II *Moving Giant*
69. James Wyeth *Waiting*
70. Peter Hurd *Predawn*
71. Franklin McMahon *Lunar Module Mission Simulator*
72. Paul Calle *Train Wreck*
73. Howard Koslow *Test Stand*
74. Theodore Hancock *Moon Buggy*
75. Theodore Hancock *Saturn Component*
76. Paul Calle *Testing the Spacesuit*
77. Paul Calle *Knight's Armor*
78. Mitchell Jamieson *Ladders*
79. Lamar Dodd *Saturn Gantry at Night*
80. James Wyeth *Fuel Storage*
81. James Wyeth *Wide Load*
82. Henry C. Pitz *Sightseeing*
83. Dong Kingman *Higher, Faster, and Farther*
84. Alfred McAdams *Recollection of Fire*
85. Alfred McAdams *Vacuum Tanks*
86. Alfred McAdams *Testing at Goddard*
87. Mitchell Jamieson *Flame Deflector*
88. Alfred McAdams *Deflector*
89. Paul Sample *Night Check*
90. Paul Sample *Liquid Fuel*
91. Paul Sample *Row*
92. Theodore Hancock *The Sounds of Engines*
93. Theodore Hancock *Wind Tunnel*
94. George Weymouth *Barrow*
95. Theodore Hancock *Alabama Landscape*
96. Theodore Hancock *Marshall*
97. James Wyeth *Support*

20.
Hugh Laidman
ALONG THE BEACH AFTER HURRICANE
Felt pen on paper, 12 × 18″

The Cape Kennedy area is well within the path of hurricanes, and in 1963 a hurricane caused damage along the coast, although the installations themselves were not affected.

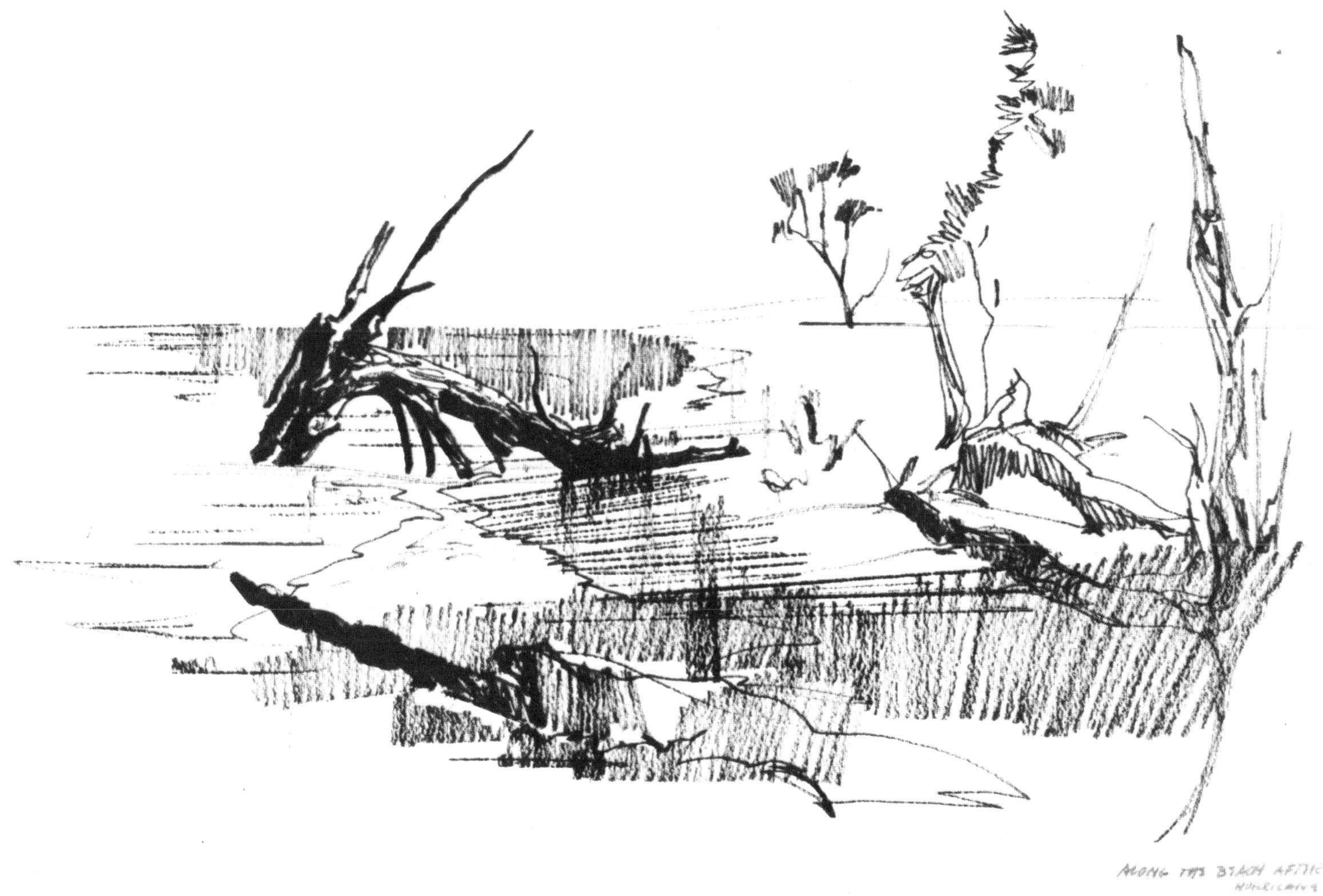

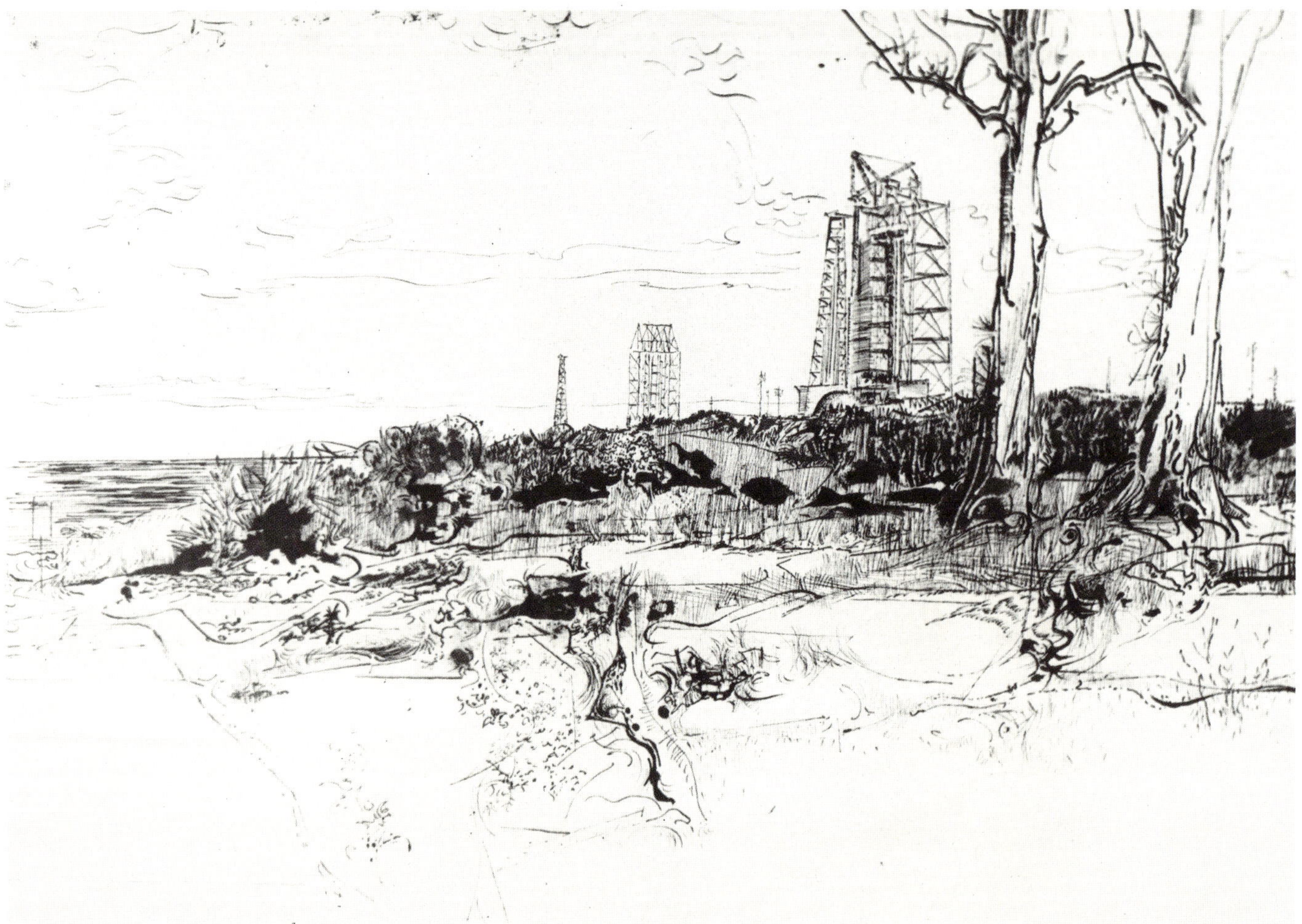

21. *Alfred McAdams.* SOUTH ALONG THE SHORE. Ink on paper, 19½ × 26¼″

22.
John W. McCoy II
FIRST LIGHT
Watercolor on paper, 21 × 30″

No artist who has seen a rocket on its pad at night, on the eve of a major launch, can ever forget the scene. It is one of the most dramatic and beautiful sights of our time. A ring of searchlights transfixes the spacecraft at the top (in this case Mercury–Atlas 9). The view is usually toward the east, and the first pale light of dawn, as here, rises out of the Atlantic behind.

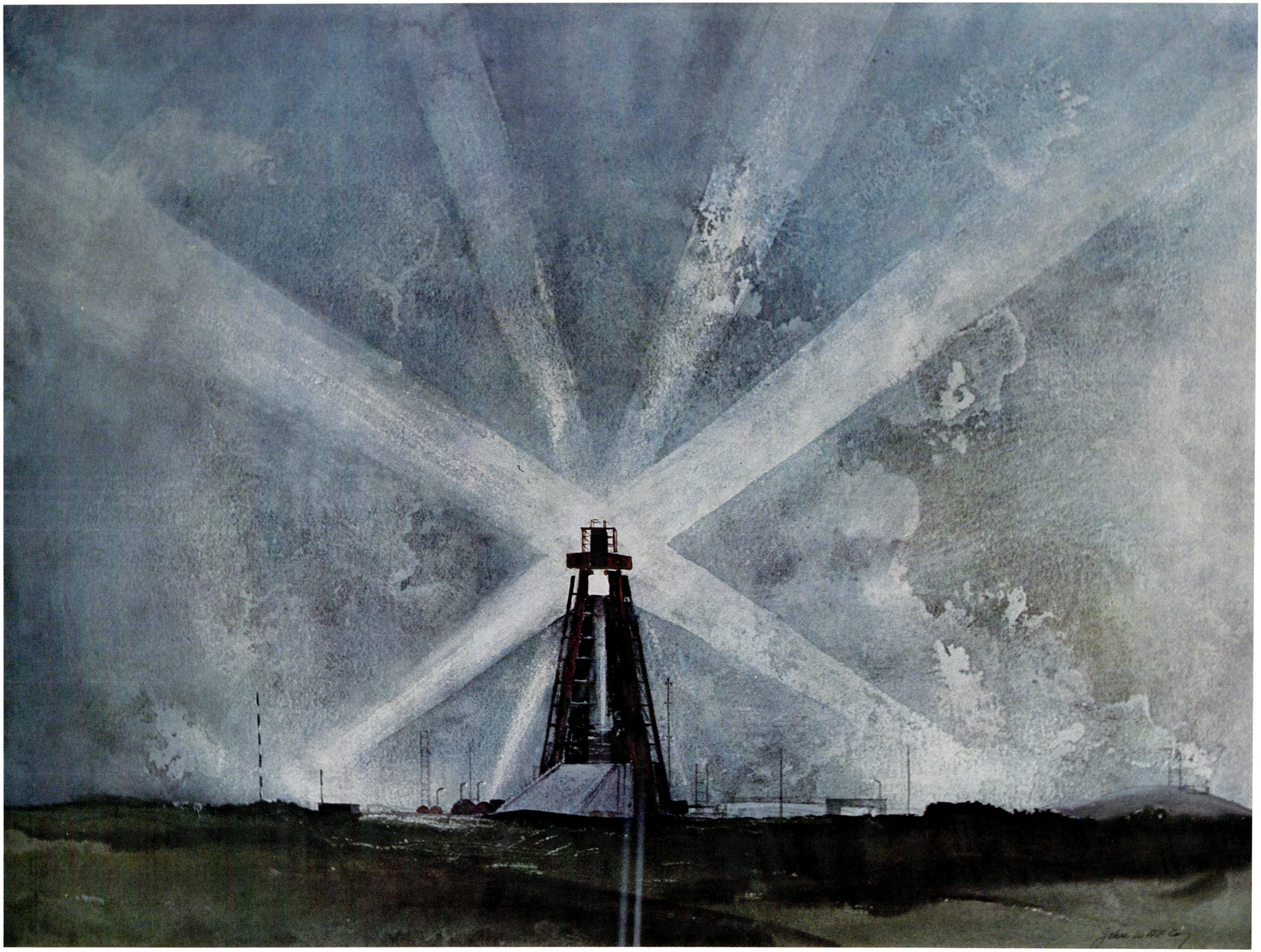

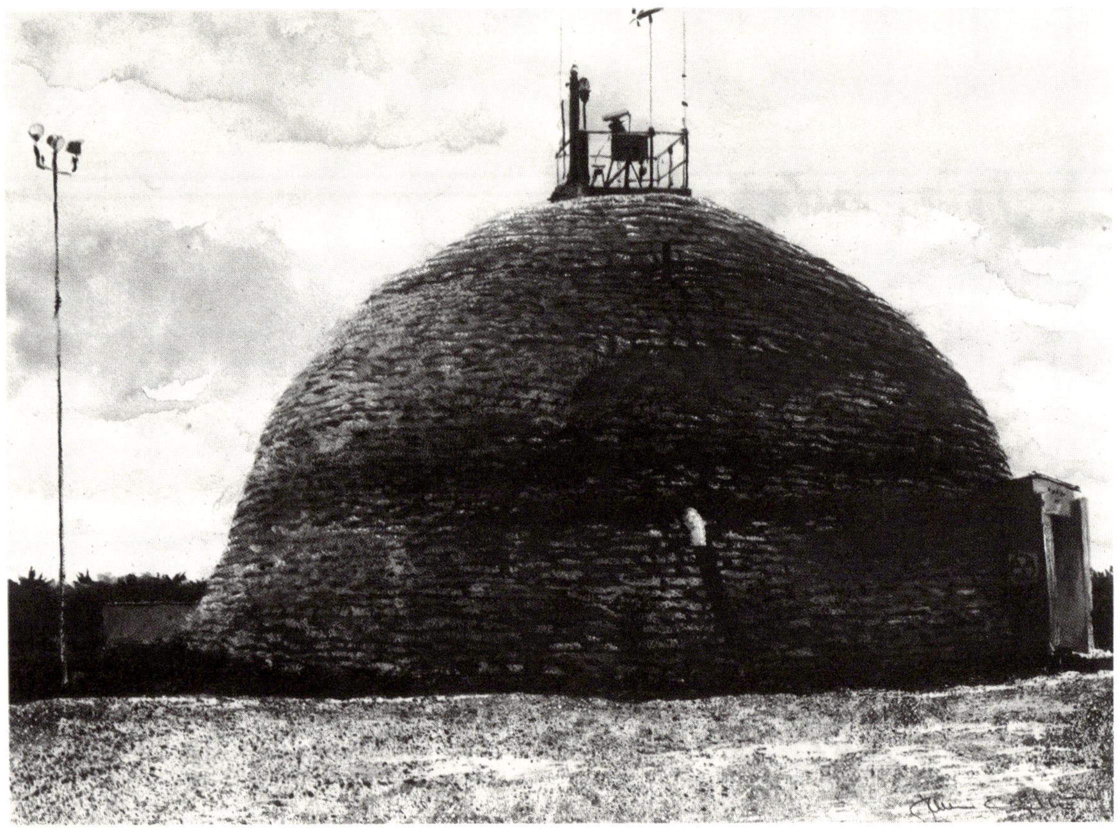

23. *James Wyeth*. BLOCKHOUSE 34. Watercolor on paper, 18×25″

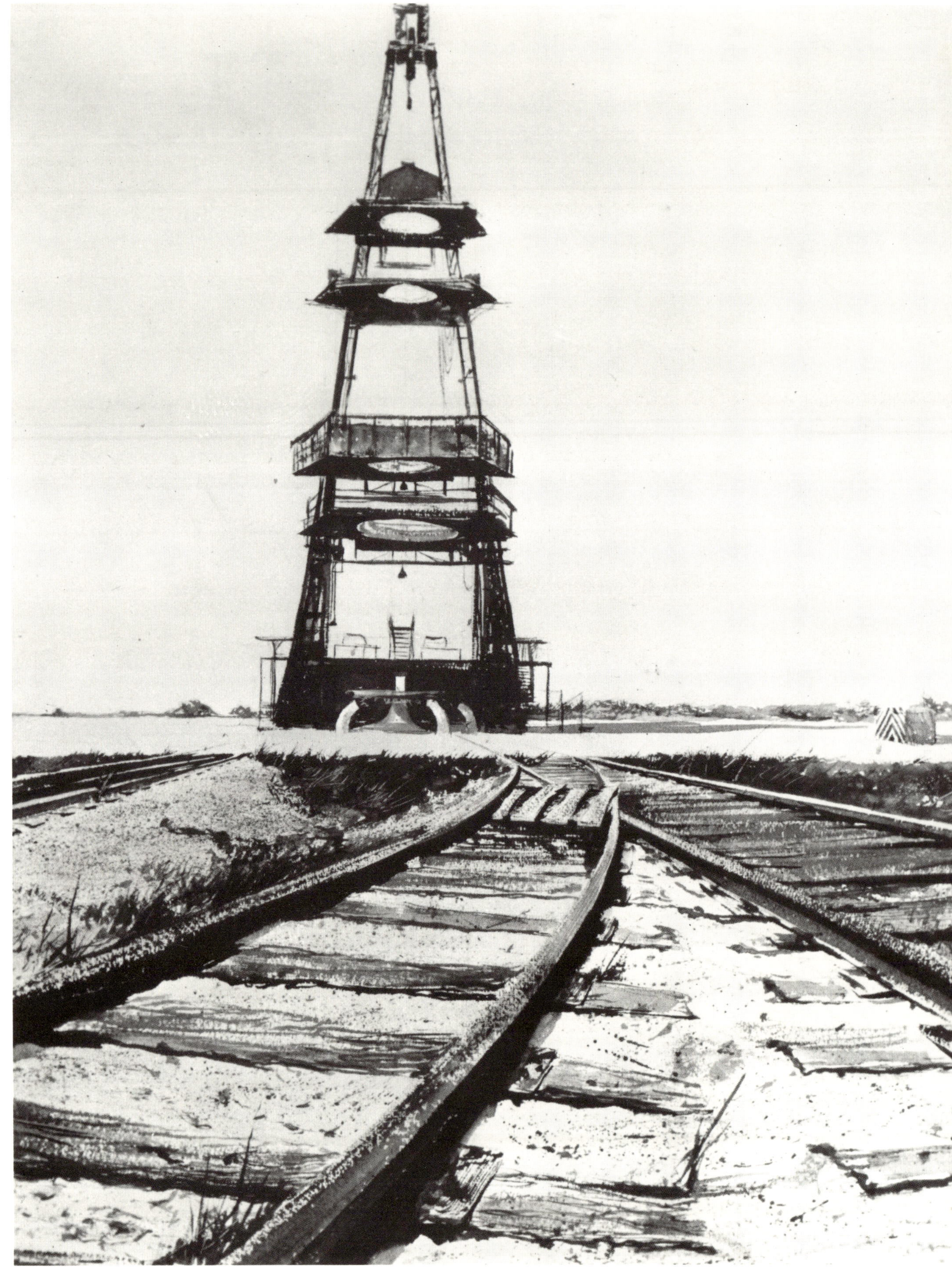

25.
Robert McCall
SECURING THE HATCH
Felt pen on paper, 17×14″

In retrospect it is difficult to imagine how small-scale, personal, and handicrafted many parts of the Mercury program were. Here Gordon Cooper is being bolted into Faith 7 on top of the Atlas rocket.

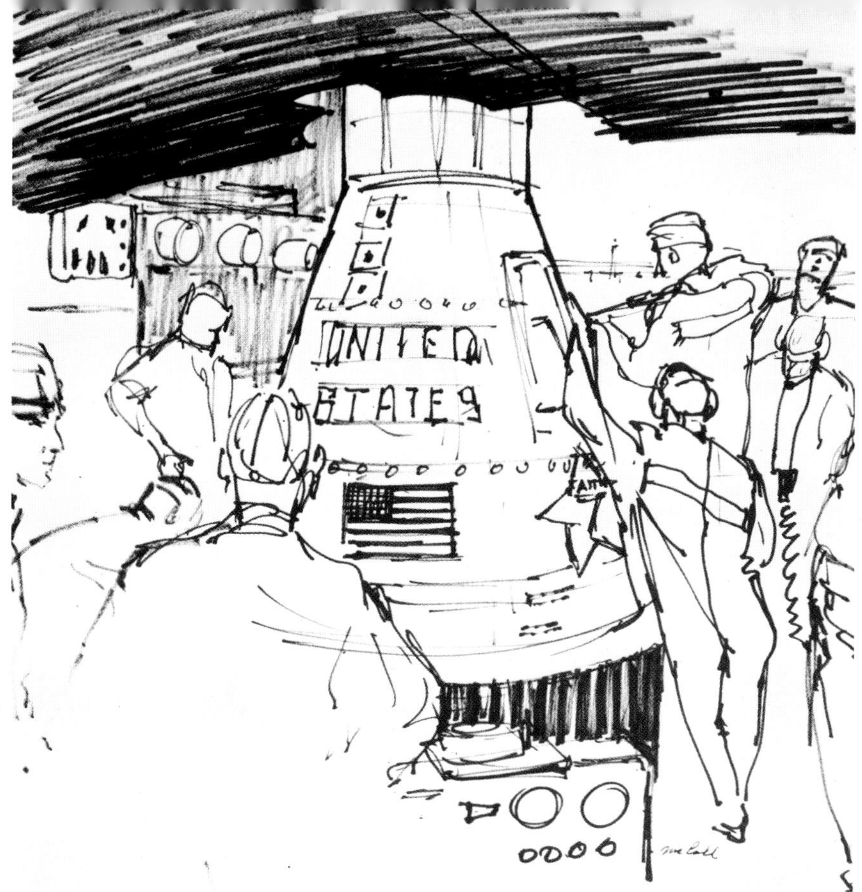

24.
George Weymouth
THE START
Watercolor on paper, 35×26″

When the first group of artists visited Cape Kennedy in 1963, the complex from which Alan Shepard, America's first astronaut, was launched in 1961 was already falling into ruins. Derived from the design of oil rigs, the structure was elemental but efficient. The flame deflector was a mandarin's hat a few feet high, and the clean room on top of the gantry was a canvas-enclosed cubicle surrounding the Mercury spacecraft.

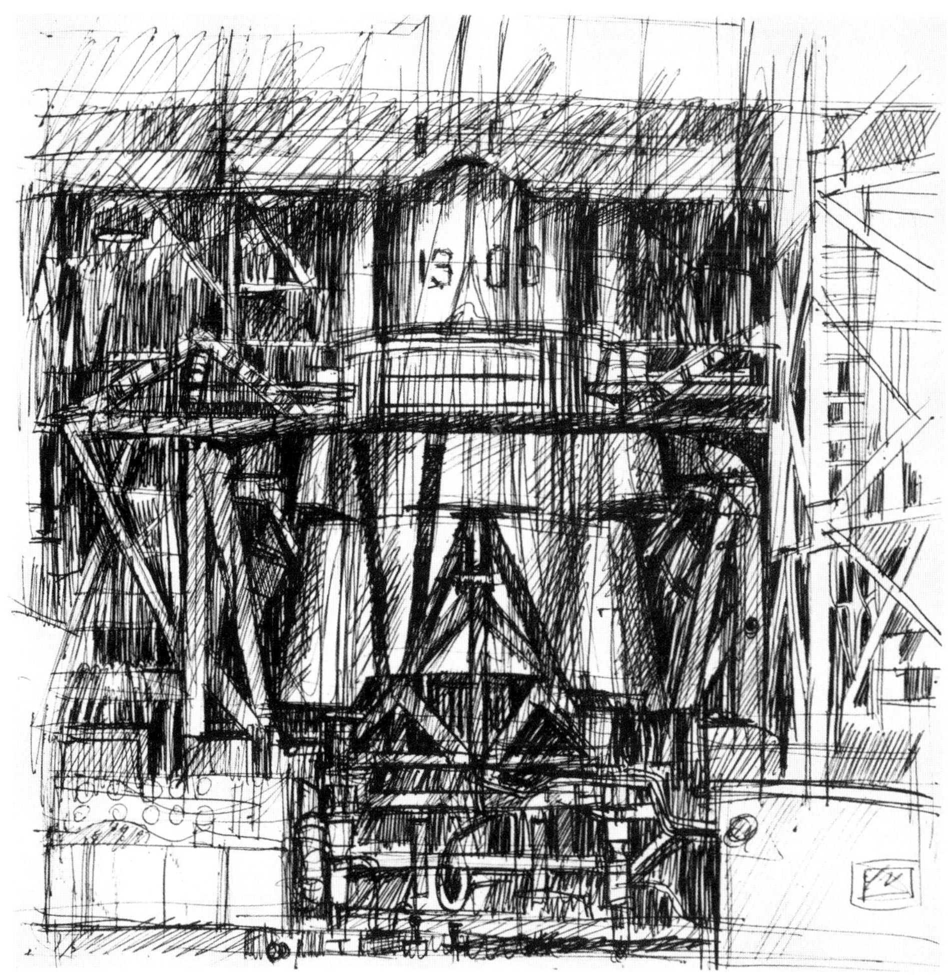

26. *Robert Shore.* ATLAS HELD. Pen and ink on paper, 14×11″

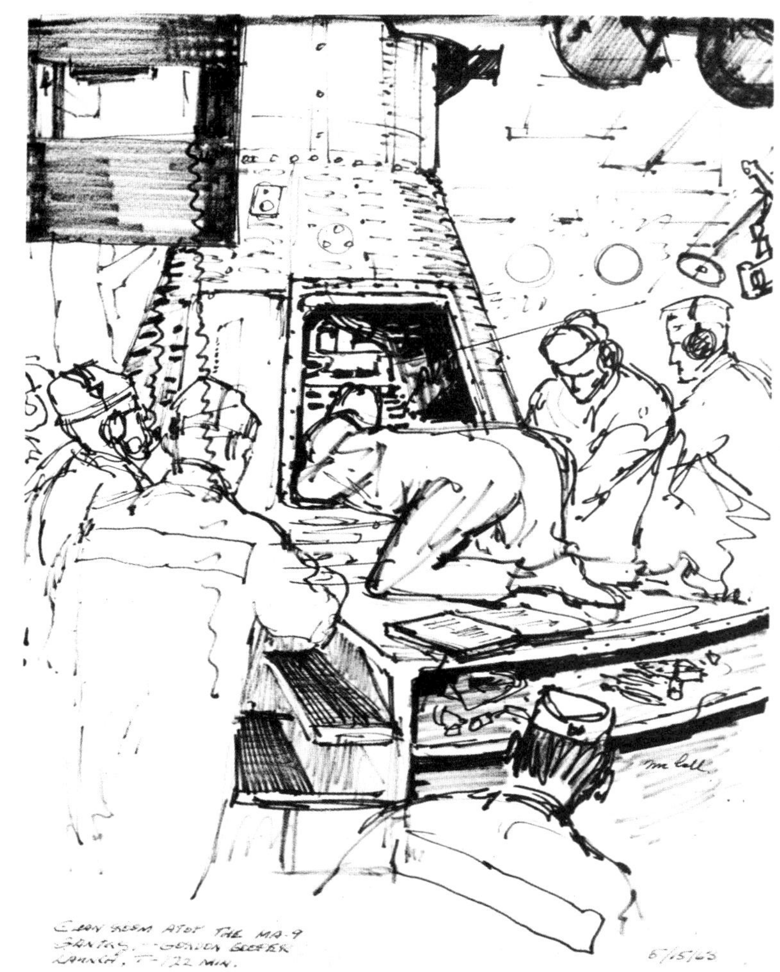

27. *Robert McCall.* GANTRY WHITE ROOM. Felt pen on paper, 17×14″

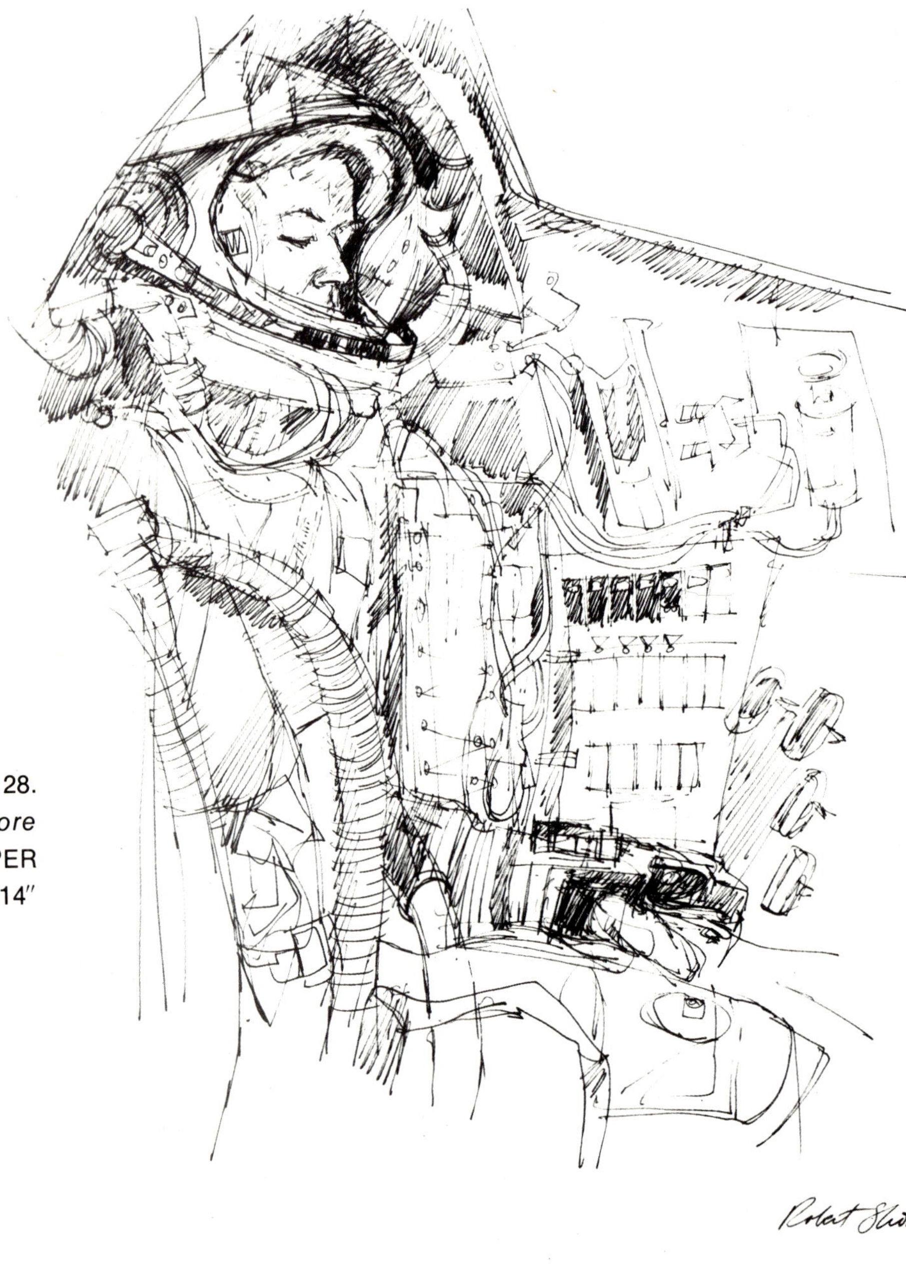

28.
Robert Shore
GORDON COOPER
Pen and ink on paper, 17×14″

29.
Robert McCall
MA9—SUIT-UP
Felt pen on paper, 17×14″

An astronaut of the Mercury program had very little freedom of movement. It is difficult to believe that Gordon Cooper remained in this strait jacket for twenty-two orbits.

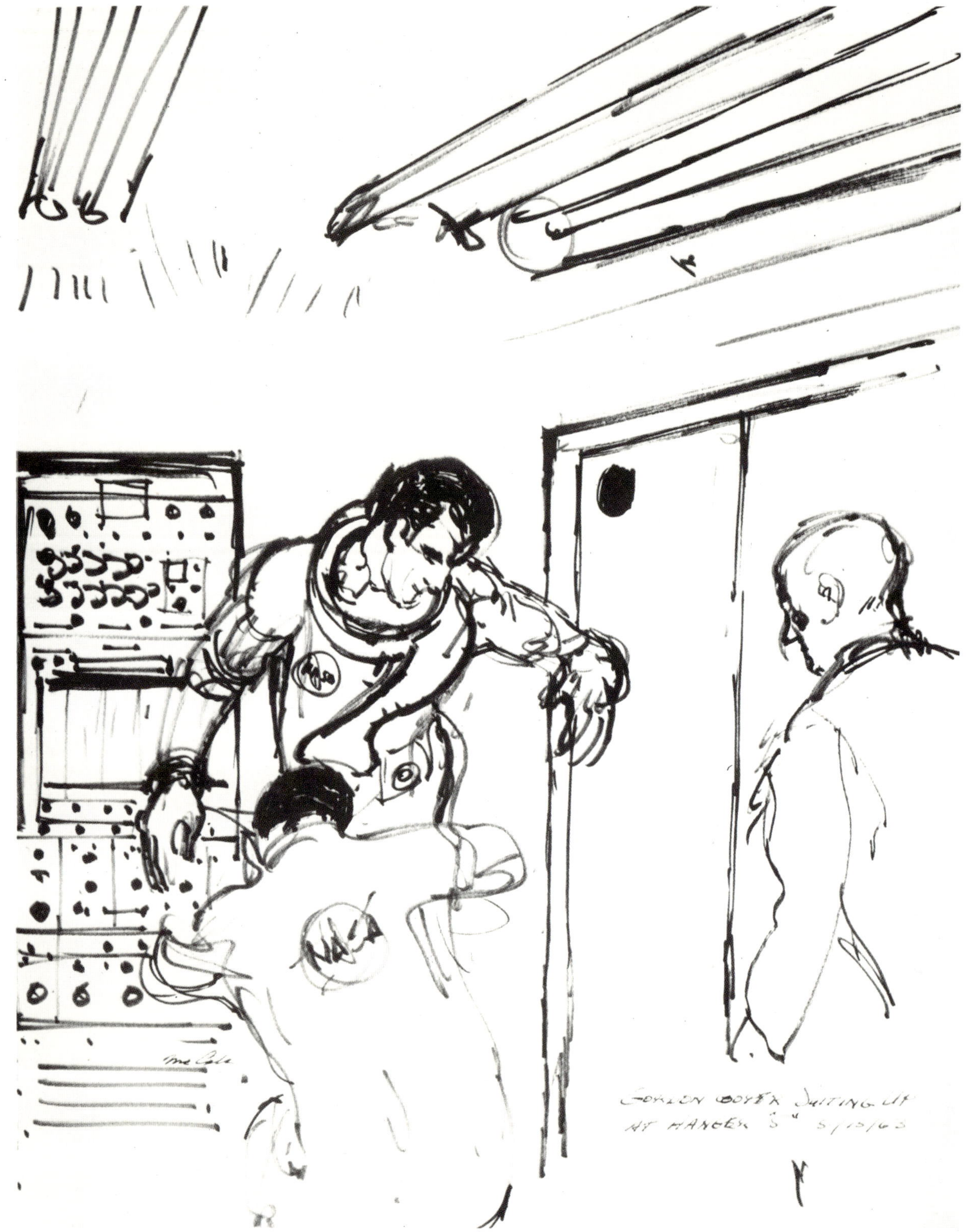

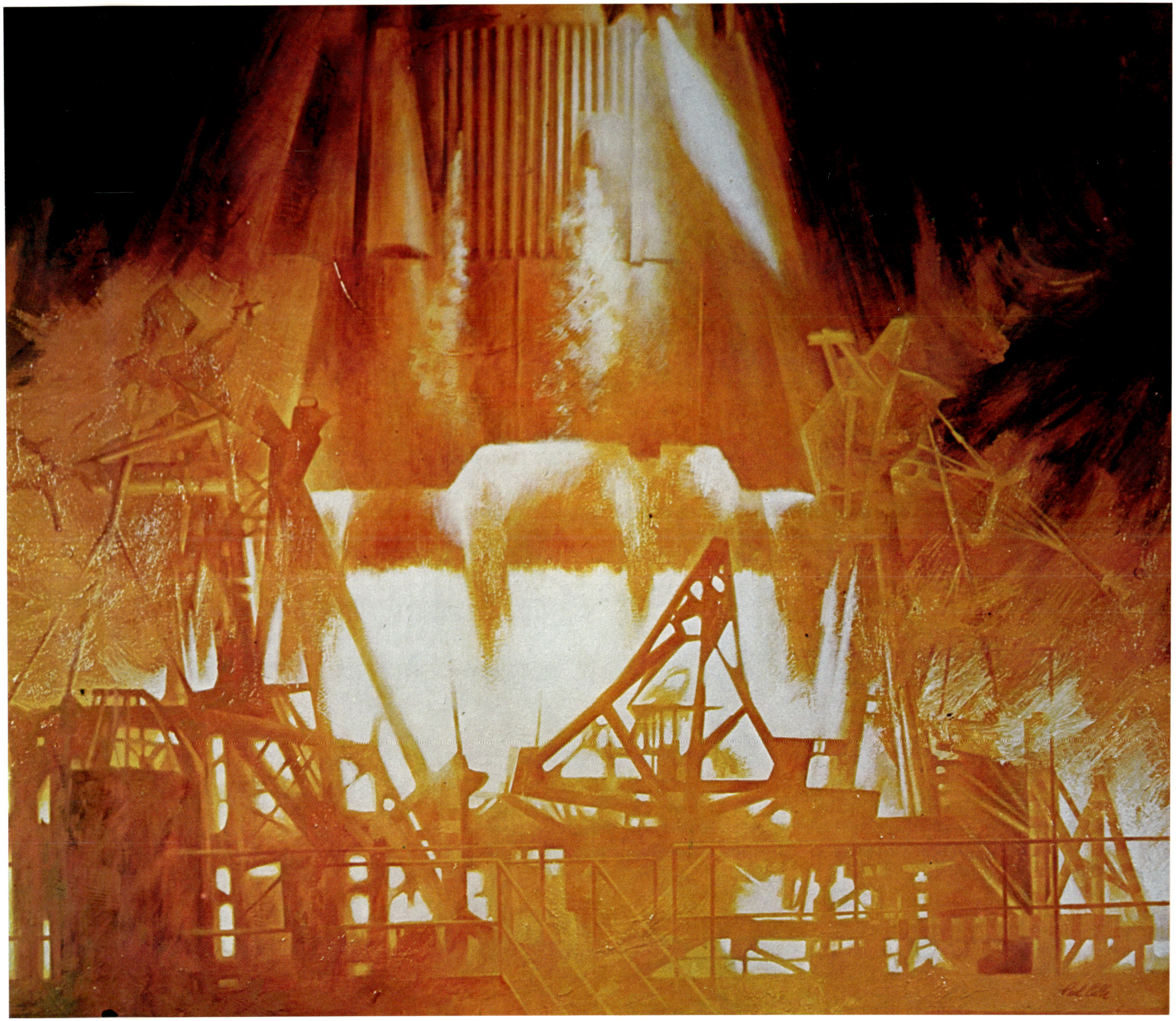

30.
Paul Calle
POWER
Oil on panel, 50⅜ × 58¼″

Those who see a launch for the first time are usually surprised that it bears so little resemblance to the image they are accustomed to in photographs and on television. They are unprepared for the blinding brilliance of the ball of flame that erupts from the rocket and for the sound waves (so strong that the ground shakes) that cascade over the flat marshland.

31. *Hugh Laidman*. TRACKING ANTENNA. Felt pen on paper, 12×22″

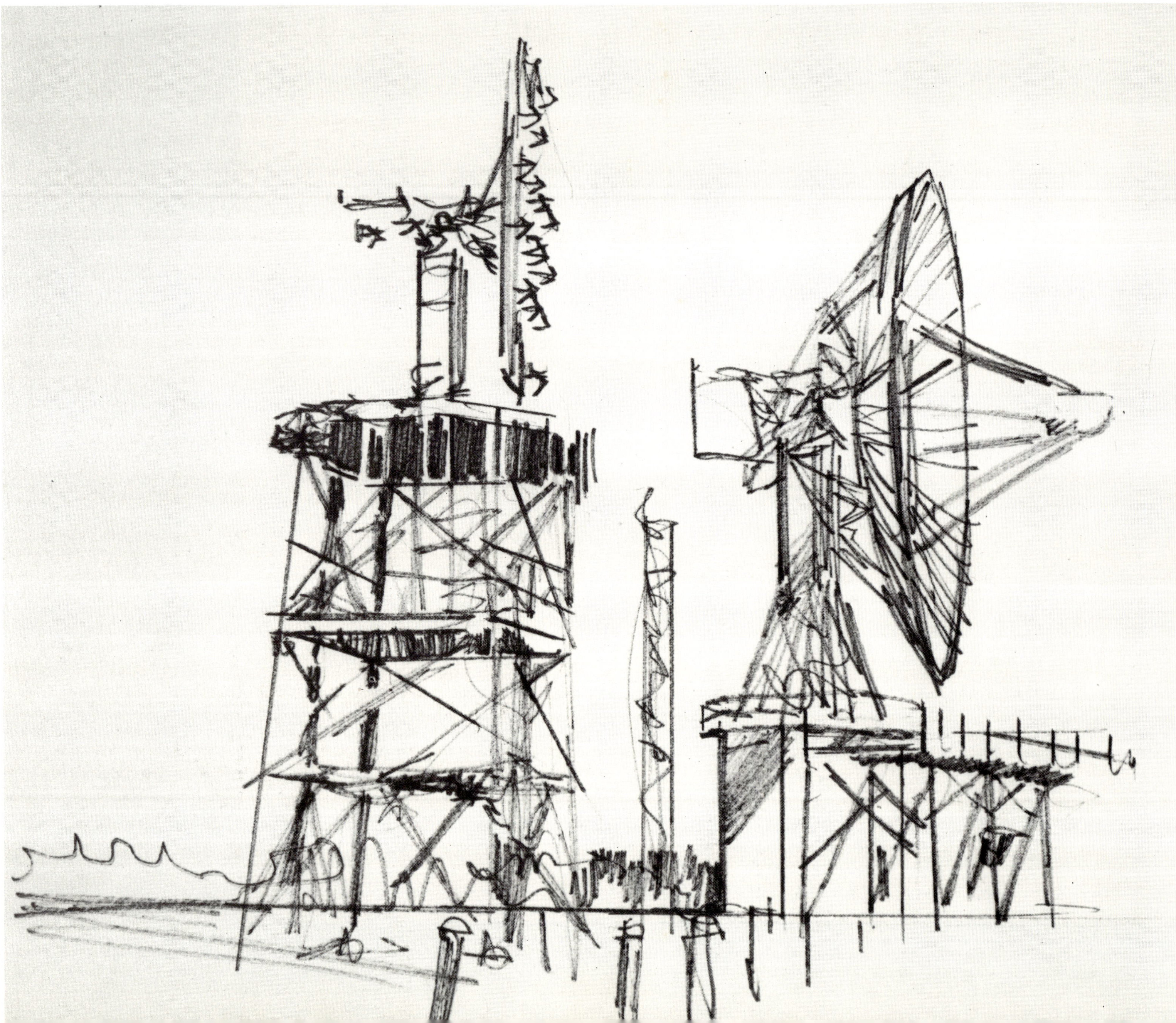

32.
Vincent Cavallaro
T PLUS 30 SECONDS
Oil on canvas, 60×50″

The sound waves and brilliance of a launch can perhaps never be expressed in either words or paint, but the afterimage in the imagination of an artist can convey at least the emotional impact.

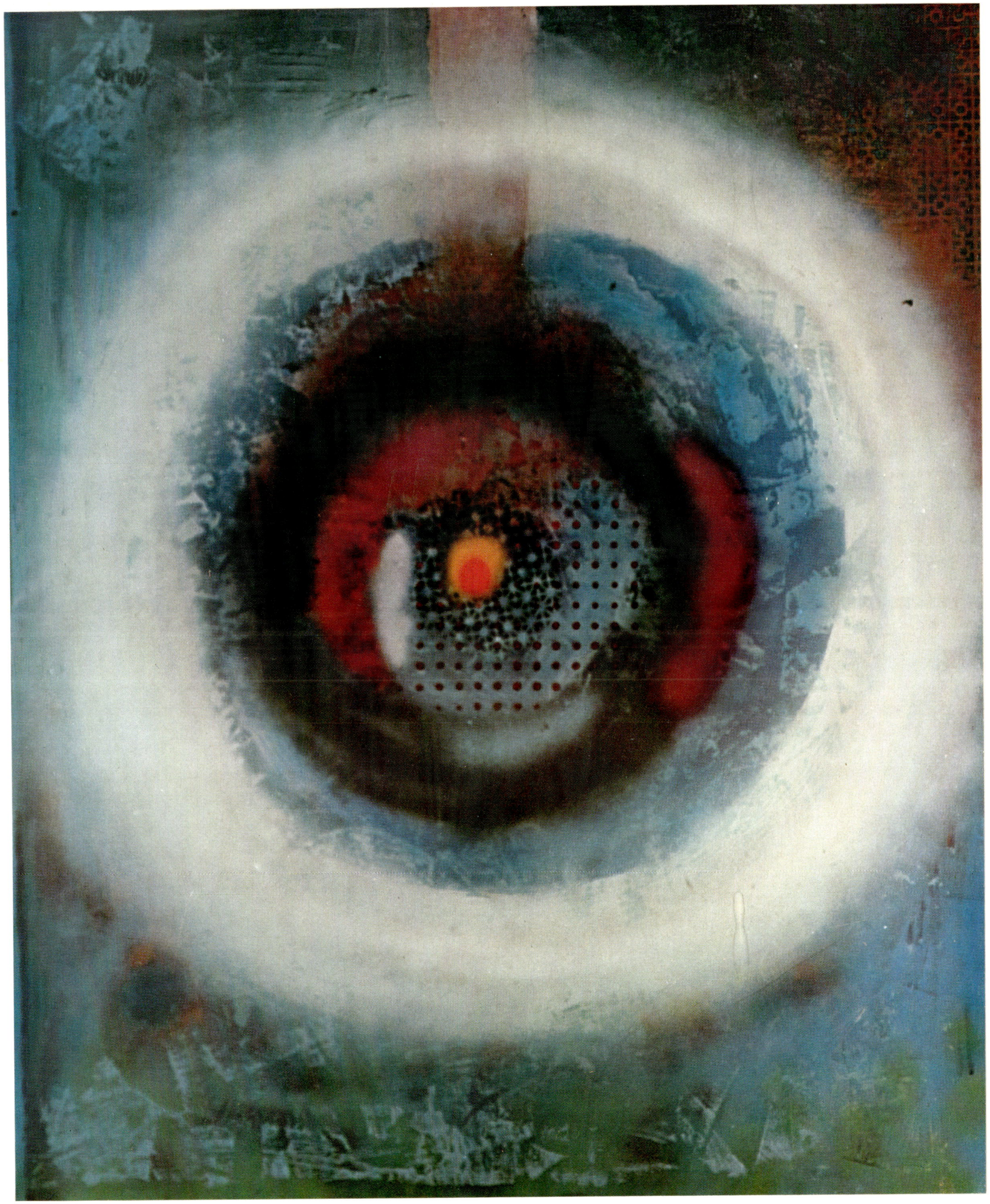

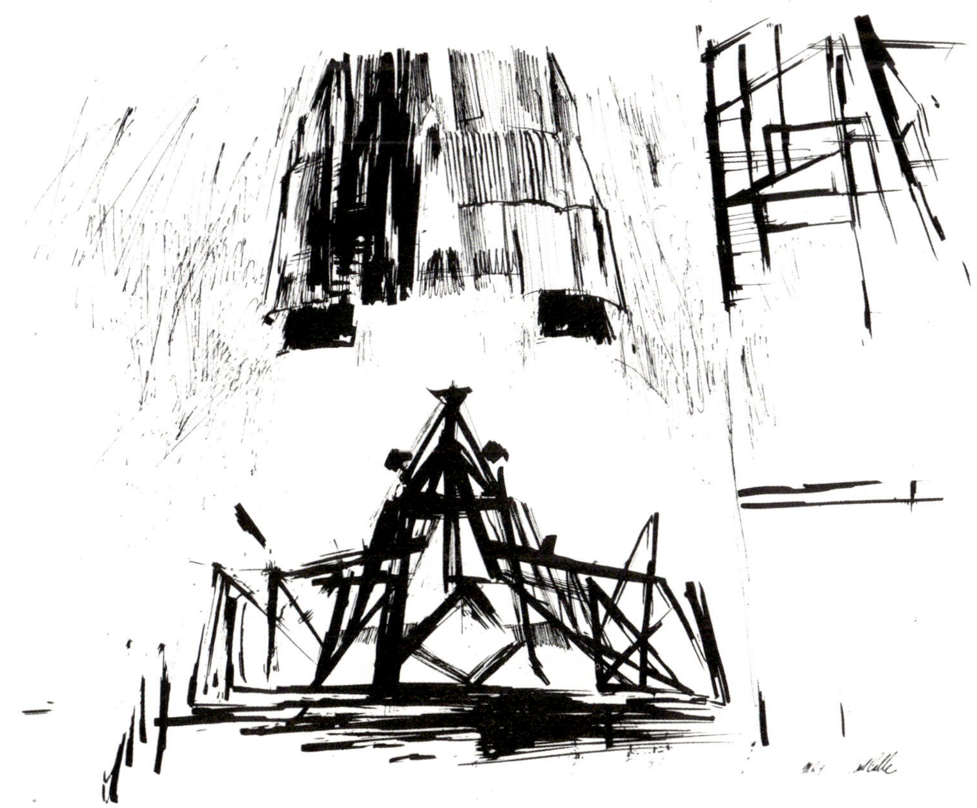

33. *Paul Calle*. BLASTOFF. Pen and ink on paper, 22×30″

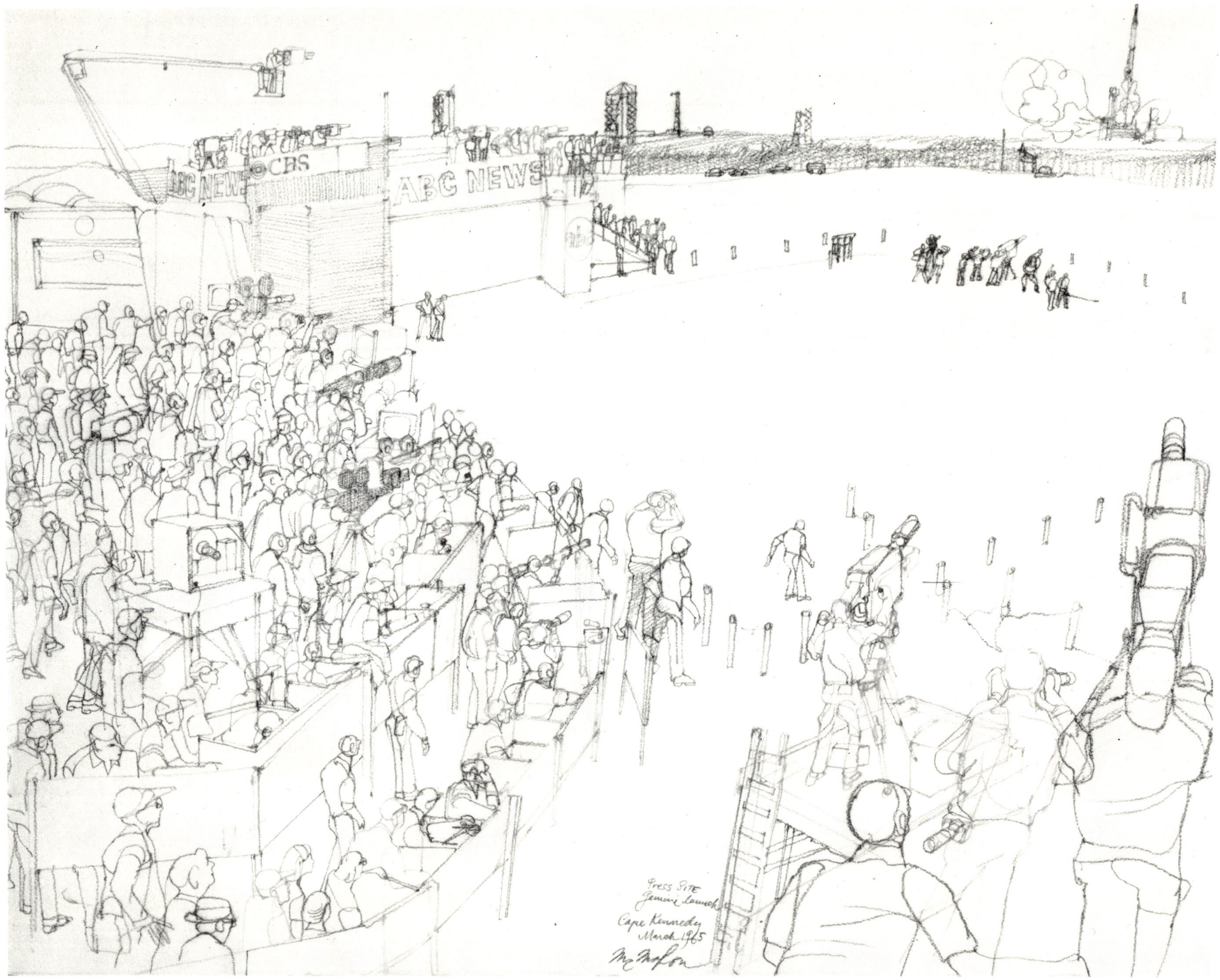

34.
Franklin McMahon
PRESS SITE
Pencil on paper, 22×30″

The press, television, and photographers during the earlier launches were corralled into an enclosure equipped with a vast number of telephone lines. Individual companies built temporary command posts and viewing platforms. Public-address systems and closed-circuit television gave them second-by-second information as the countdown proceeded.

35.
James Wyeth
GEMINI LAUNCH PAD
Watercolor on paper, 27×34″

Those responsible for the launch during earlier programs huddled in beehive-shaped fortresses a few hundred yards from the pad. Protected by thick layers of concrete in case of accidental explosions, the crew watched through periscopes and closed-circuit television. In Wyeth's painting the commonplace and the wonderful new world of technology coexist by necessity. Beside the domed control center, a workman has propped his bicycle, and weeds grow among the concrete slabs.

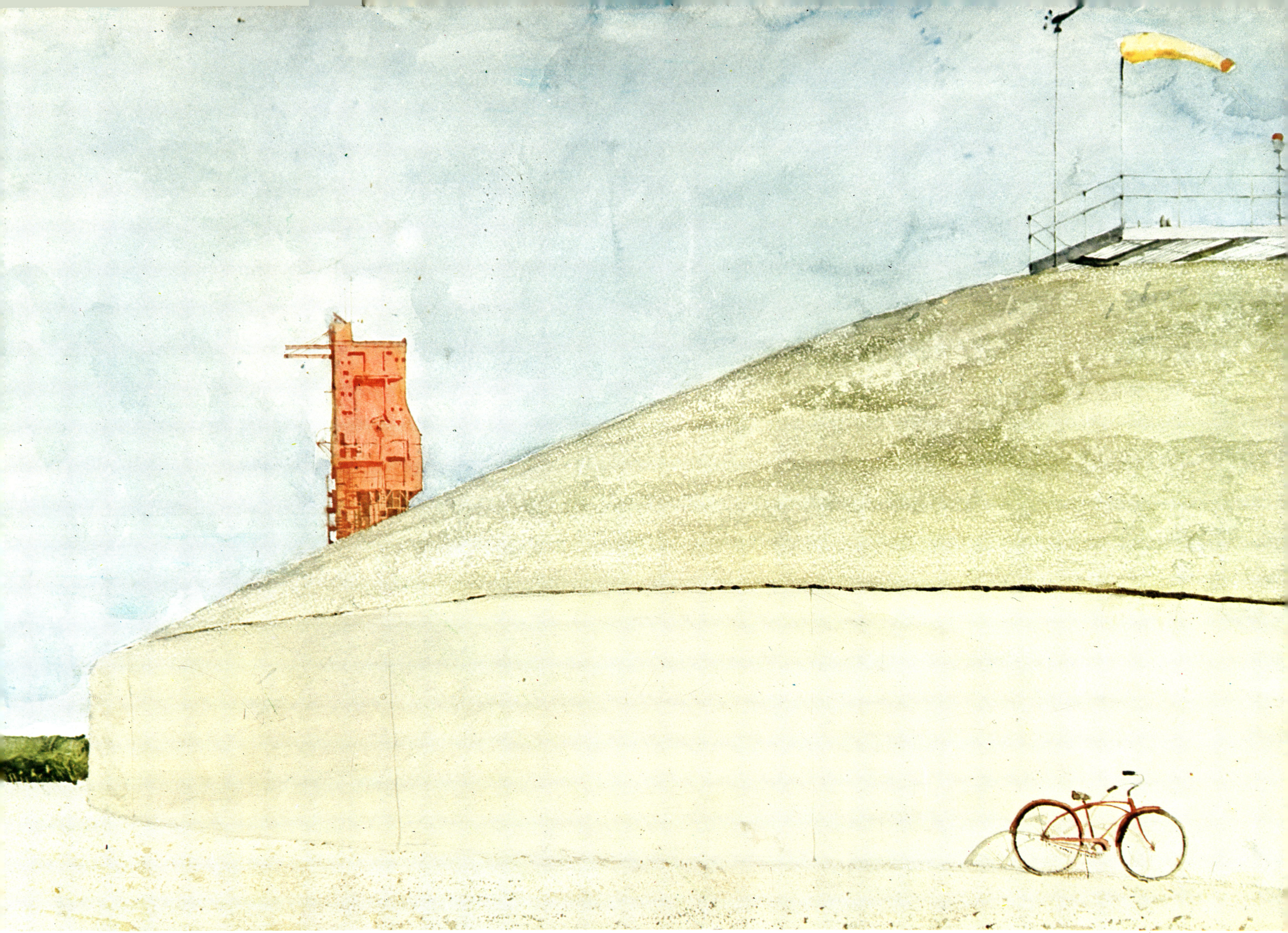

36.
Mitchell Jamieson
INSIDE BLOCKHOUSE
Ink on acetate, 7×9″

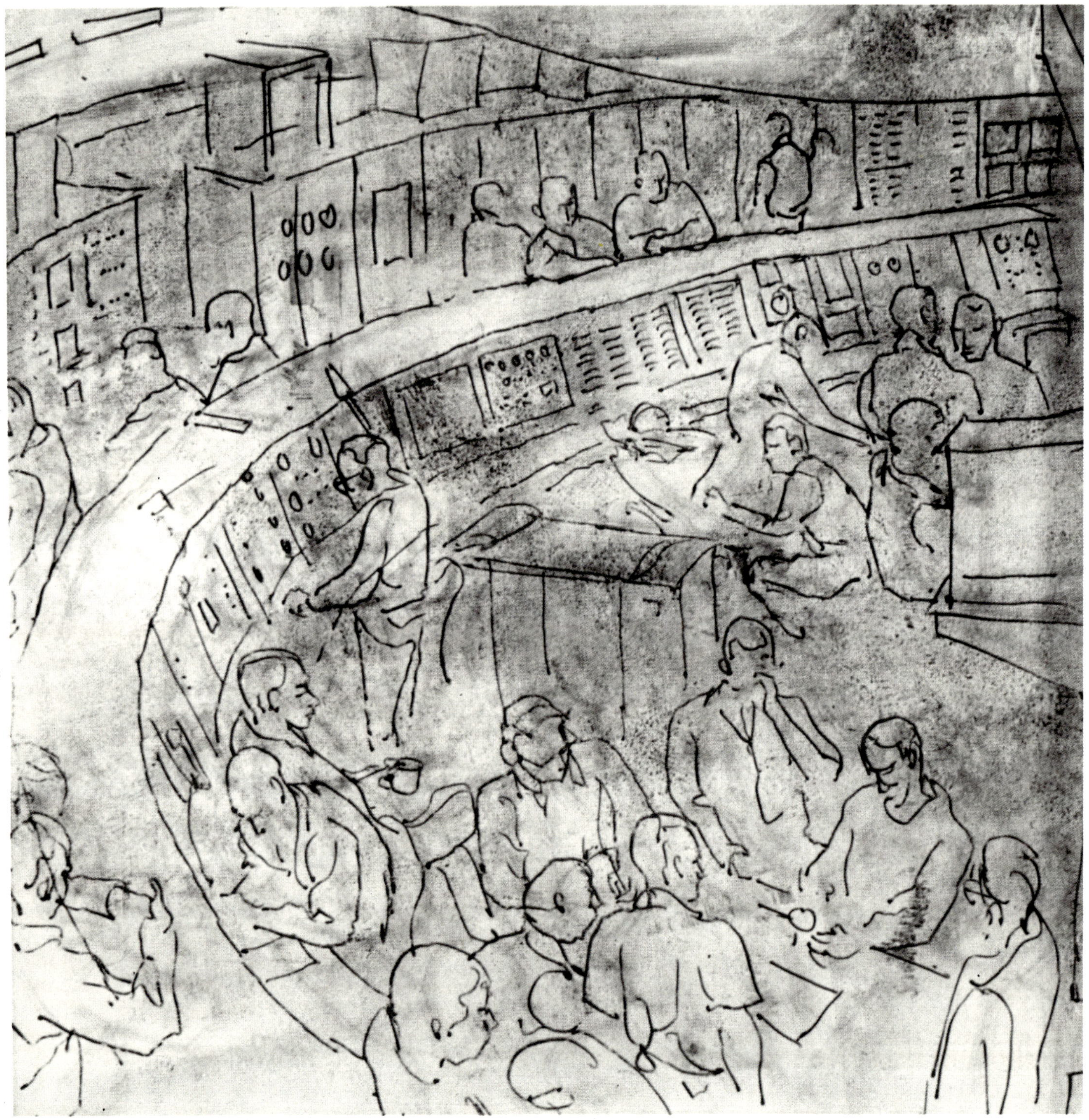

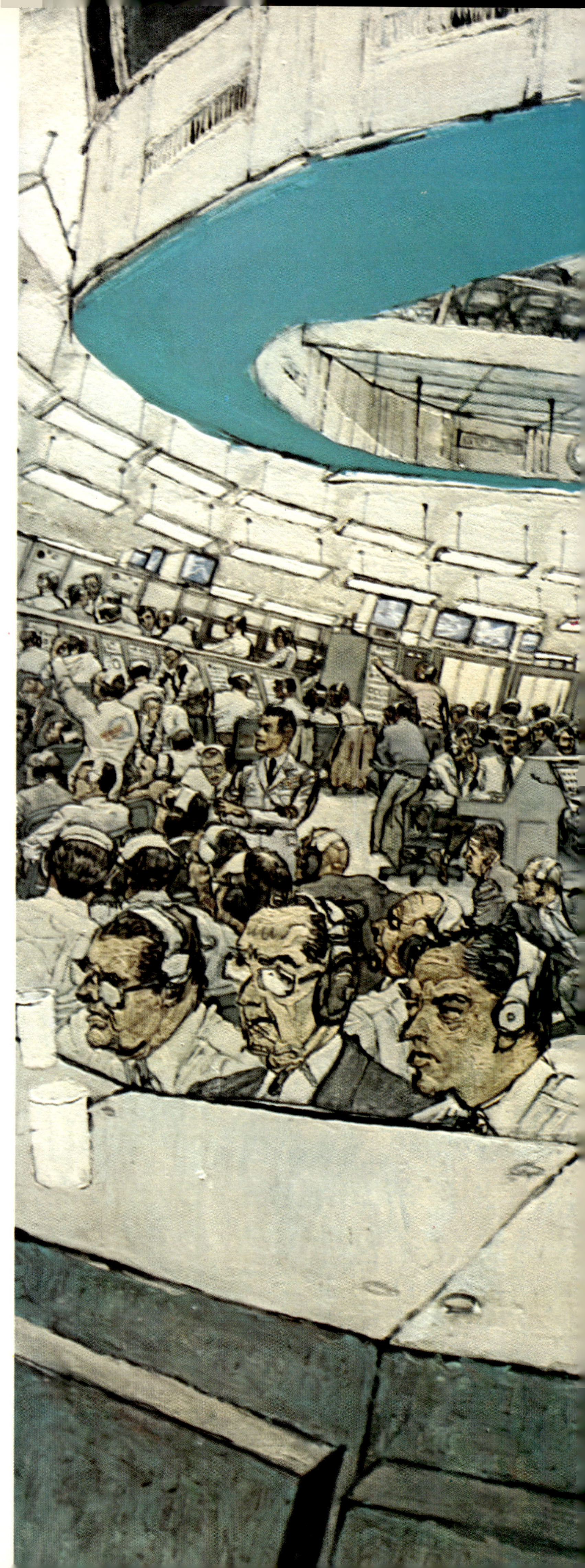

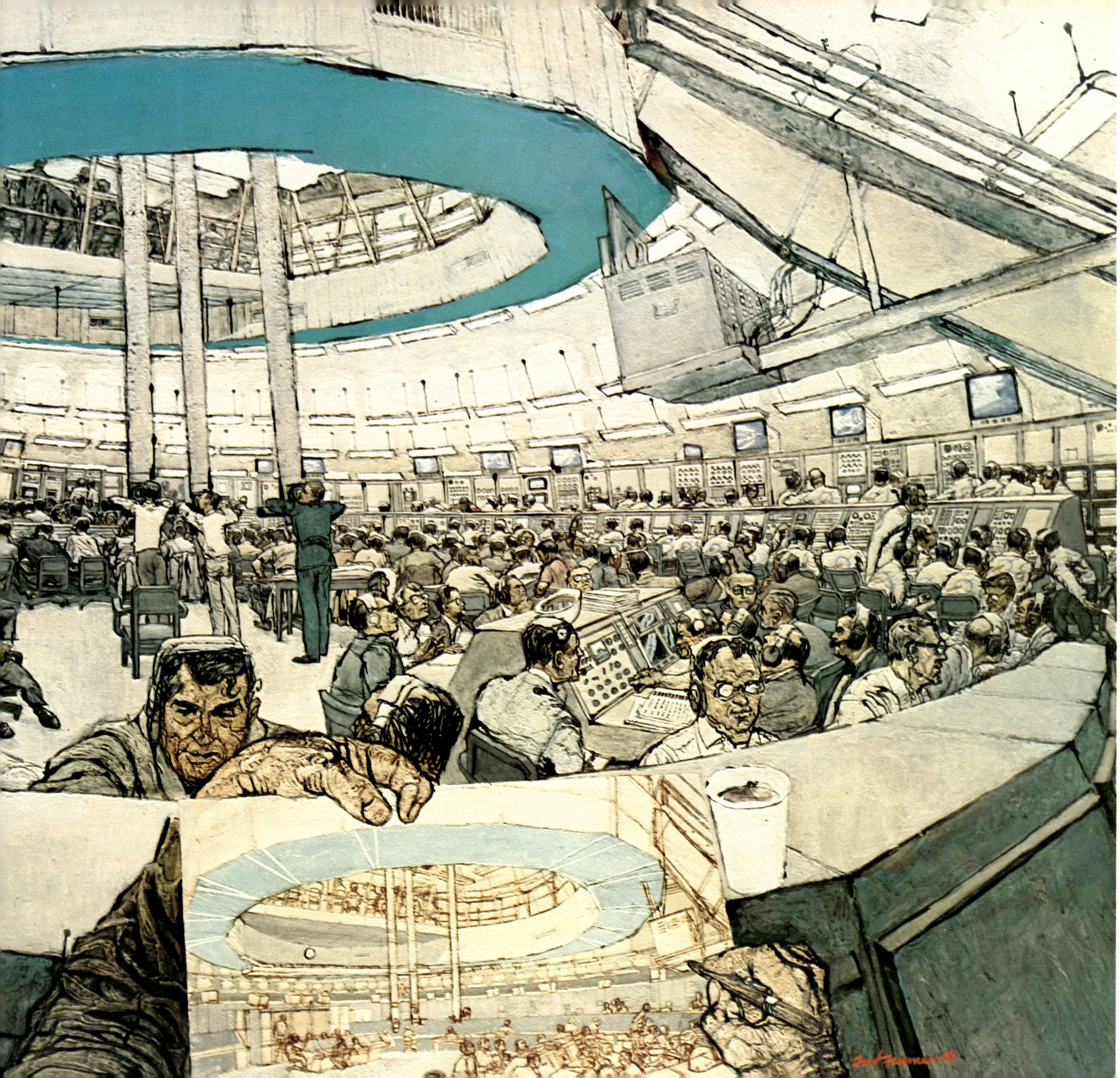

37.
Fred Freeman
SATURN BLOCKHOUSE
Acrylic on canvas, 3×4′

Under twelve feet of concrete, the launch team huddles like a family of troglodytes. Each specialist watches his console equipped with a closed-circuit television or a ticker tape of facts and figures. In the center, three crewmen watch through periscopes. The man leaning on the parapet watches the artist at work, and in the visitors' gallery near the ceiling VIPs watch the watchers.

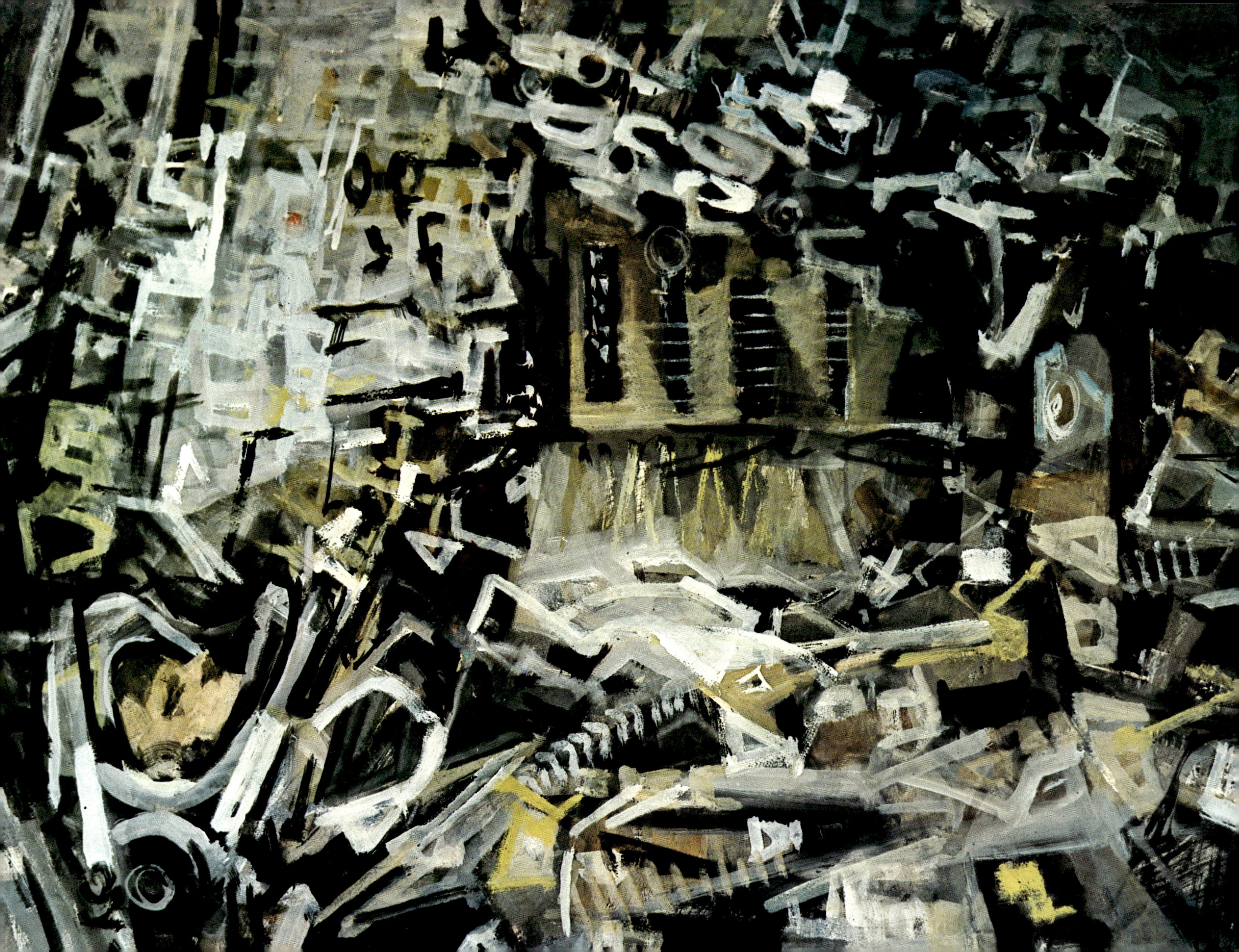

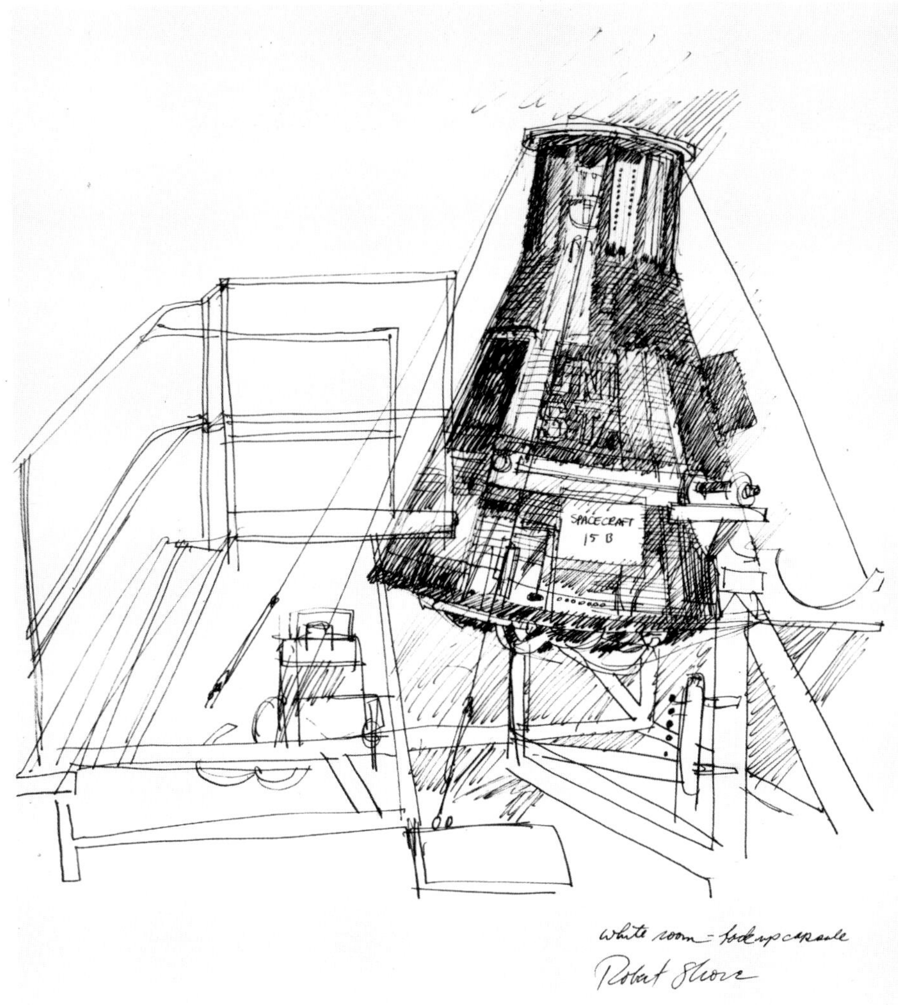

39. *Robert Shore.* BACK-UP SPACECRAFT. Pen and ink on paper, 14×11″

38.
Lamar Dodd
MAX Q
Gouache on paper, 27×36″

The Mercury spacecraft, one is apt to forget, was a tiny, fragile, corrugated container into which the astronaut was strapped. Surrounding him was a maze of controls, switches, instrument panels, and warning lights, hanging like stalactites from the roof of his paper-thin vehicle. Spacecraft, in the rush of technological advances, are becoming much more roomy, and soon it will be hard to imagine the feelings of utter isolation that the first astronauts must have felt in their phone-booth-sized projectiles.

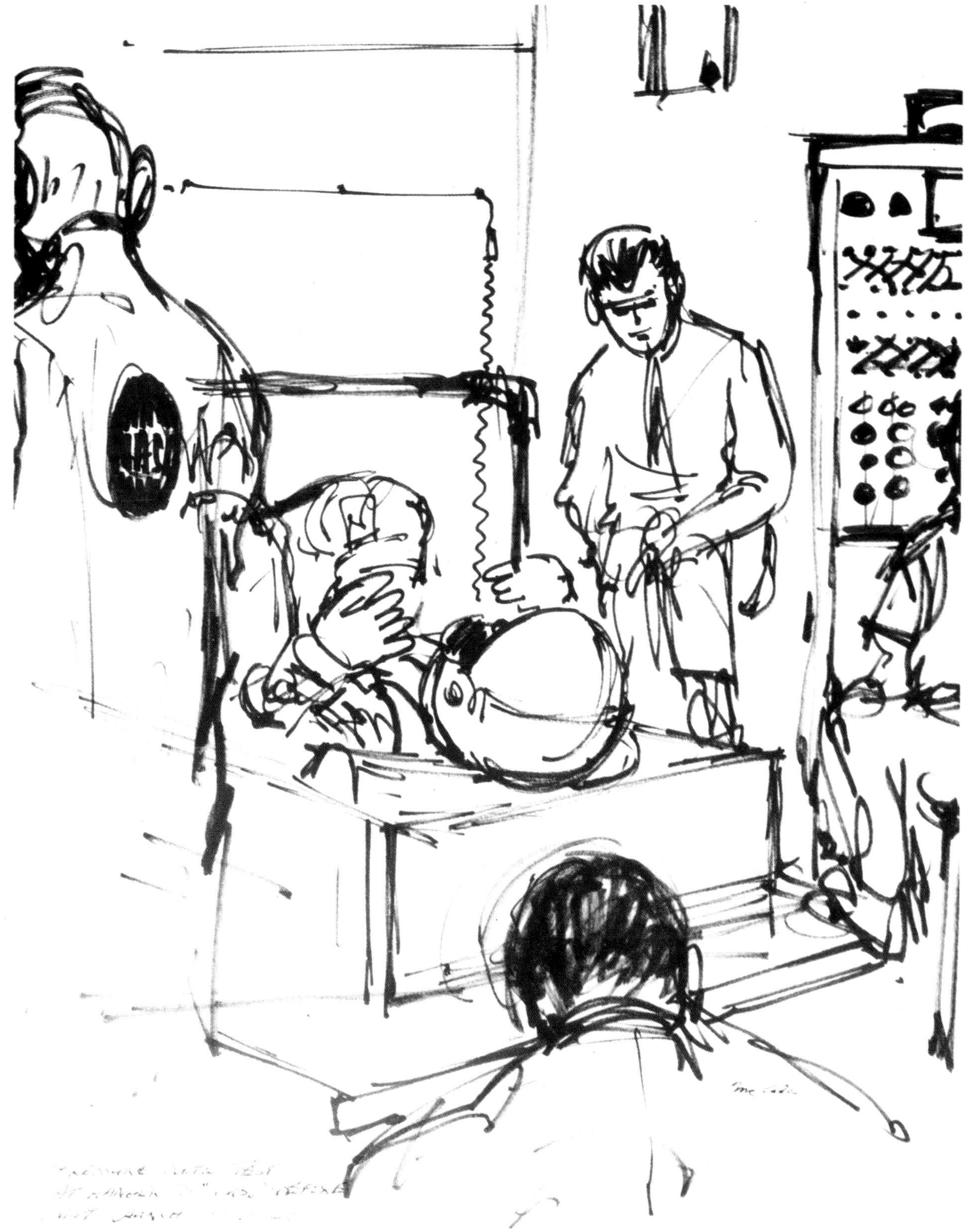

40. *Robert McCall.* MERCURY SUIT TEST. Felt pen on paper, 17×14″

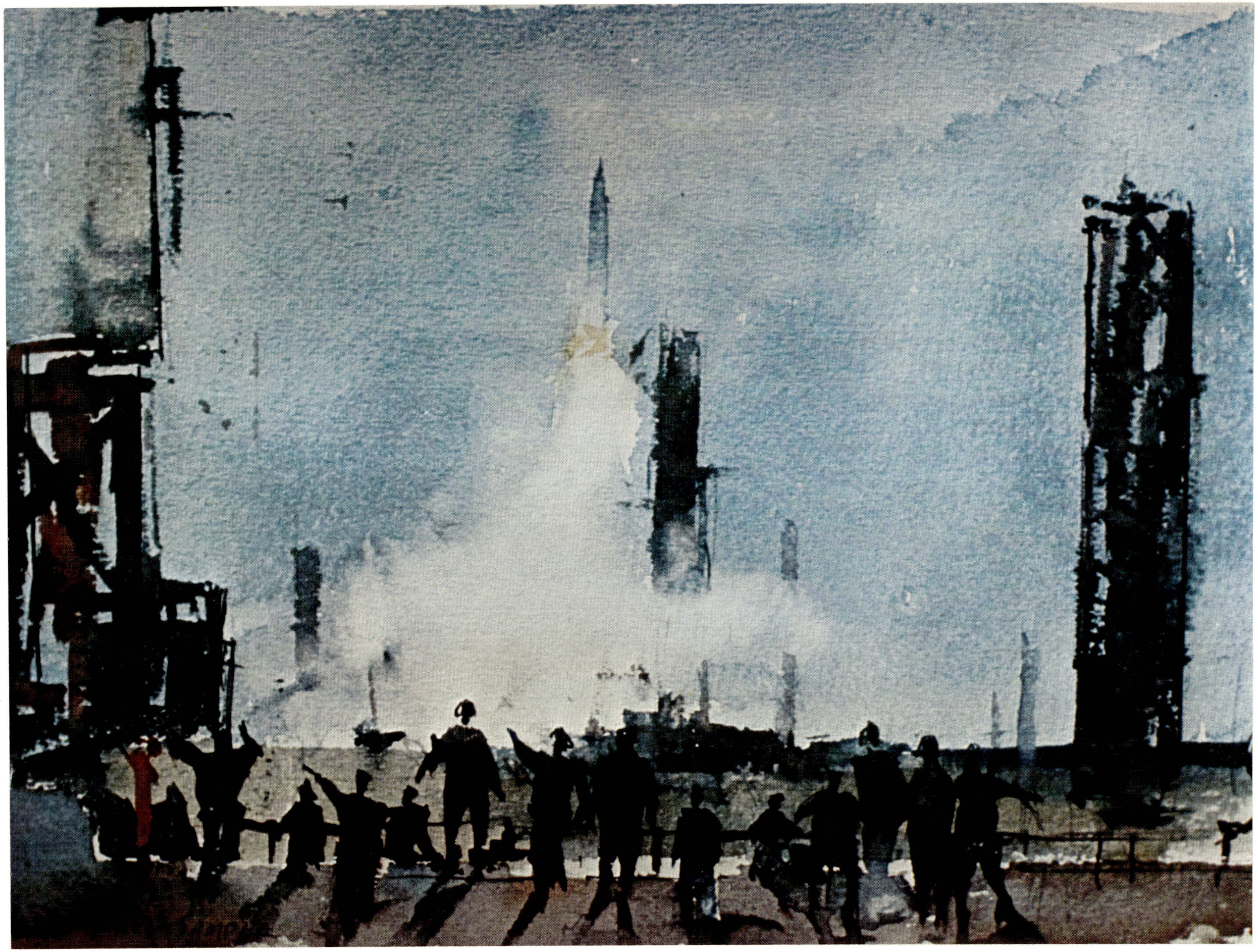

42.
Fred Freeman
SATURN IB
Watercolor on paper, 18 × 12″

41. *Paul Sample*. NEIGHBORS. Watercolor on paper, 6½ × 10″

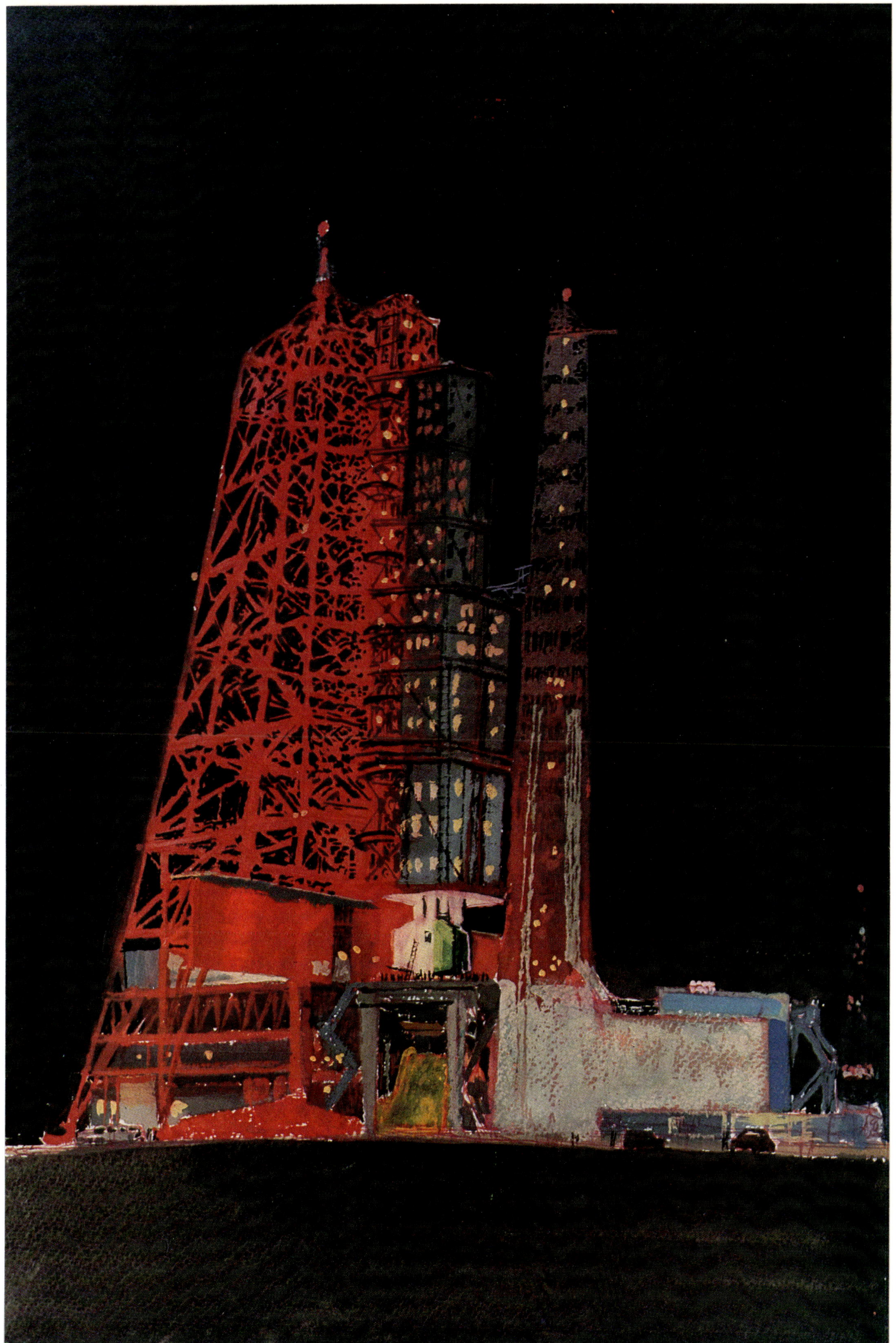

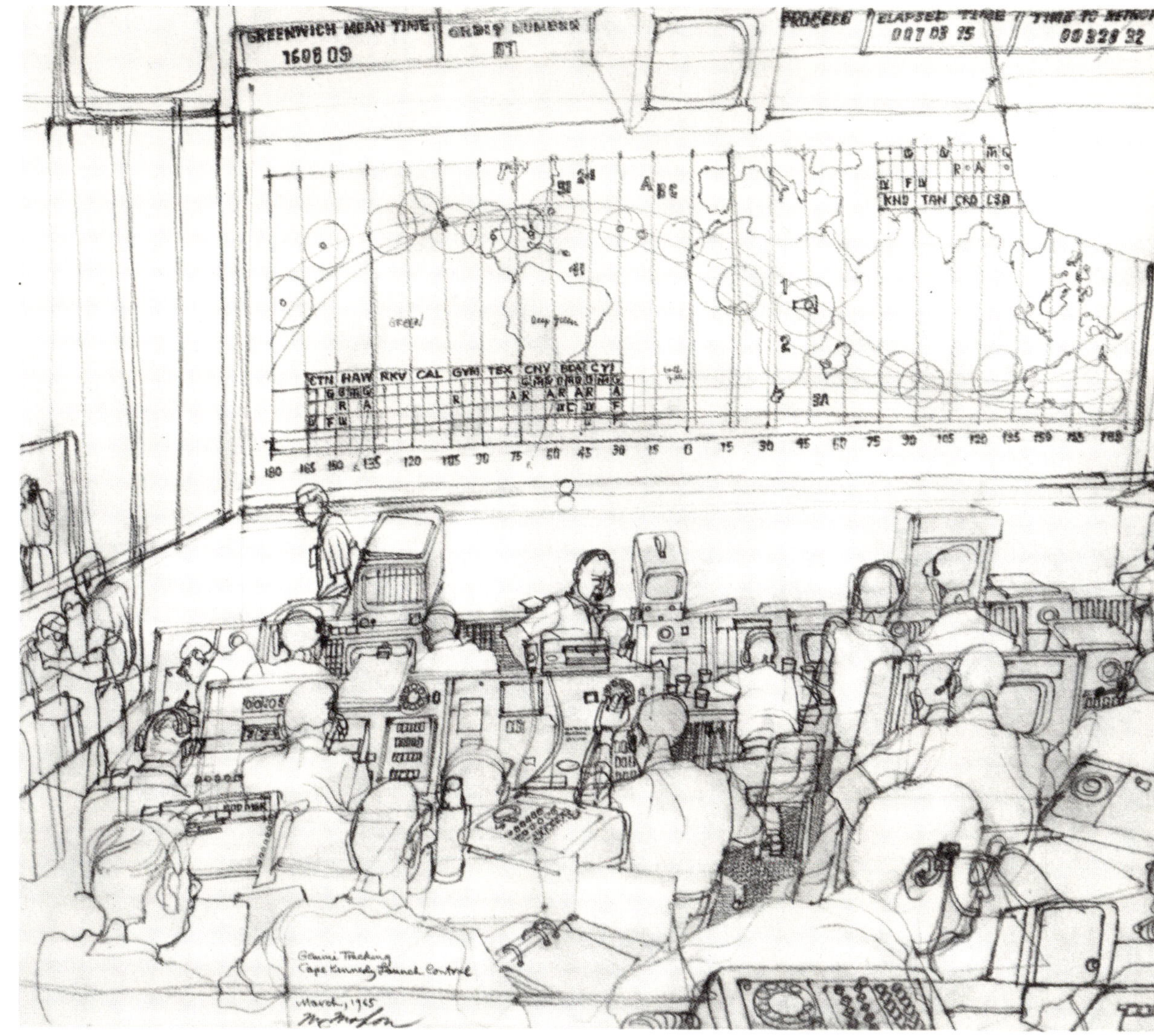

43.
Franklin McMahon
ORBIT NUMBER 1
Pencil on paper, 22×30″

A few seconds after the rocket is clear of its pad, control passes to the Mission Control Center, which, during the Mercury program, was in an unassuming building on Cape Kennedy. Here, amid thermos jugs of coffee, the position of the spacecraft as it orbited the earth was plotted, the flight surgeons watched the pulses of their speeding patients, and the voices from outer space were relayed to the appropriate adviser or confidant.

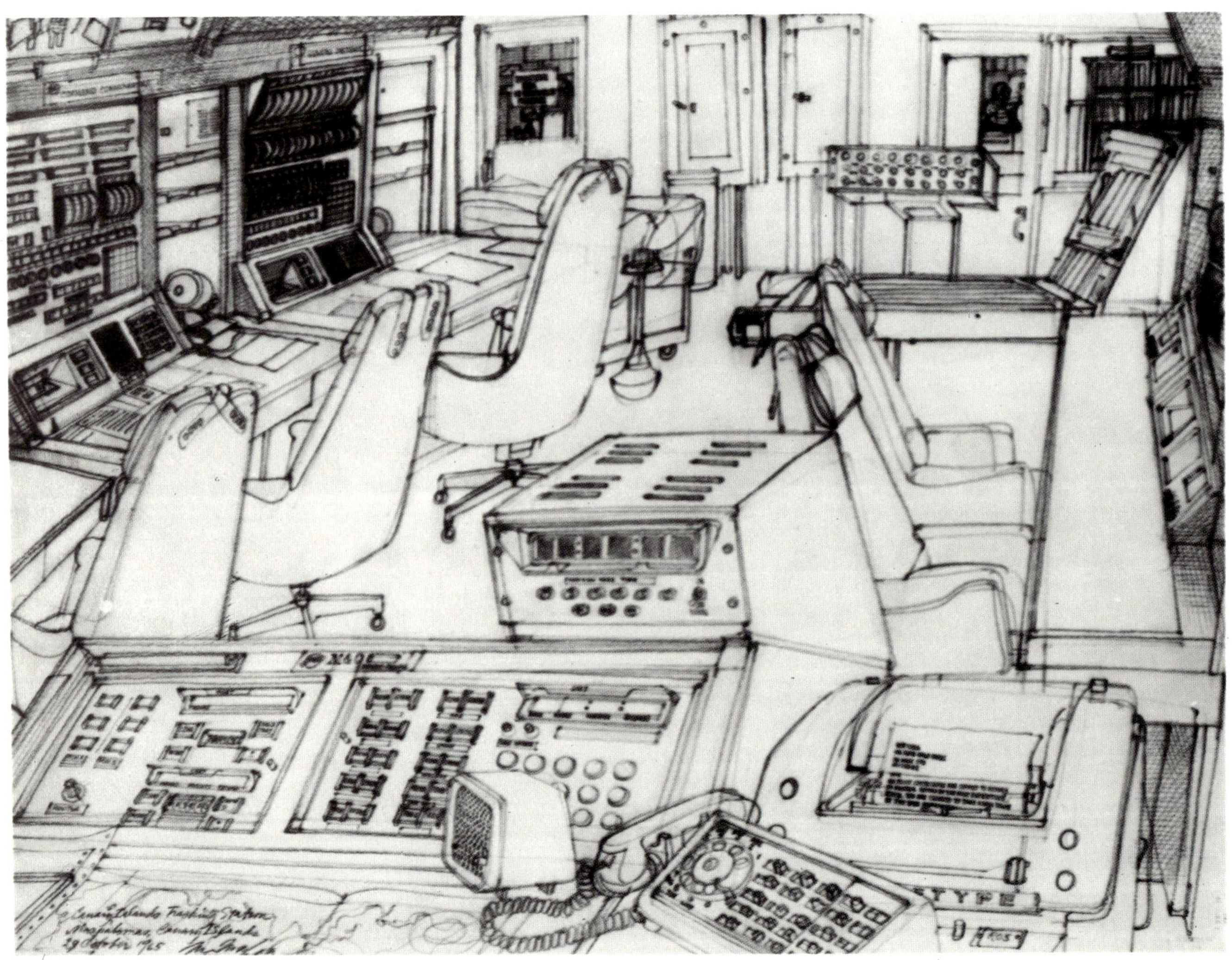

44. *Franklin McMahon.* CANARY ISLAND TRACKING STATION. Pencil on paper, 22×30″

45.
Lamar Dodd
LISTENING
Pen and ink on paper, 20×12″

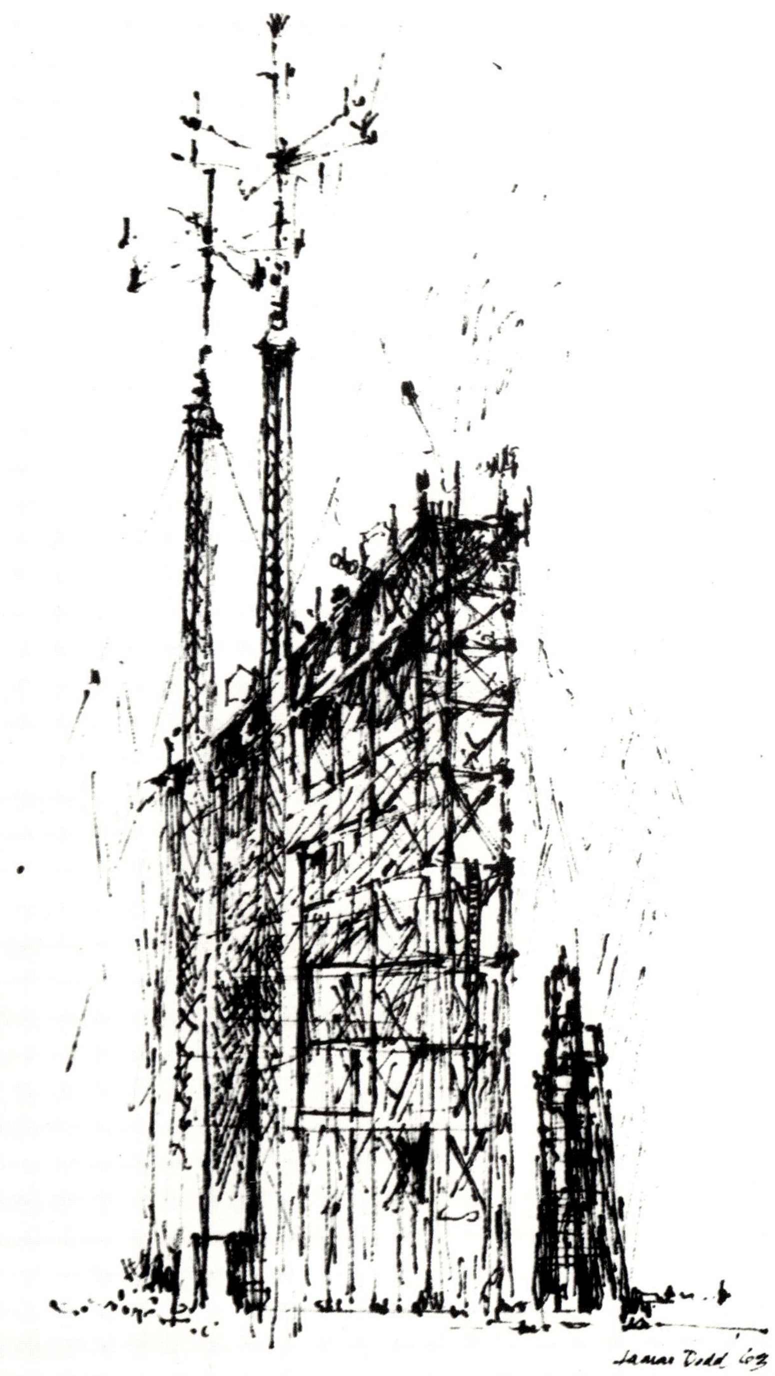

46.
Alfred McAdams
RANGE SAFETY
Watercolor on paper, 17×22″

One of the crucial moments in a launch comes right after the rocket leaves the pad, when a decision must be made as to whether everything is going according to plan. If it is not, the rocket must be destroyed in flight. Commands for the rocket to blow itself up are given over a group of antennae that among artists have become known as "The Four Horsemen." No sculptor could imagine or construct more striking and original designs.

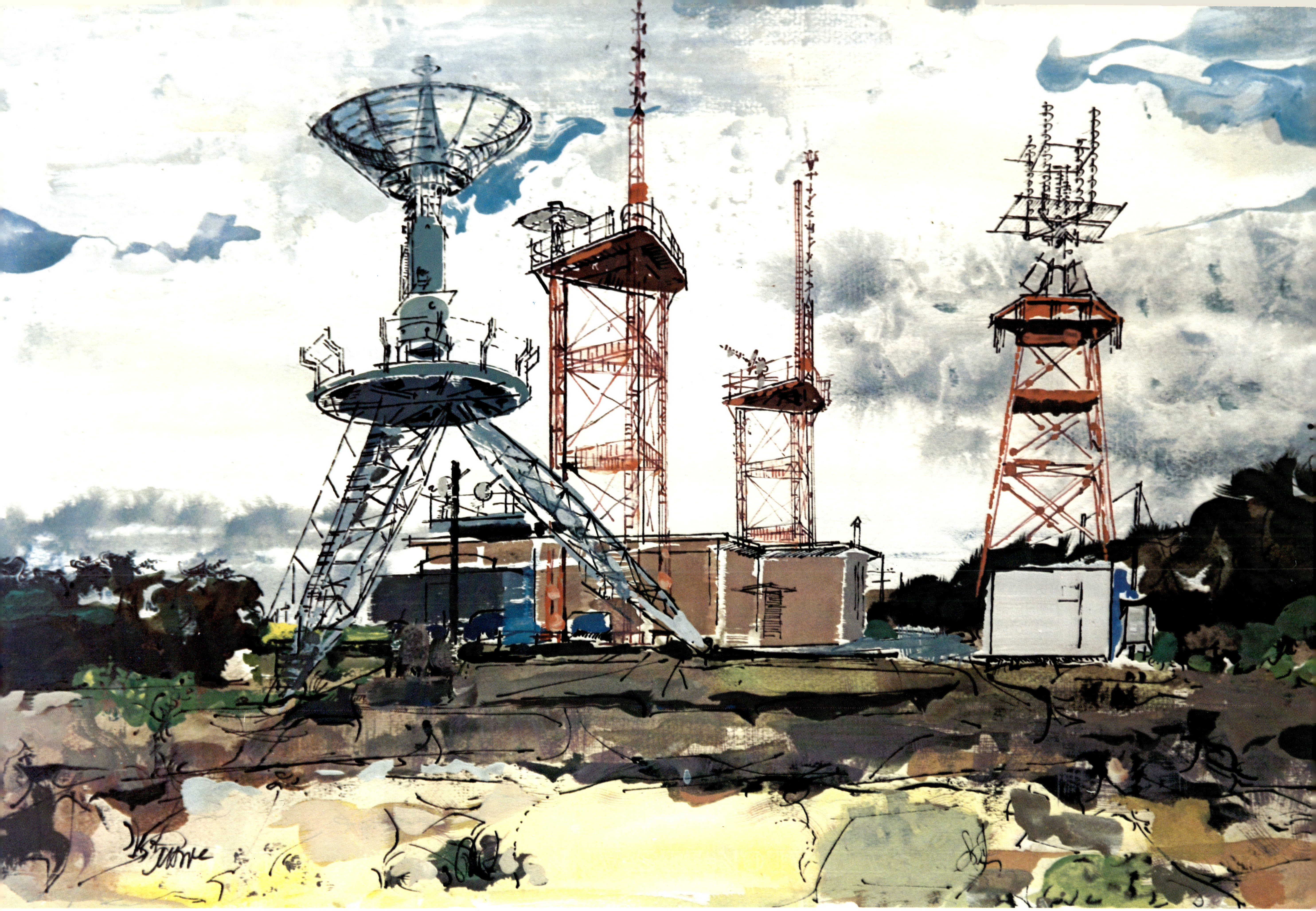

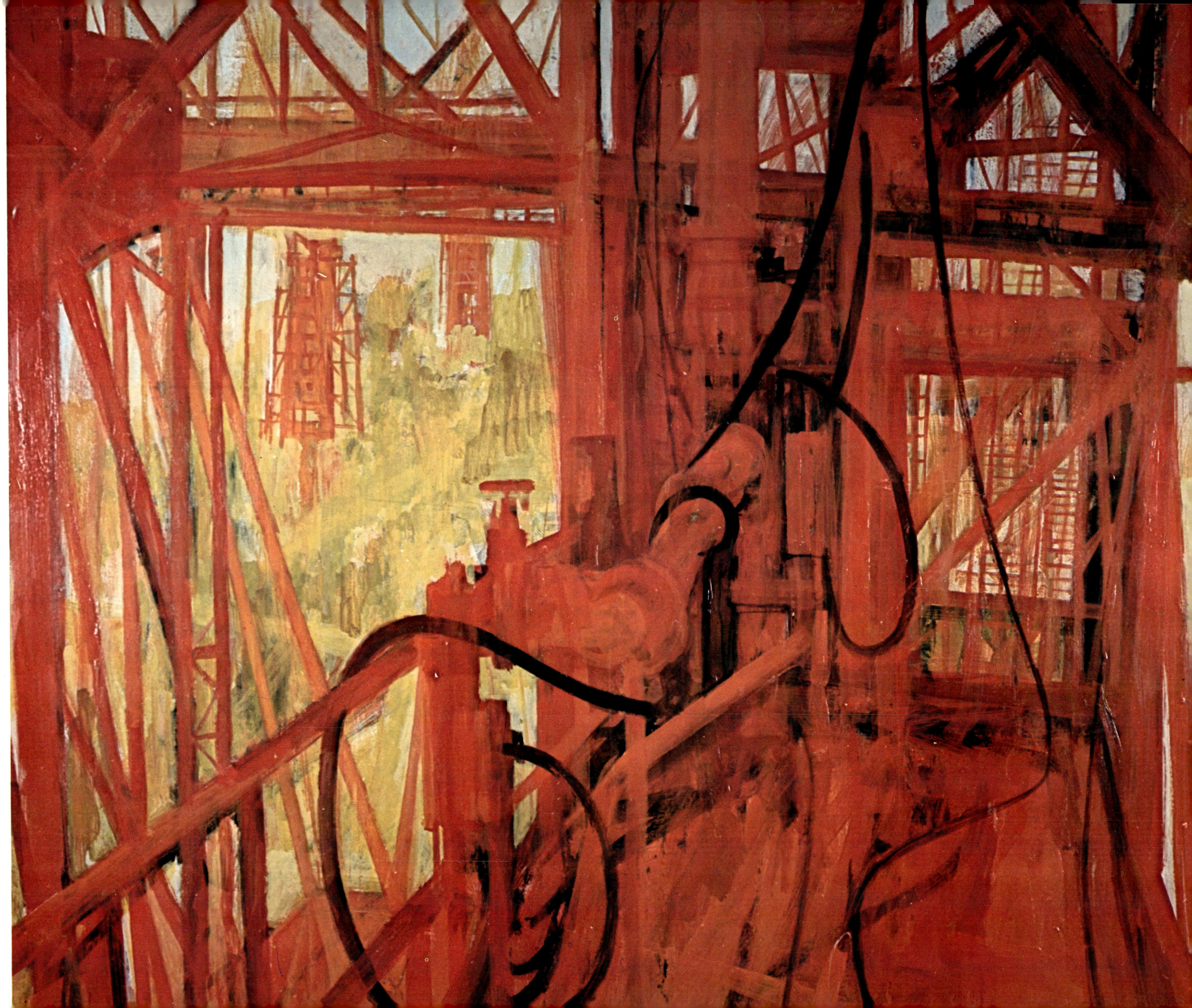

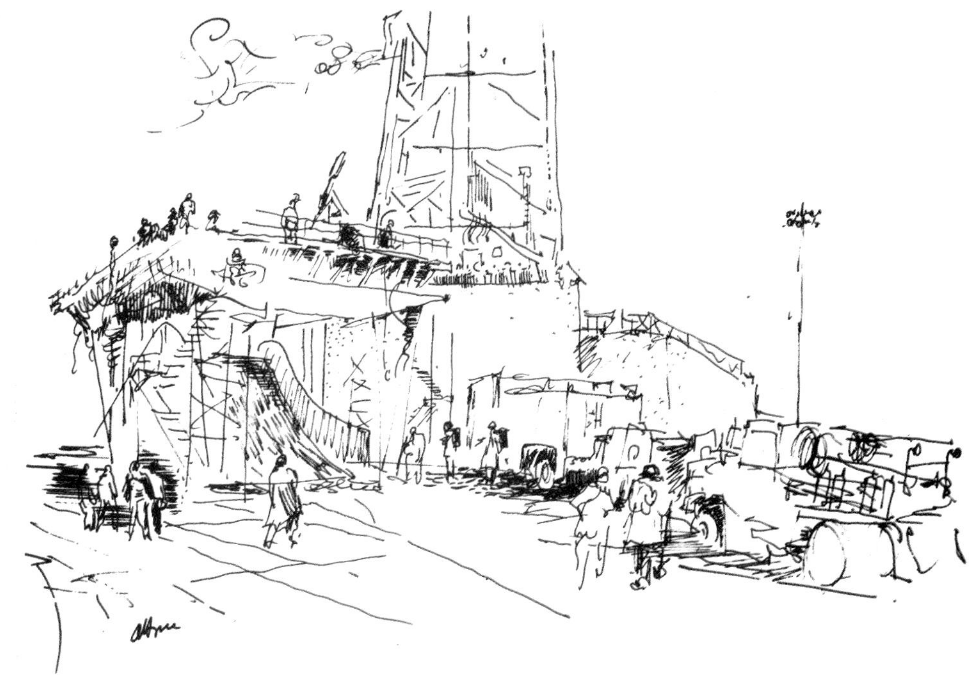

49.
Alfred McAdams
FIRE DAMAGE
Ink on paper, 15 × 22″

47.
Robert Shore. FROM THE INSIDE
Oil on Masonite, 42 × 48″

Several artists have become intrigued by the maze of girders, pipes, catwalks, ladders, and wires that crisscross the interior of a gantry. This view is from one of the higher platforms of an early Saturn I gantry. Before the Vehicle Assembly Building was constructed, many of the technicians worked in the open, perched on these spindly platforms.

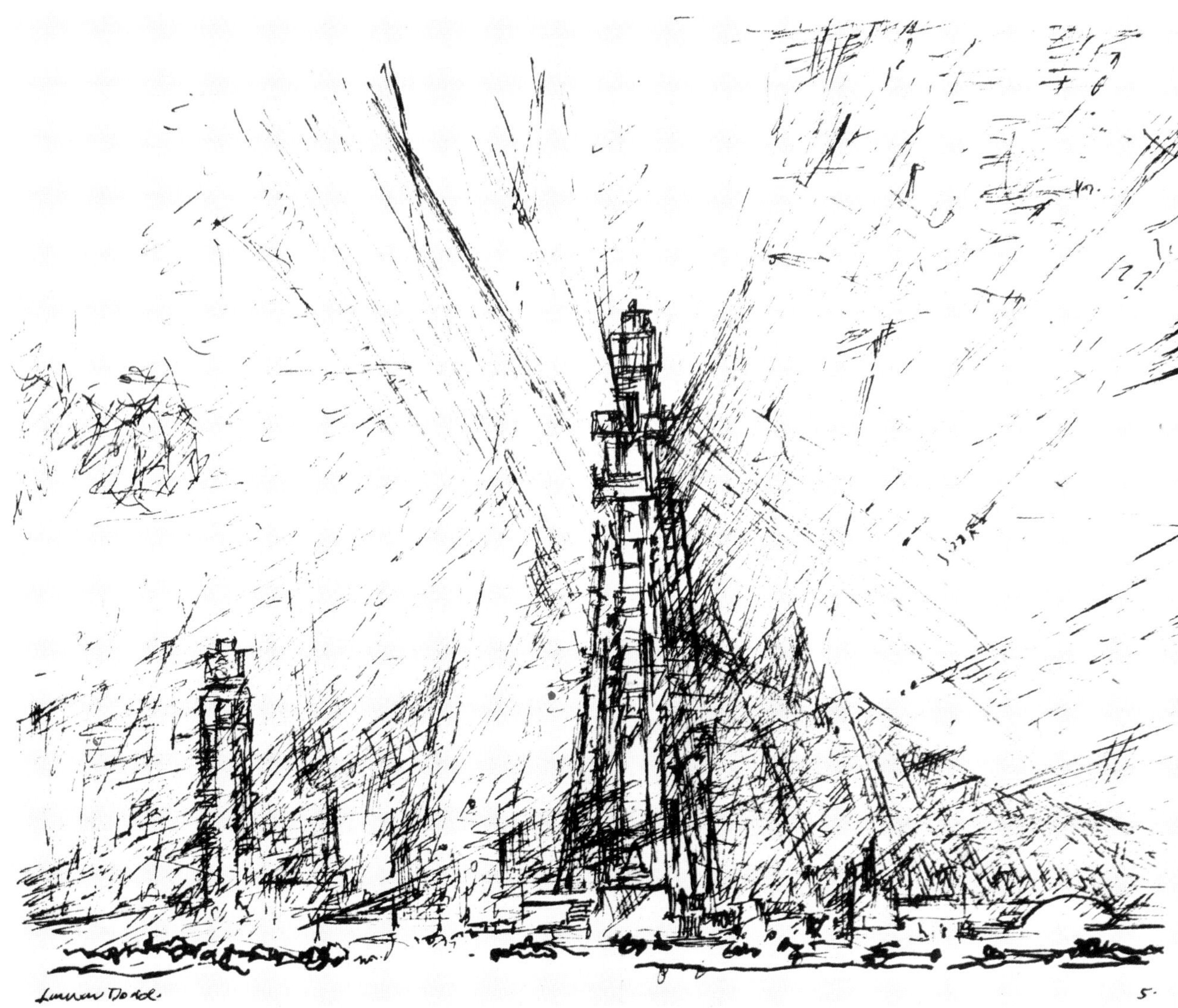

48.
Lamar Dodd
LAUNCH PAD
Pen and ink on paper, 11 × 14″

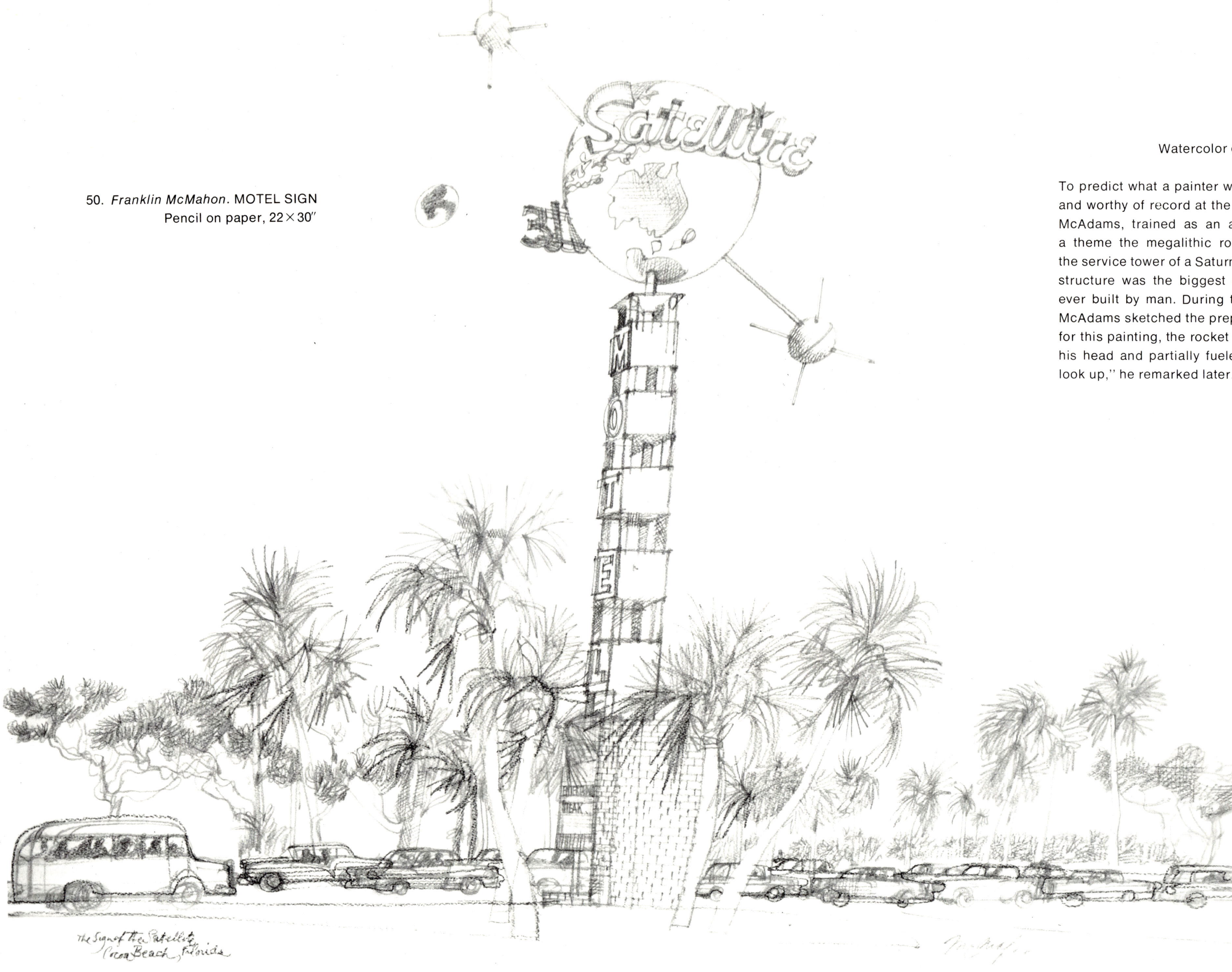

50. *Franklin McMahon*. MOTEL SIGN
Pencil on paper, 22×30″

51.
Alfred McAdams
ROLLBACK
Watercolor on paper, 21×27″

To predict what a painter will find interesting and worthy of record at the Cape is not easy. McAdams, trained as an architect, took as a theme the megalithic rolling stock under the service tower of a Saturn I. At this time the structure was the biggest object on wheels ever built by man. During the evening when McAdams sketched the preparatory drawings for this painting, the rocket was in place over his head and partially fueled. "I tried not to look up," he remarked later.

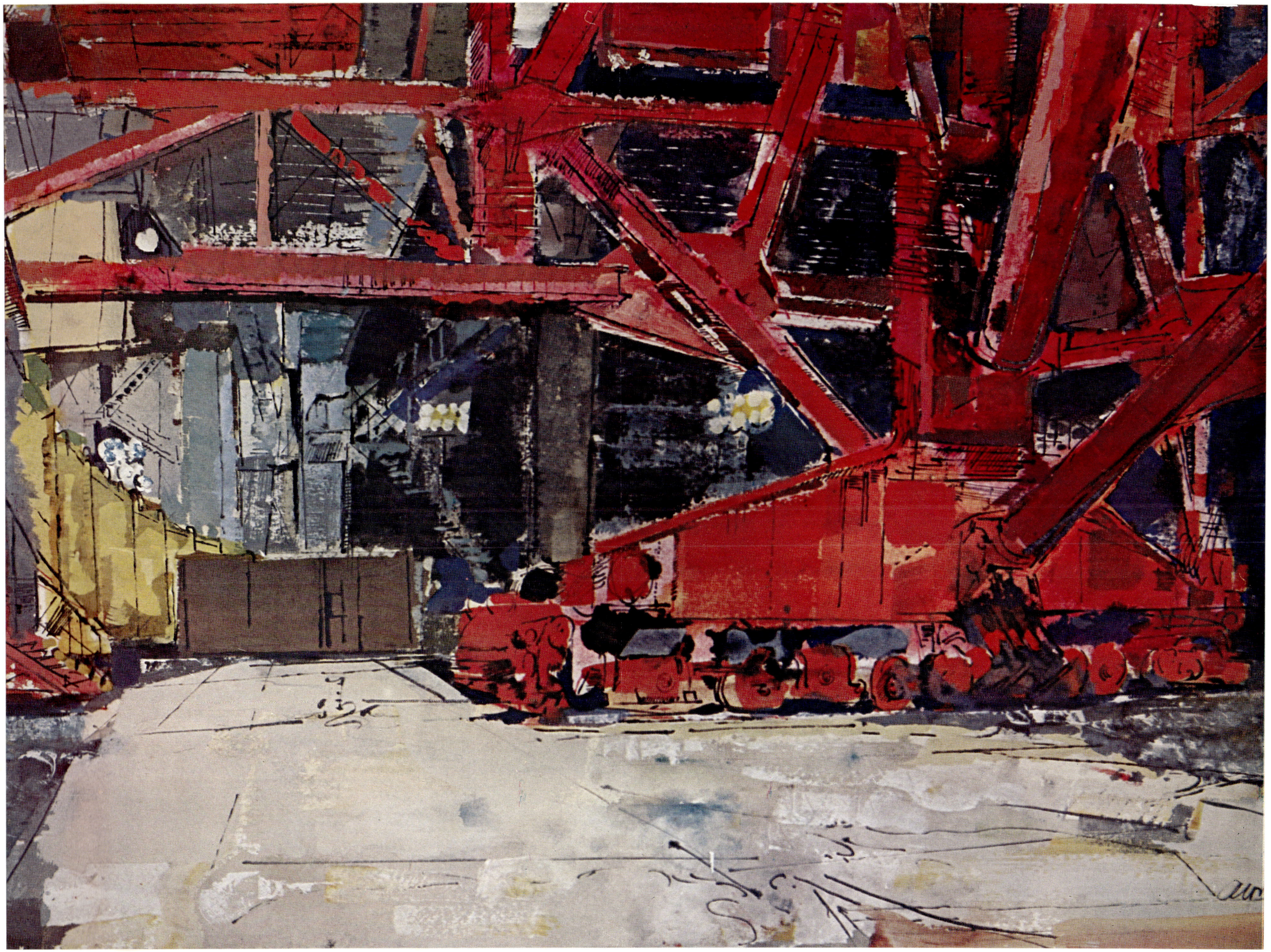

54.
Lamar Dodd
WHITE ROOM
Gouache on paper, 27×36″

The White Room for the Mercury spacecraft was a modest enclosure, the purpose of which was to prevent dust and other particles from lodging in the spacecraft. In weightless space, dust particles become dangerous saboteurs, ready to float into the eyes of an astronaut or the workings of an instrument. Technicians, therefore, working in this room wear lint-free garments, caps, and gloves, and the most heinous offense in this hushed environment is to drop a screw and lose it.

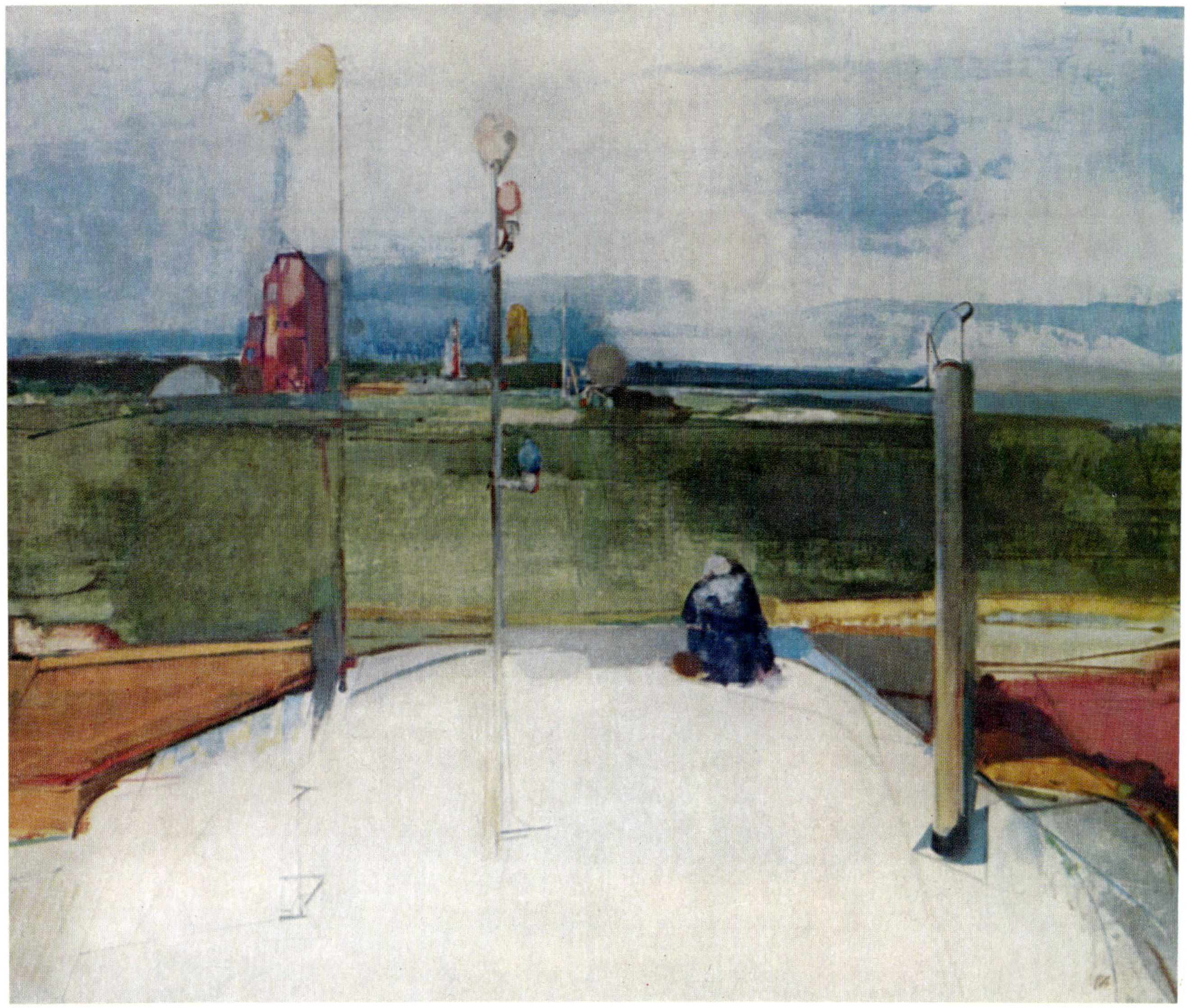

52. *Alfred McAdams.* CAPE KENNEDY 1964. Oil on canvas, 43×51″

53.
Robert McCall
ROLLBACK
Felt pen on paper, 18×12″

On top of the Mercury spacecraft there was a rocket mounted on a lattice support. In case of an accident during launch, this rocket could pull the spacecraft free and parachute it into the sea.

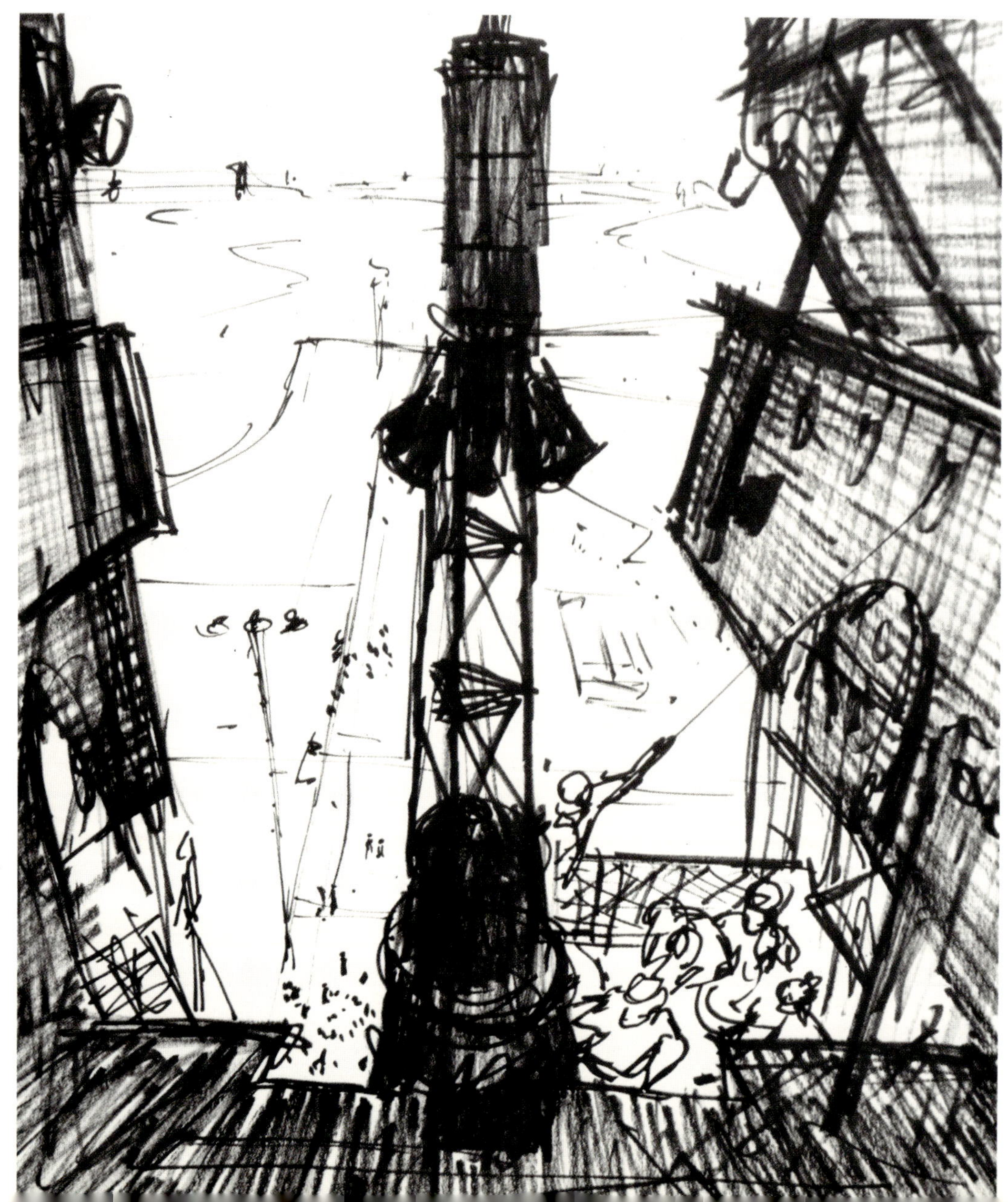

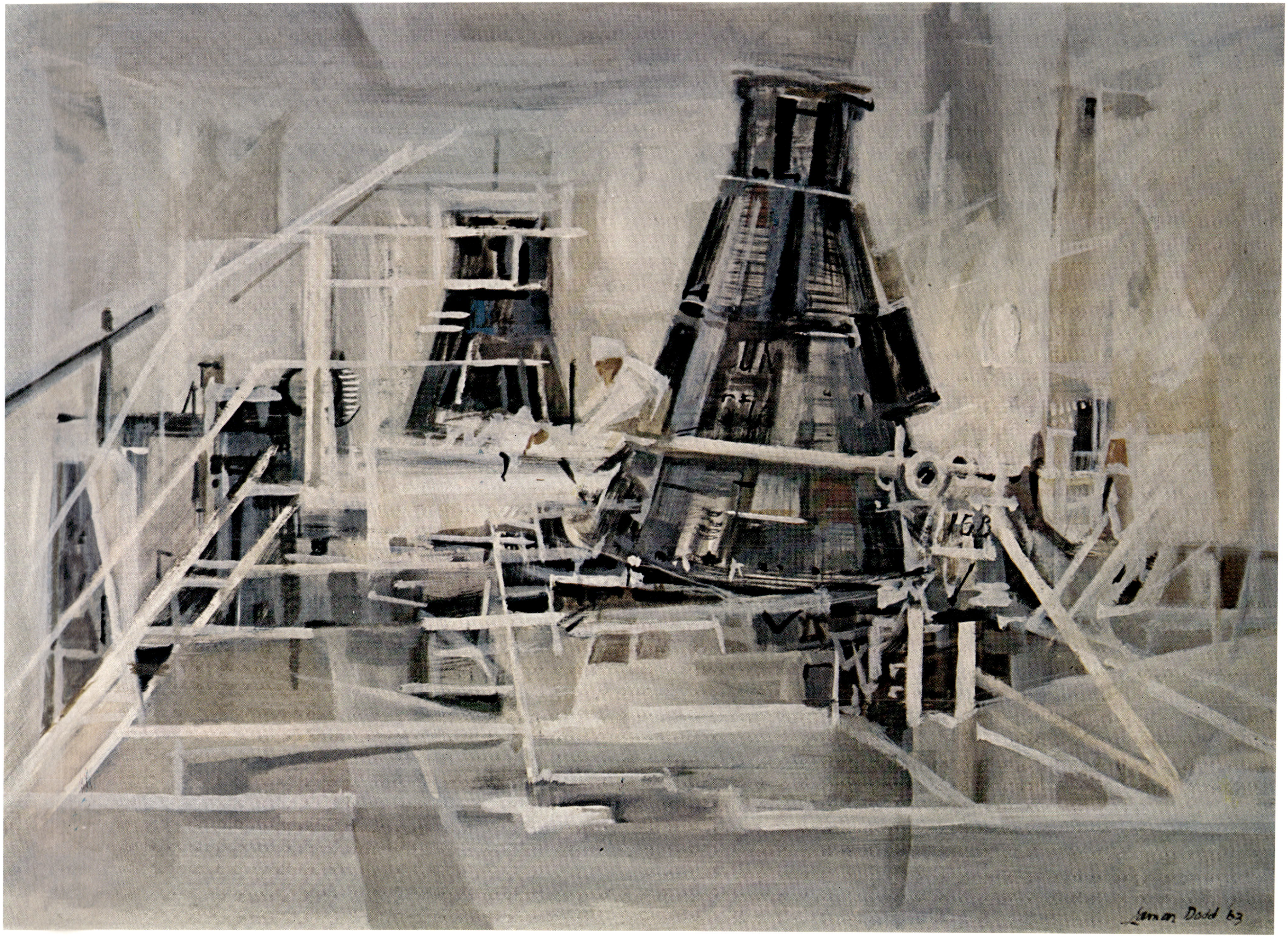
Lamar Dodd '63

55.
Fred Freeman
ROADSIDE VIEWING STAND
Ink and watercolor on paper, 11 × 17″

Most photographers and pressmen waited and watched from the official press site, where batteries of cameras, wonderful in their precision and weirdly complex in their designs, were trained on the distant pad. During early launches, some crews preferred a roadside location where a clearing in the underbrush permitted a good view.

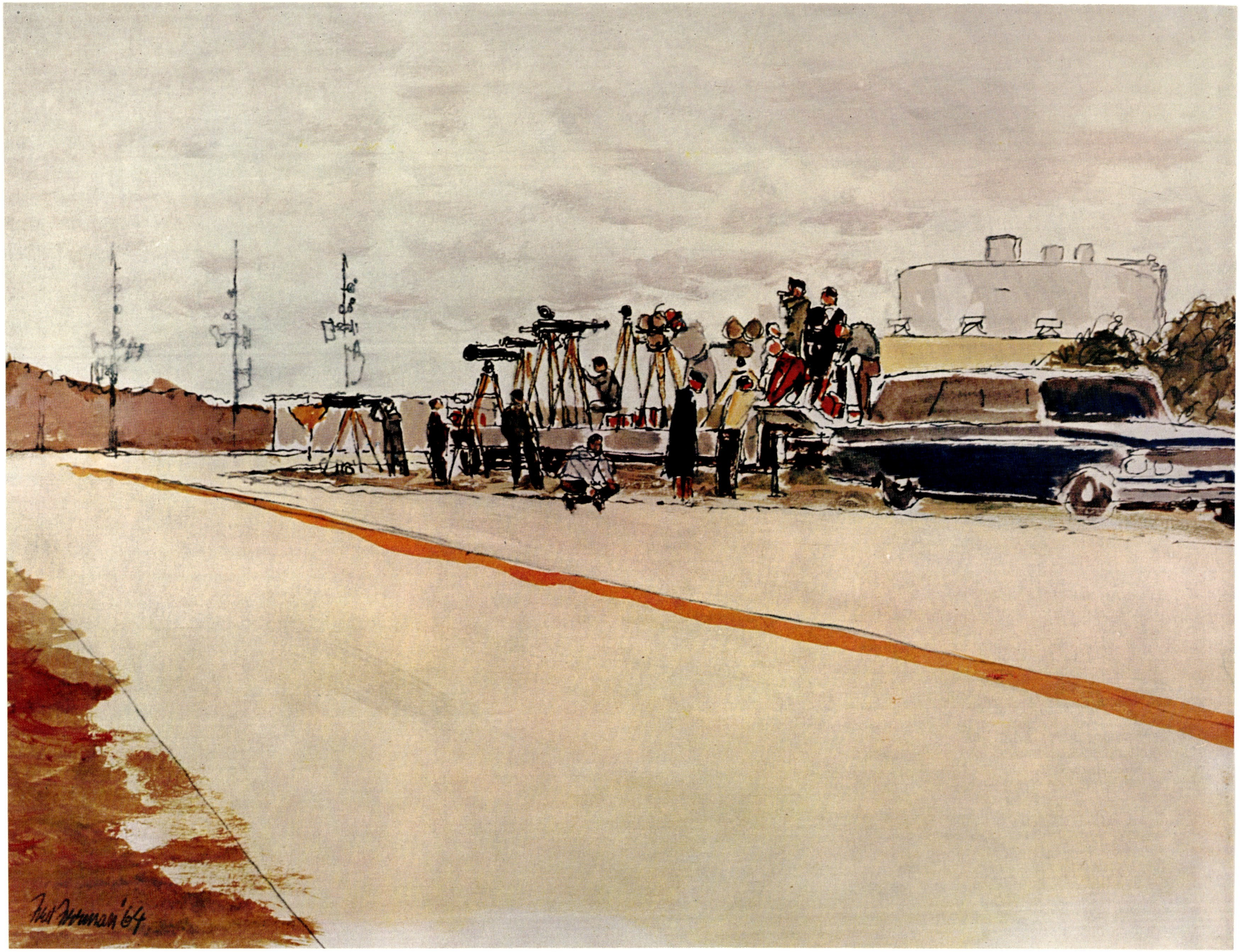

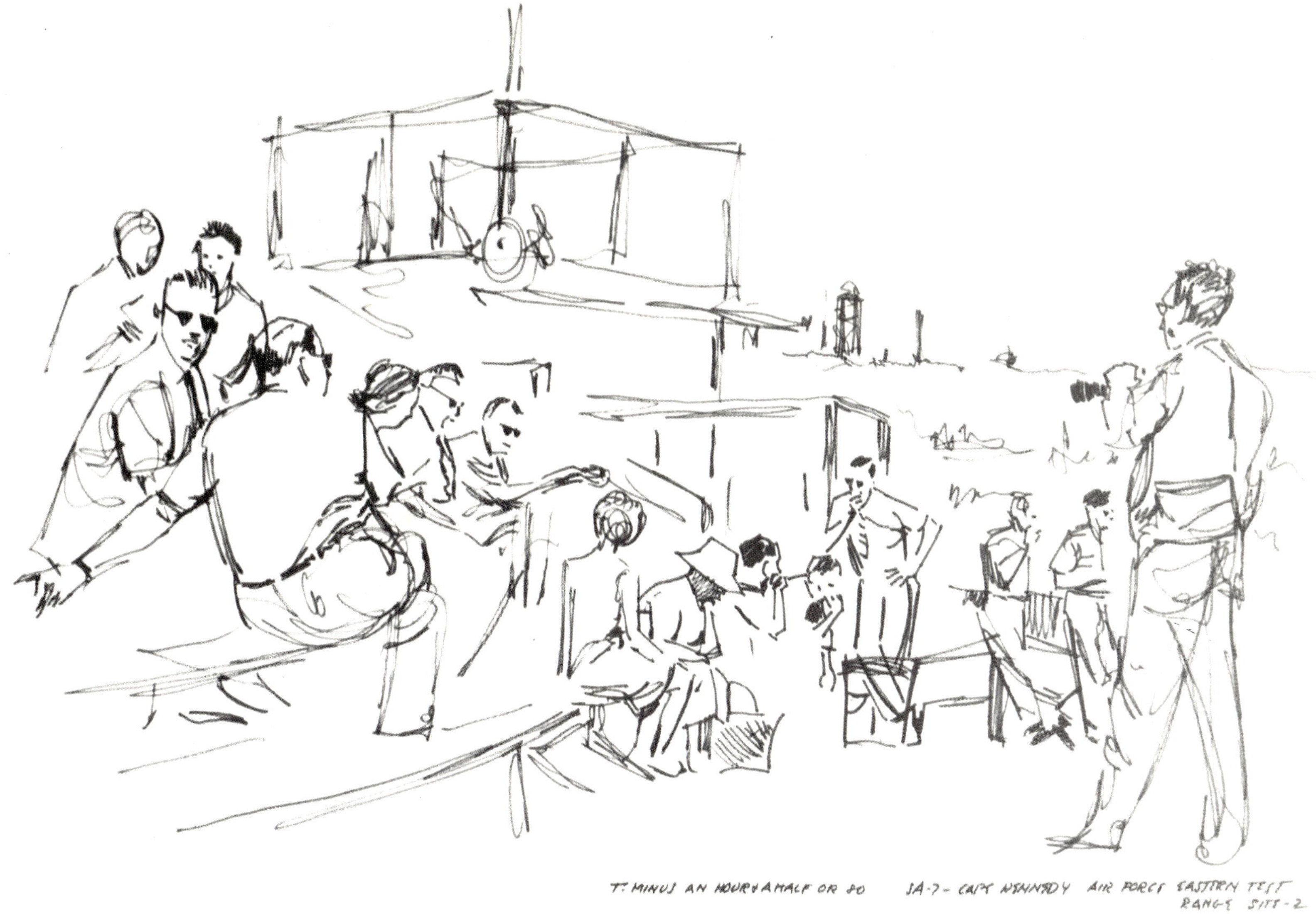

56.
Hugh Laidman
T MINUS 80
Felt pen on paper, 12 × 18″

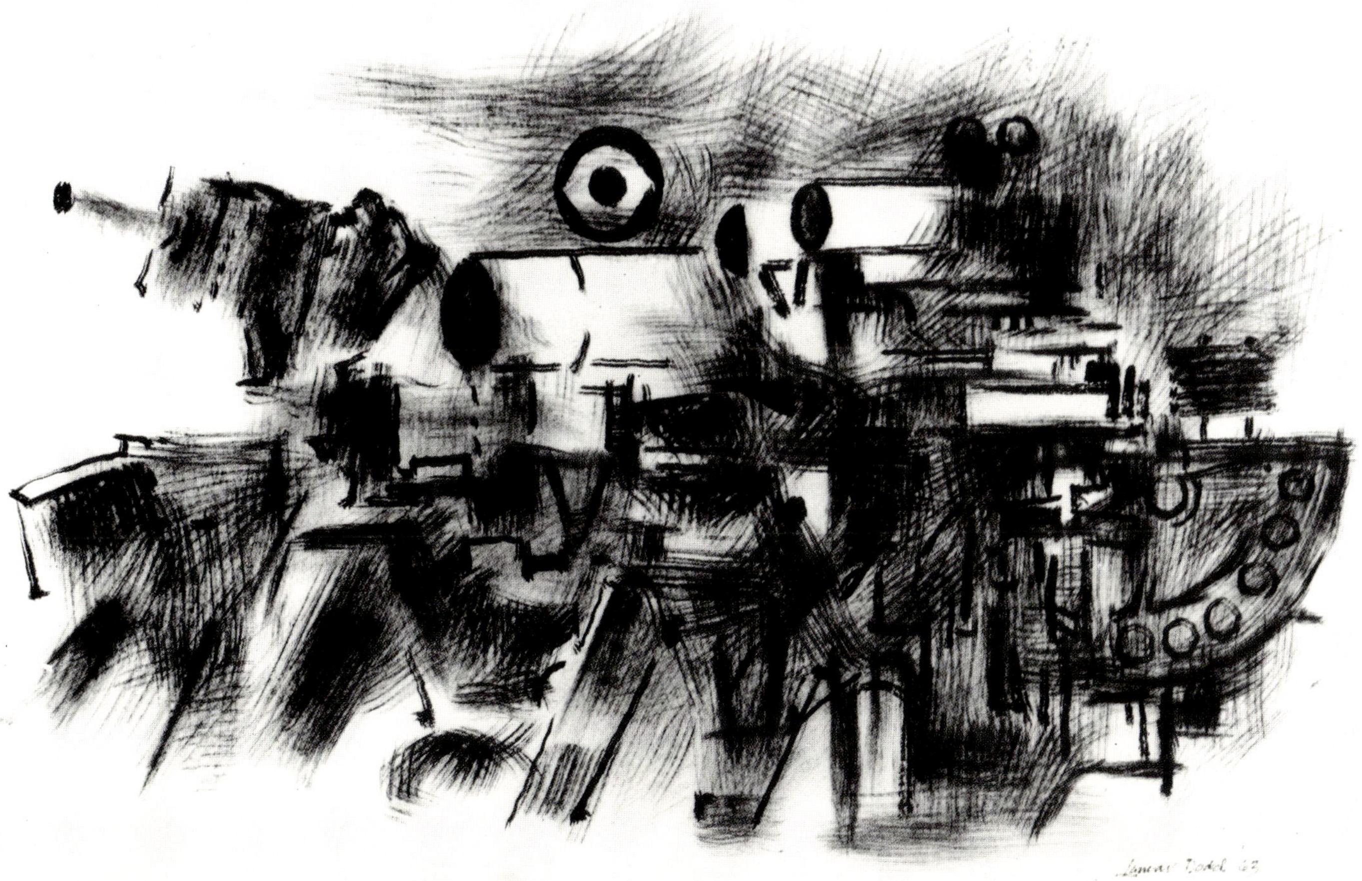

57.
Lamar Dodd
WATCHING
Ink on paper, 21 × 30″

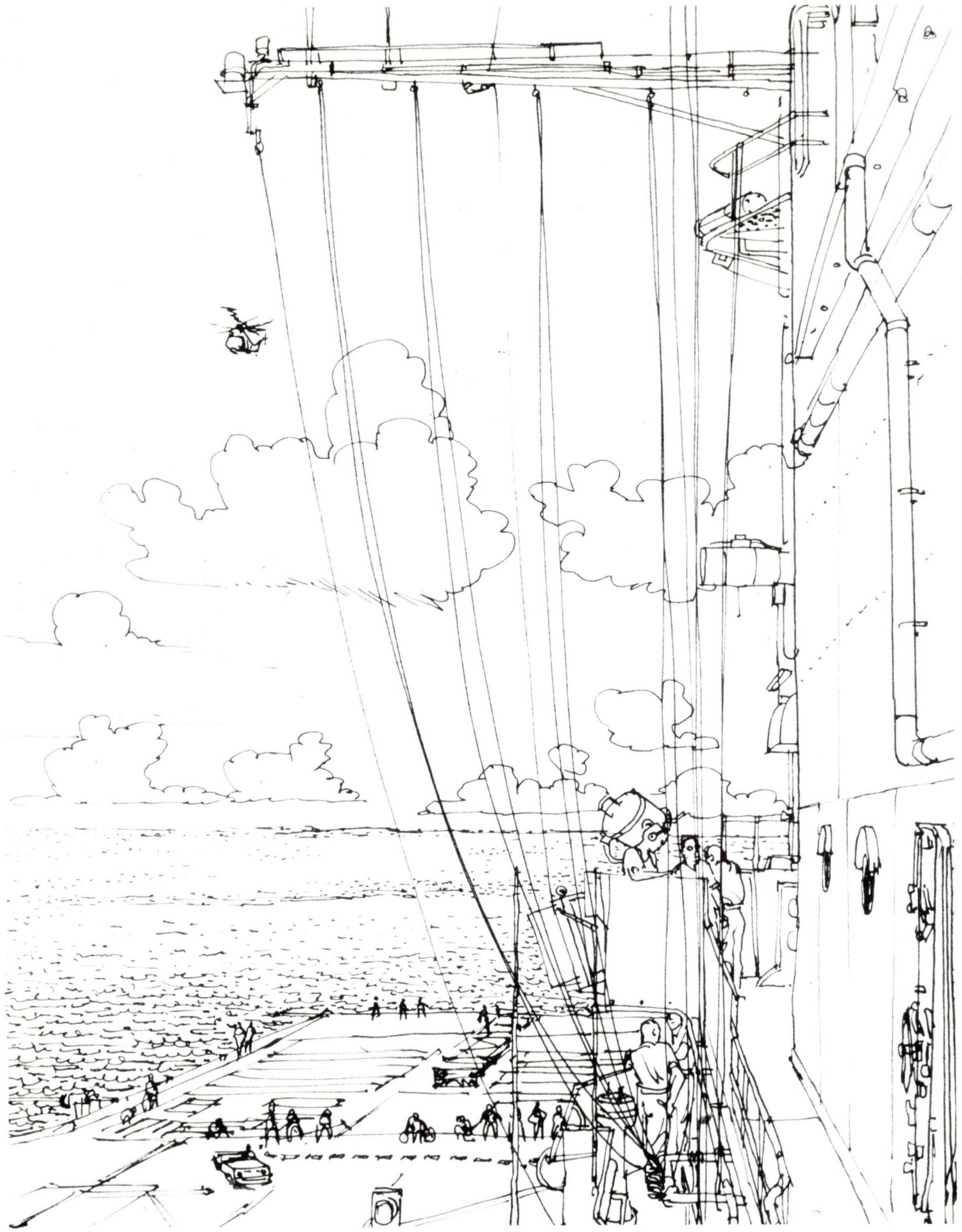

58.
Robert McCall
WAITING FOR GT-5
Pen and ink on paper, 14×11″

60.
Mitchell Jamieson
FIRST SIGHTING
Acrylic on wood panel, 27×51″

Gordon Cooper, manually controlling his Mercury spacecraft, after breaking all records for man in space, landed very nearly on the deck of the aircraft carrier Kearsage in the Pacific. Jamieson said that the occasion was one of the most exciting and dramatic events he had ever witnessed. The first indication of the astronaut's return was a thunderclap from the sonic boom as the craft entered the atmosphere. The second indication was sighting the orange and white parachute coming through the cloud coverage. "When it was sighted," Jamieson said, "everyone on board went wild, cheering, shouting, clapping their hands, and running toward the side of the flight deck nearest to the distant speck in the sky."

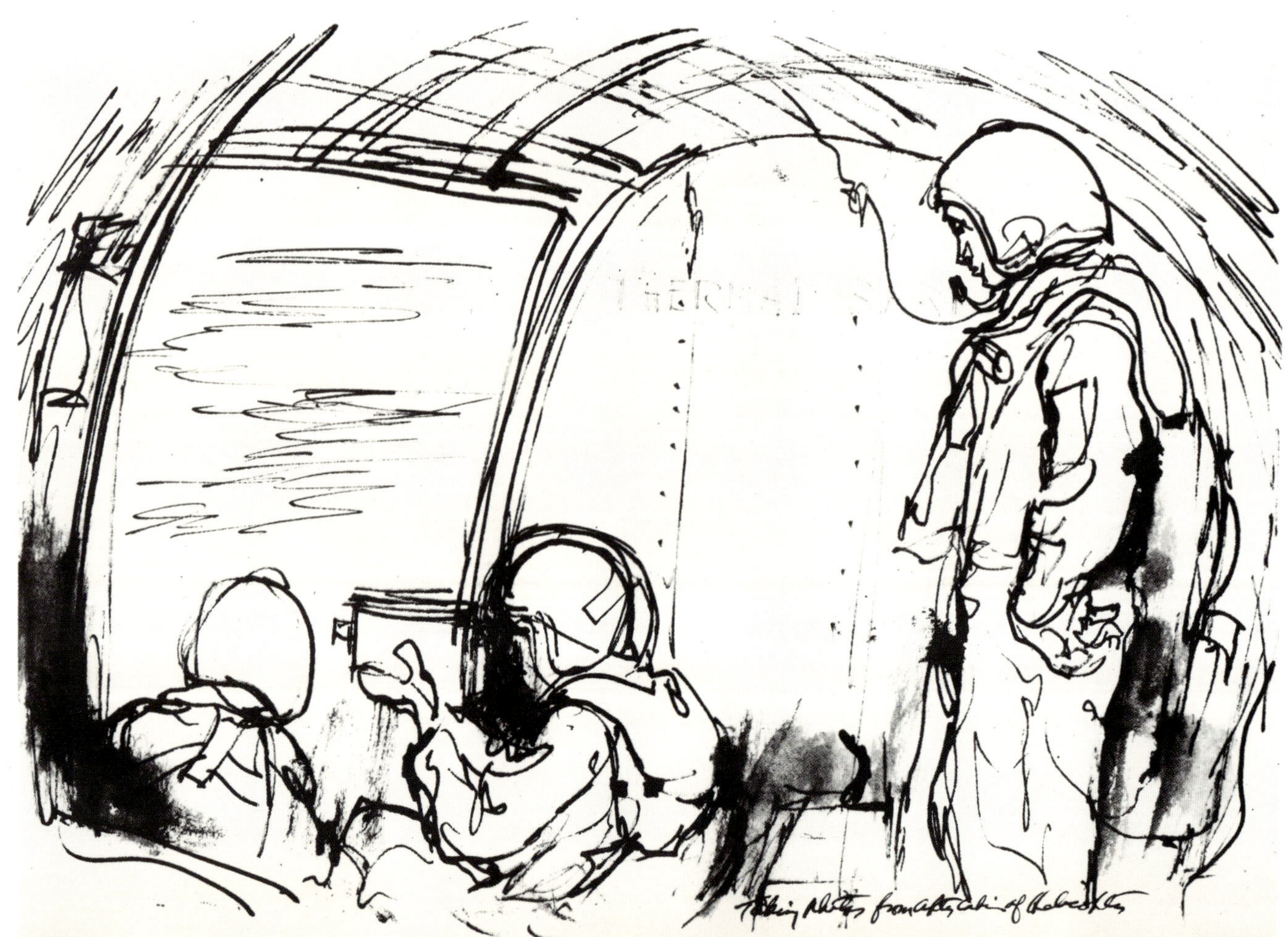

59.
Mitchell Jamieson
HELICOPTER CREWMEN
Ink on paper, 9×12″

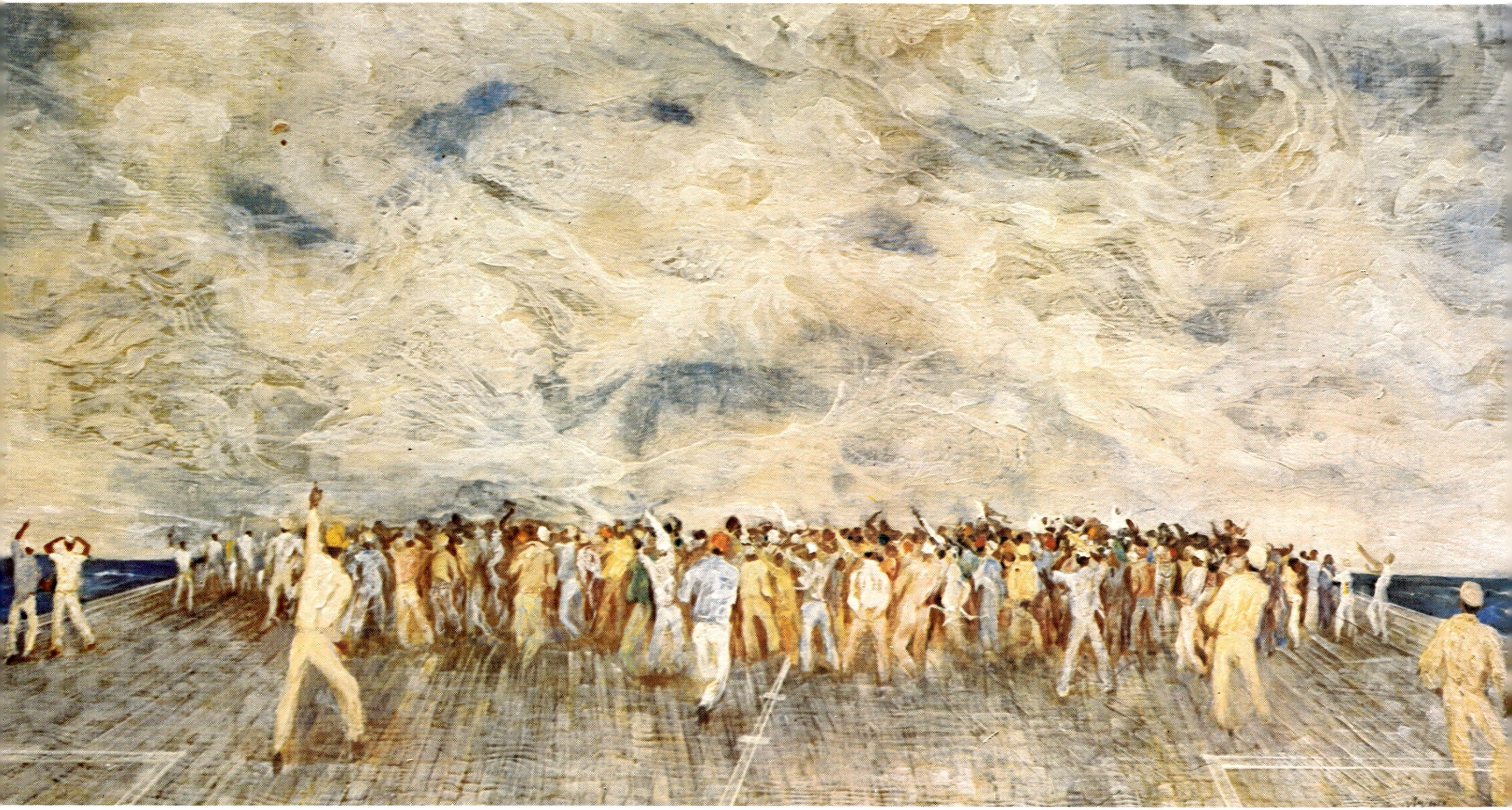

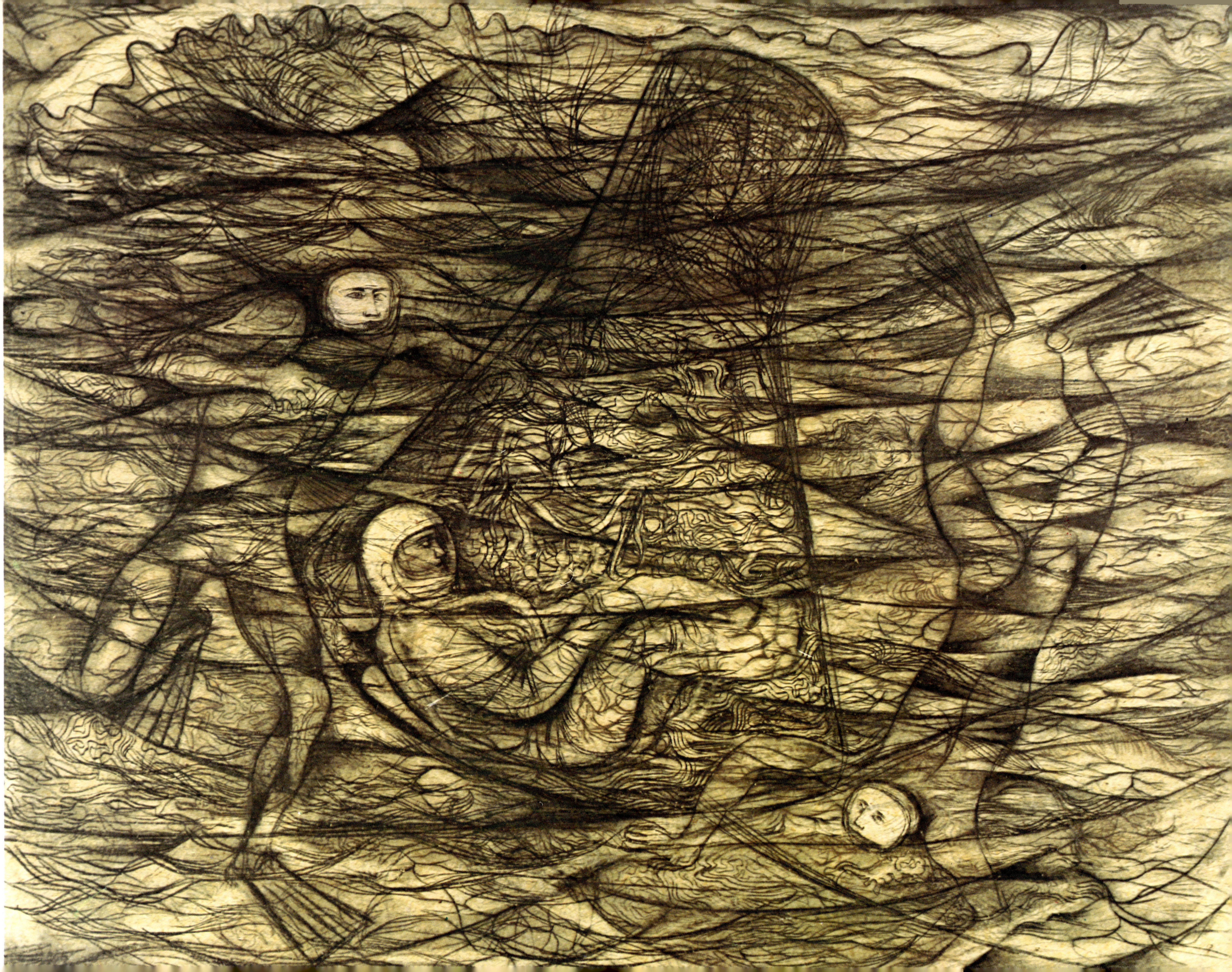

61.
Mitchell Jamieson
PACIFIC RECOVERY
Ink on board, 38×44″

Man, home from outer space where without protection he would die within a few seconds, sinks into the water, where without other kinds of protection he would drown.

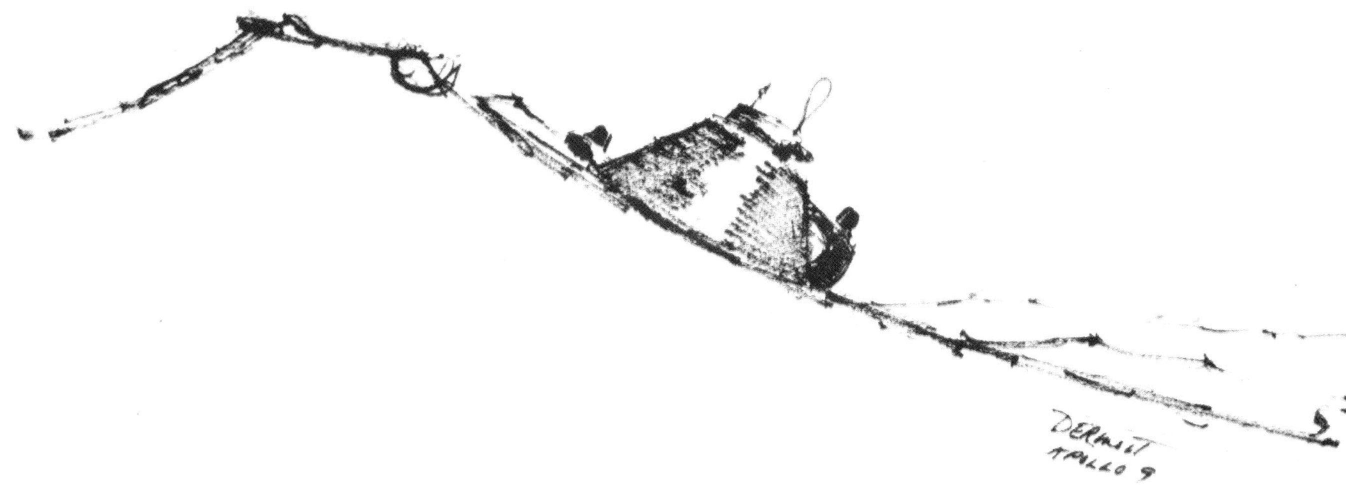

63.
Leonard Dermott
WAITING FOR PICKUP
Ink on paper, 14×20″

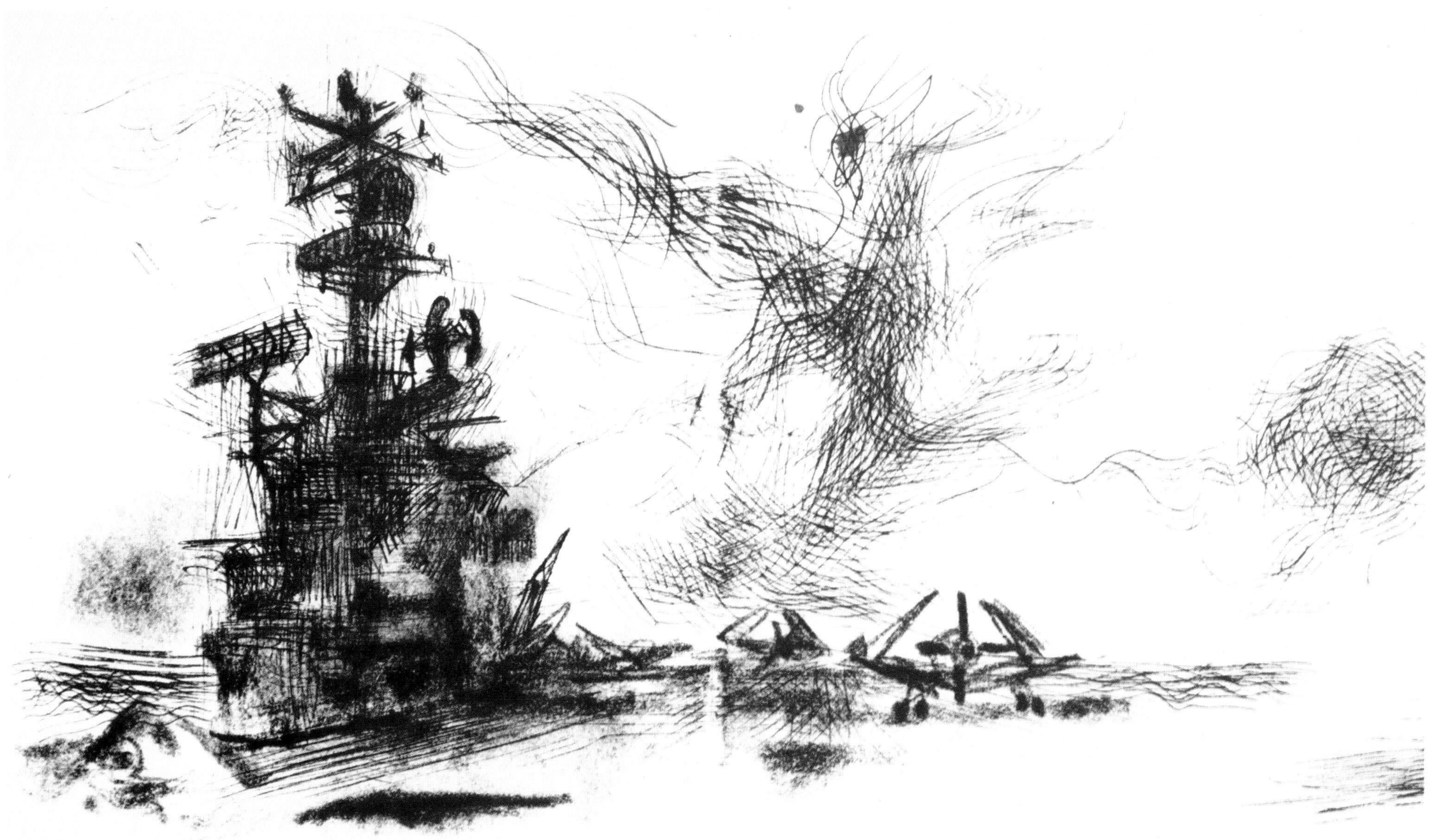

62.
Mitchell Jamieson
FLIGHT DECK
Ink on paper, 11¼×14″

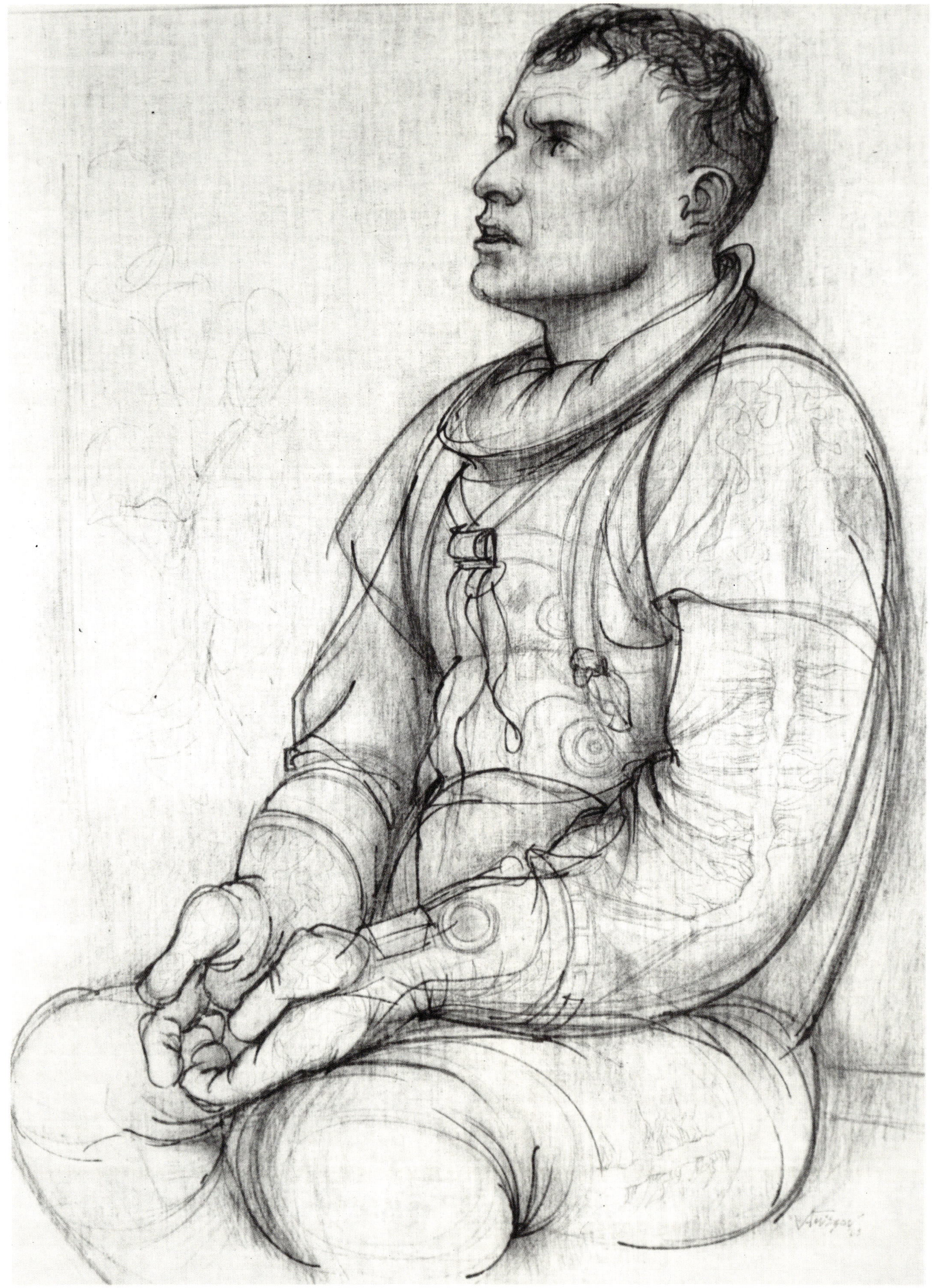

64.
Mitchell Jamieson
AFTERTHOUGHTS
Ink and chalk on paper, 40×30″

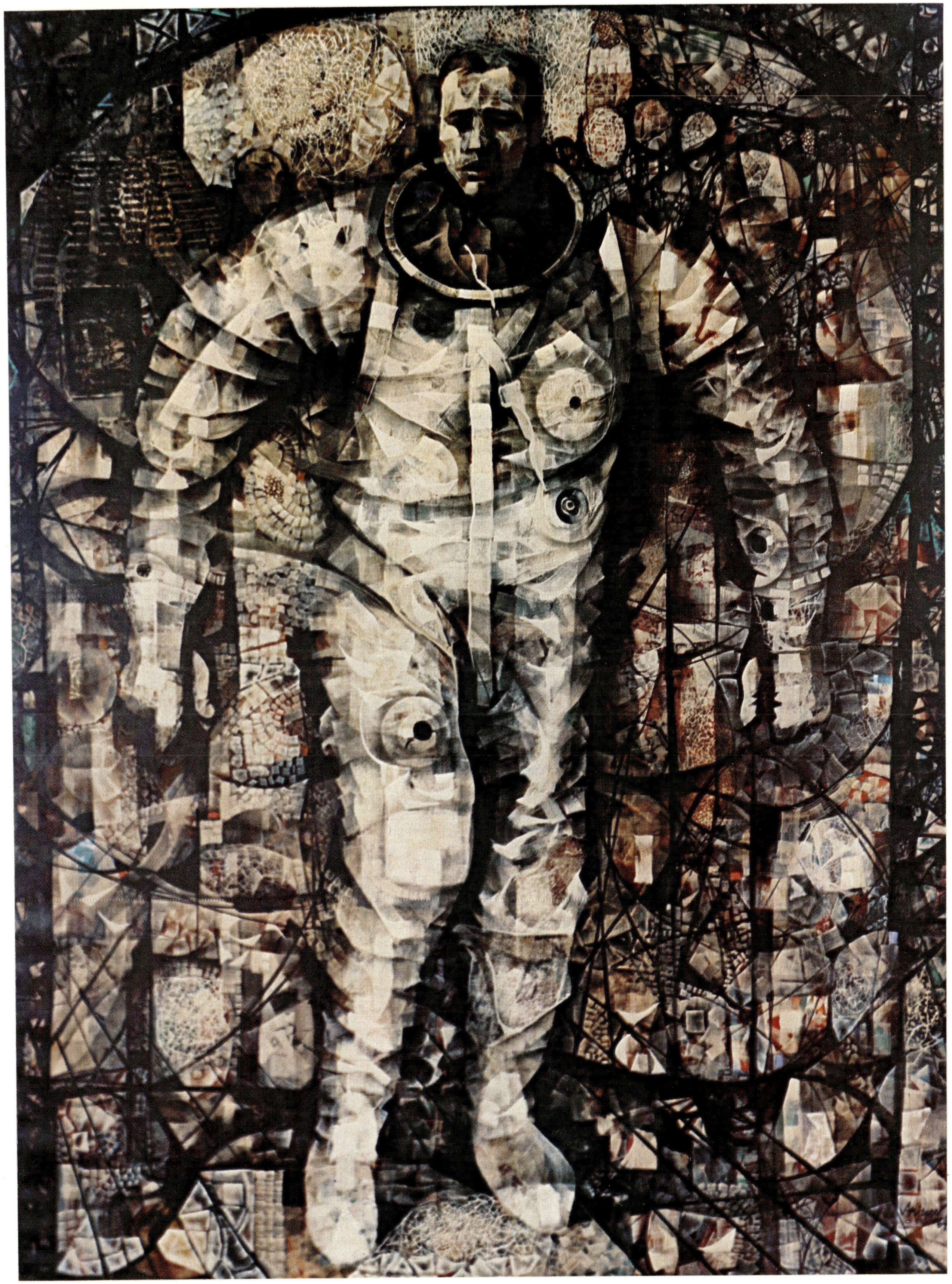

65.
Mitchell Jamieson
FIRST STEPS
Acrylic on canvas, 79×61″

Mitchell Jamieson said that when Gordon Cooper stepped out of his scorched and fragile Mercury spacecraft, having been in space for more than twenty-two orbits, and walked onto the deck of the aircraft carrier, he suddenly seemed to be much larger than life. He seemed to represent not NASA or the United States, but all mankind, which was opening the door to a new dimension and a new future. He seemed like one of the gods of Olympus, who descends to earth and mingles with mortals, but whose real home is in the skies.

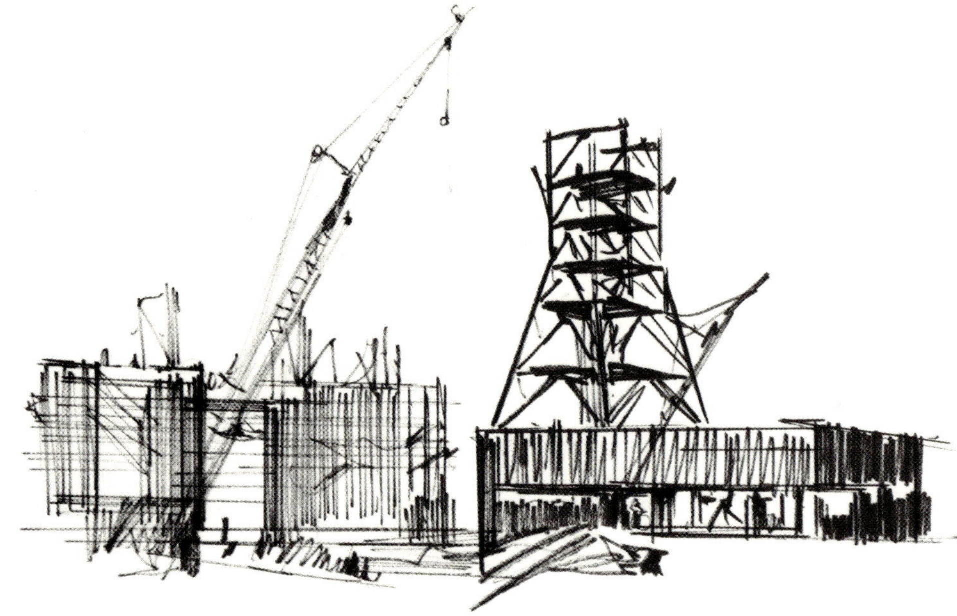

66. *Hugh Laidman*. MILA UNDER CONSTRUCTION
Felt pen on paper, 12×18″

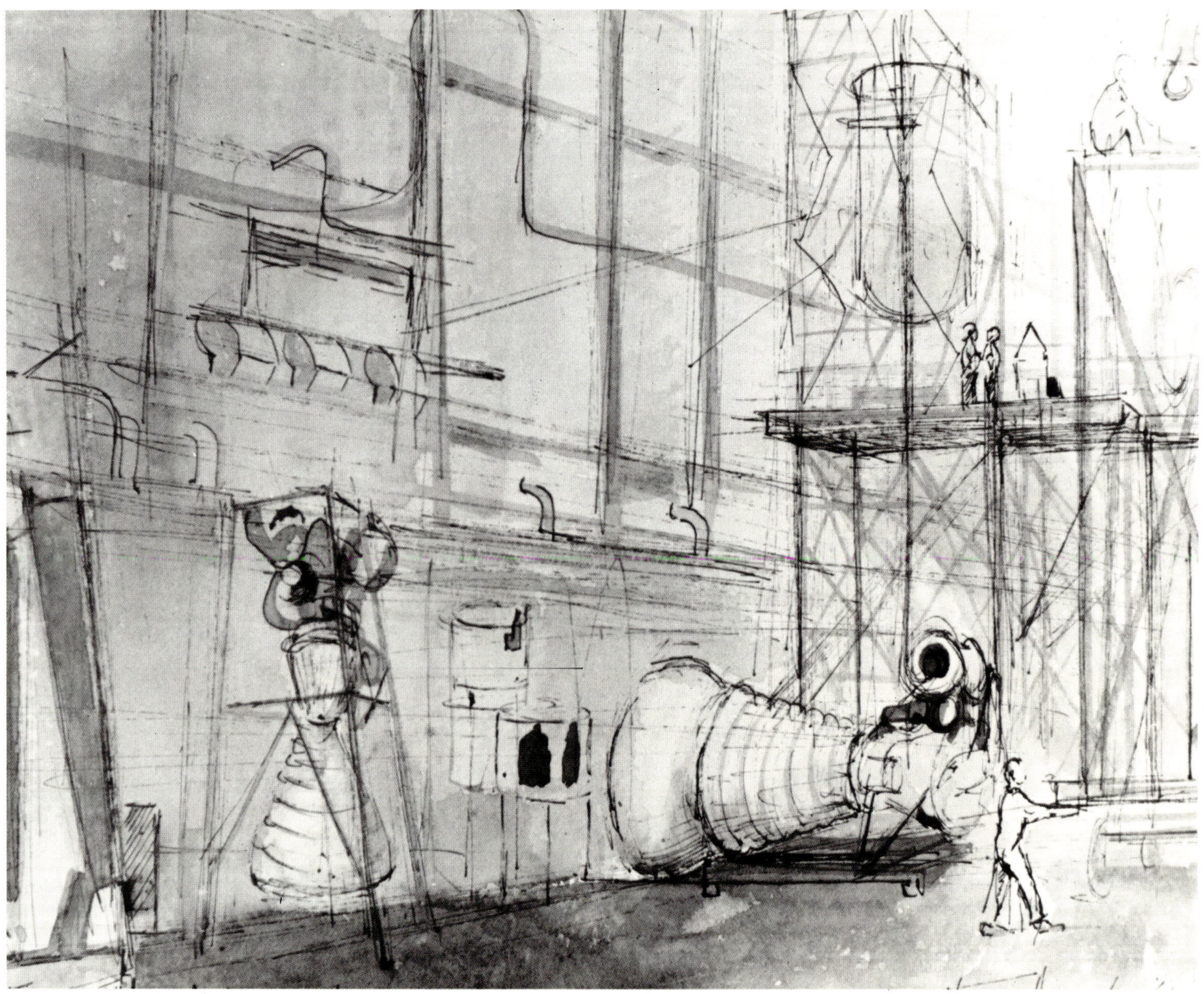

67. *Theodore Hancock*. ENGINES AT HUNTSVILLE. Ink and watercolor on paper, 9×12″

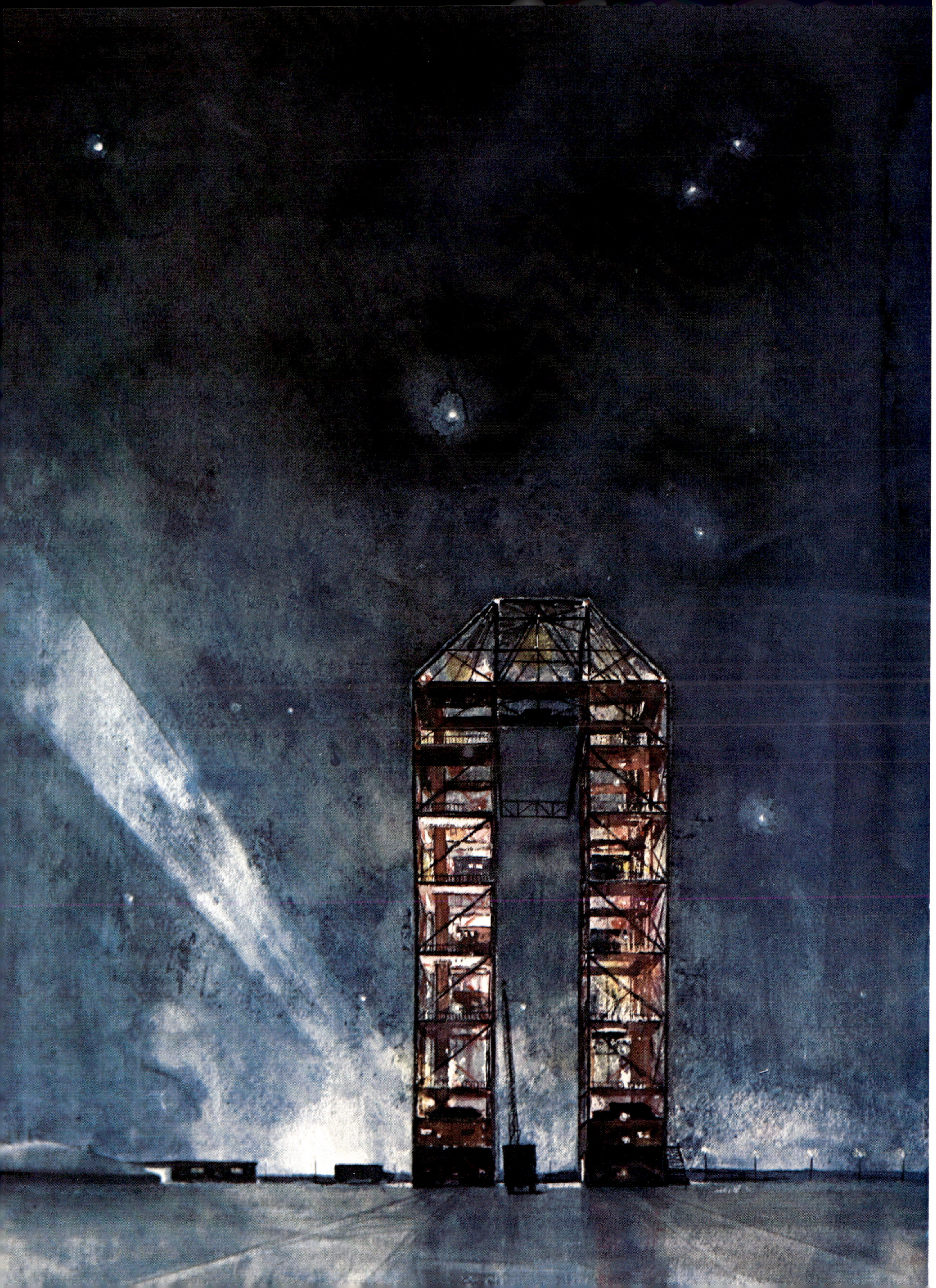

68.
John W. McCoy II
MOVING GIANT
Watercolor on paper, 40×30″

The Saturn I gantry, one of the many experimental designs for a mobile service tower, stands astride the sky like a giant gateway. At this time it ranked as the biggest structure on wheels ever built. Later, it acquired tragic fame from the fire that killed astronauts Grissom, White, and Chaffee.

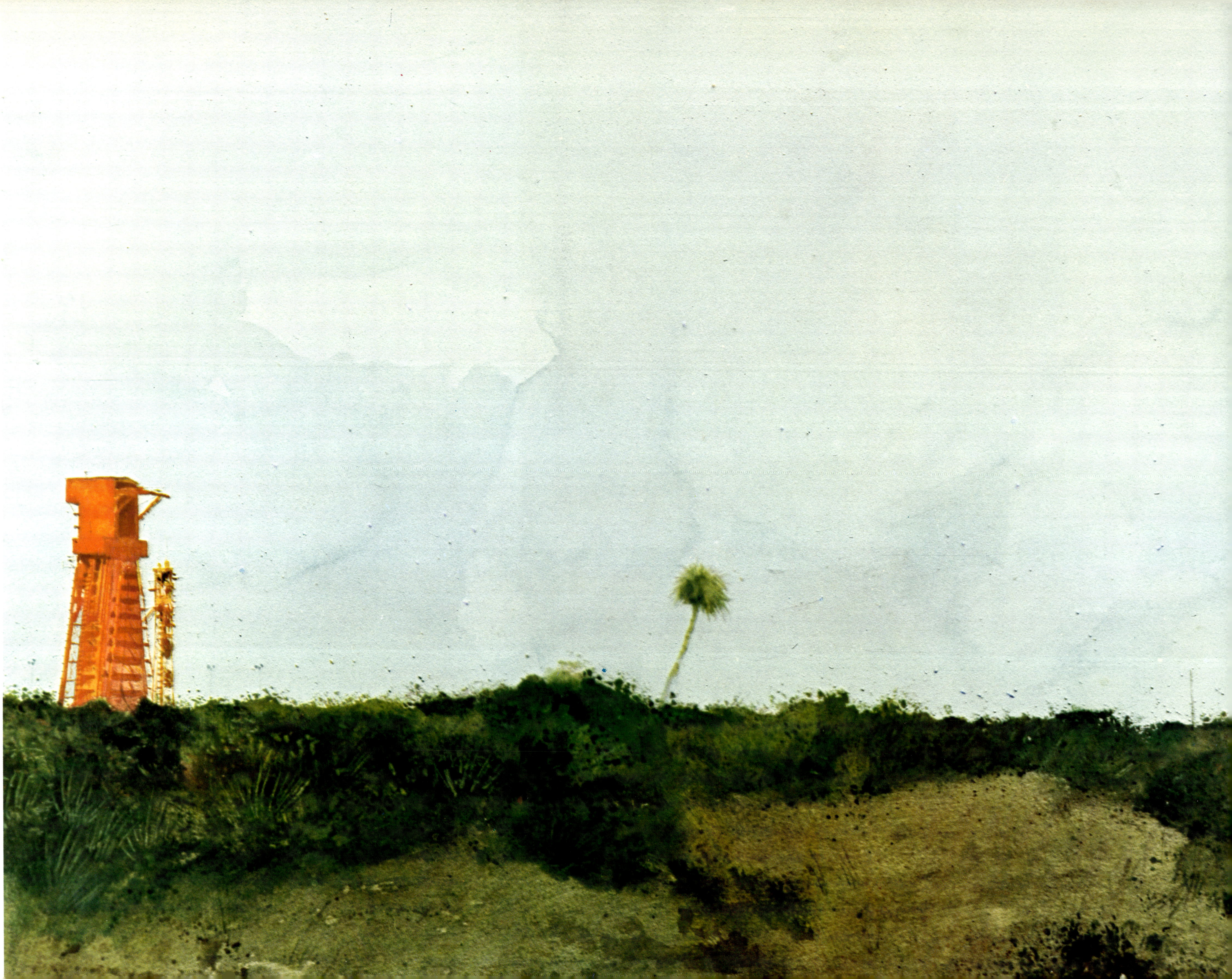

69.
James Wyeth
WAITING
Watercolor on paper, 27 × 34″

The Cape, as most visitors are surprised to find, consists mostly of dreary stretches of sandy scrub. At intervals, gantries loom above the palmetto pines. Barbed wire, cryptic signs, and occasional guardhouses divide the Cape into fiefdoms assigned to either the Navy, Air Force, or NASA.

70.
Peter Hurd
PREDAWN
Watercolor on paper, 21⅞ × 18¼″

Fortunately for the artist, gantries are painted a brilliant vermilion. At night, lit by hundreds of lights as technicians work around the clock, the launch pad (in this case Complex 14, with Mercury–Atlas 9) has been compared to a Christmas tree, a cathedral, and a frozen fireworks display. In fact, it is like none of these, but a new creation for a new mission.

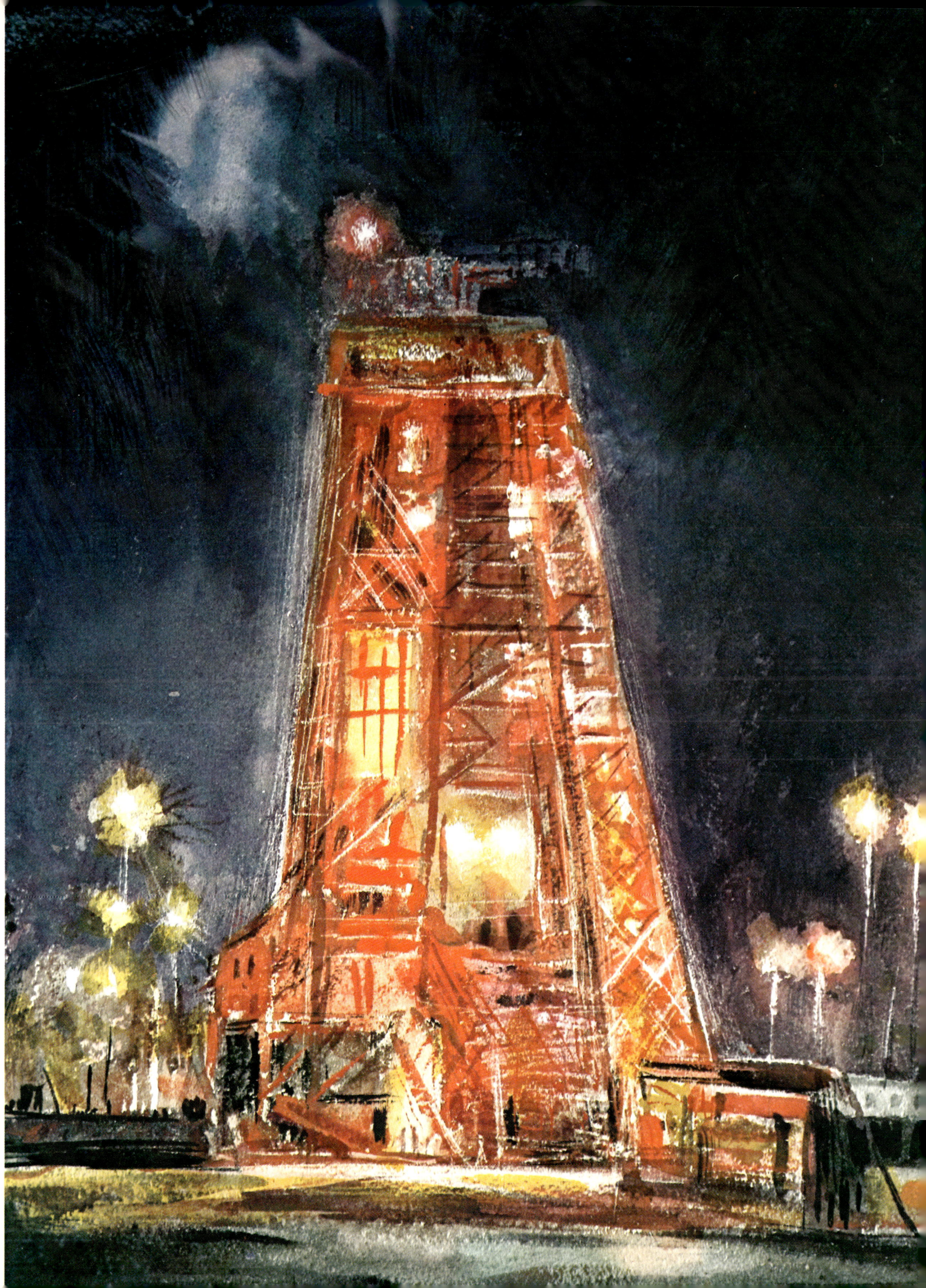

71.
Franklin McMahon
LUNAR MODULE MISSION SIMULATOR
Watercolor on paper, 30×22″

The artist always has the option of rearranging the facts of vision if he wishes. No single view of the Lunar Module Mission Simulator could show the upside-down scale model of the moon being scanned by a moving television camera in the center, the jumble of television projectors duplicating these images in the upper center, the astronaut standing at the controls of the Lunar Module in the lower center, the serried ranks of computers on the right, the hutch in the foreground with more simulating devices, and the spectator gallery at the upper right filled with spellbound tourists. This facility, known as the "Train Wreck," is at the Kennedy Space Center.

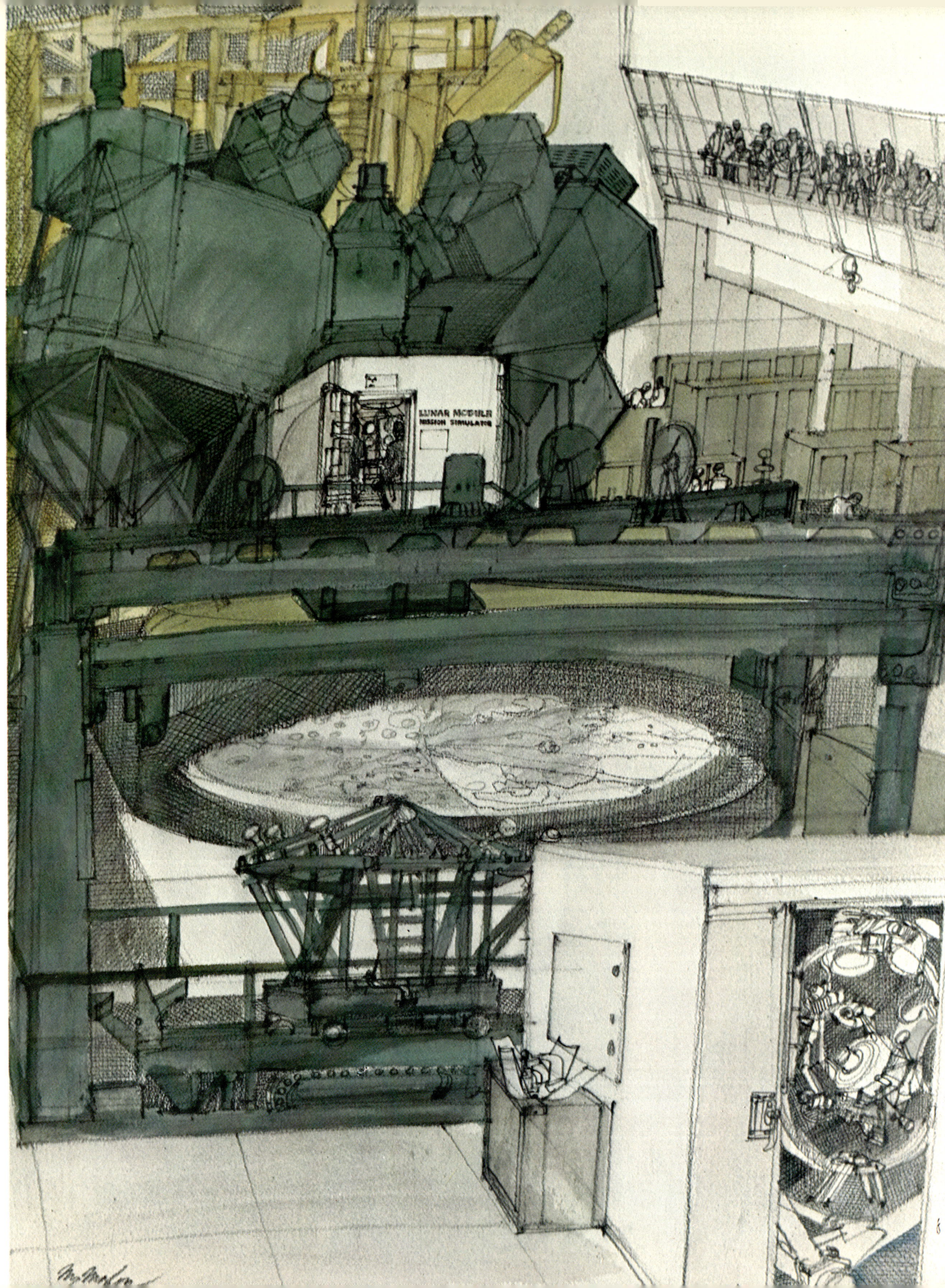

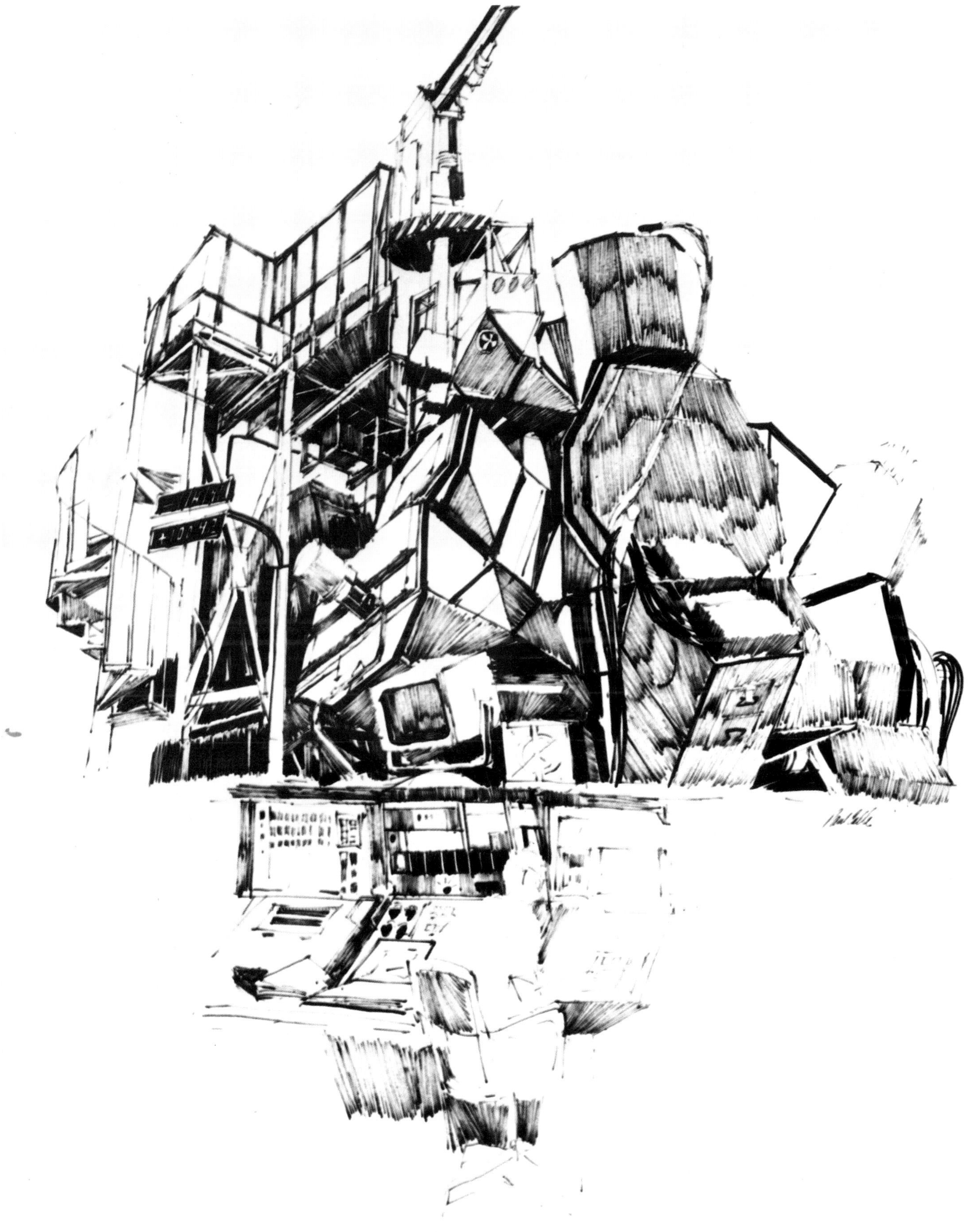

72.
Paul Calle
TRAIN WRECK
Ink on paper, 23×17″

73.
Howard Koslow
TEST STAND
Oil on panel, 5×4′

A part of NASA's operations that rarely gets in the limelight is the Marshall Space Flight Center in Huntsville, Alabama, where the Saturn moon rockets and many other vitally important pieces of equipment were designed and built. A rocket engine must be tested at full power before it is shipped to Cape Kennedy. Residents of the Huntsville area have become accustomed to the shattering roar of engines fettered in this concrete and steel stockade. Here for a few seconds at a time the engine spews out orange flames and deafening noise waves across the peaceful countryside.

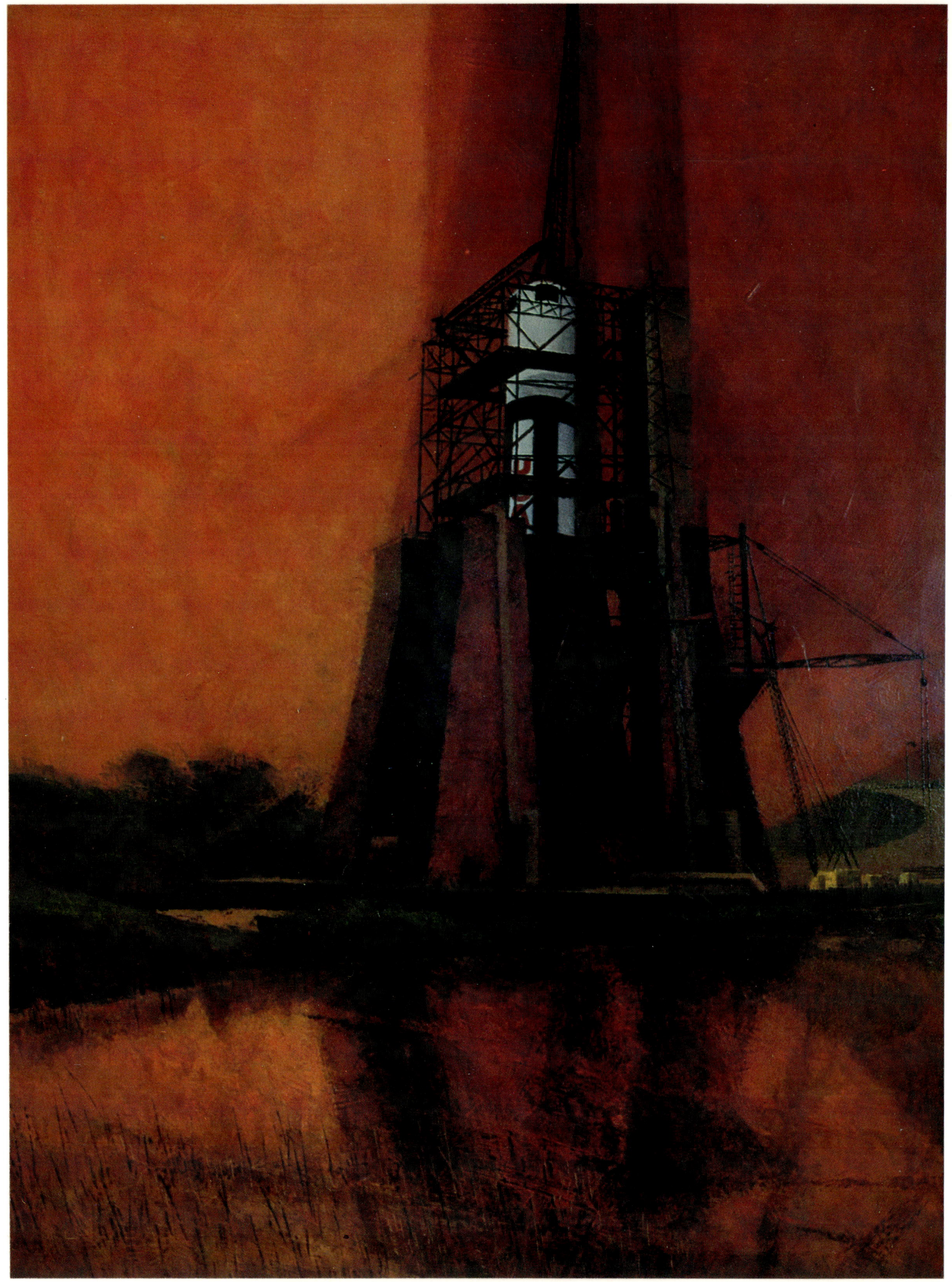

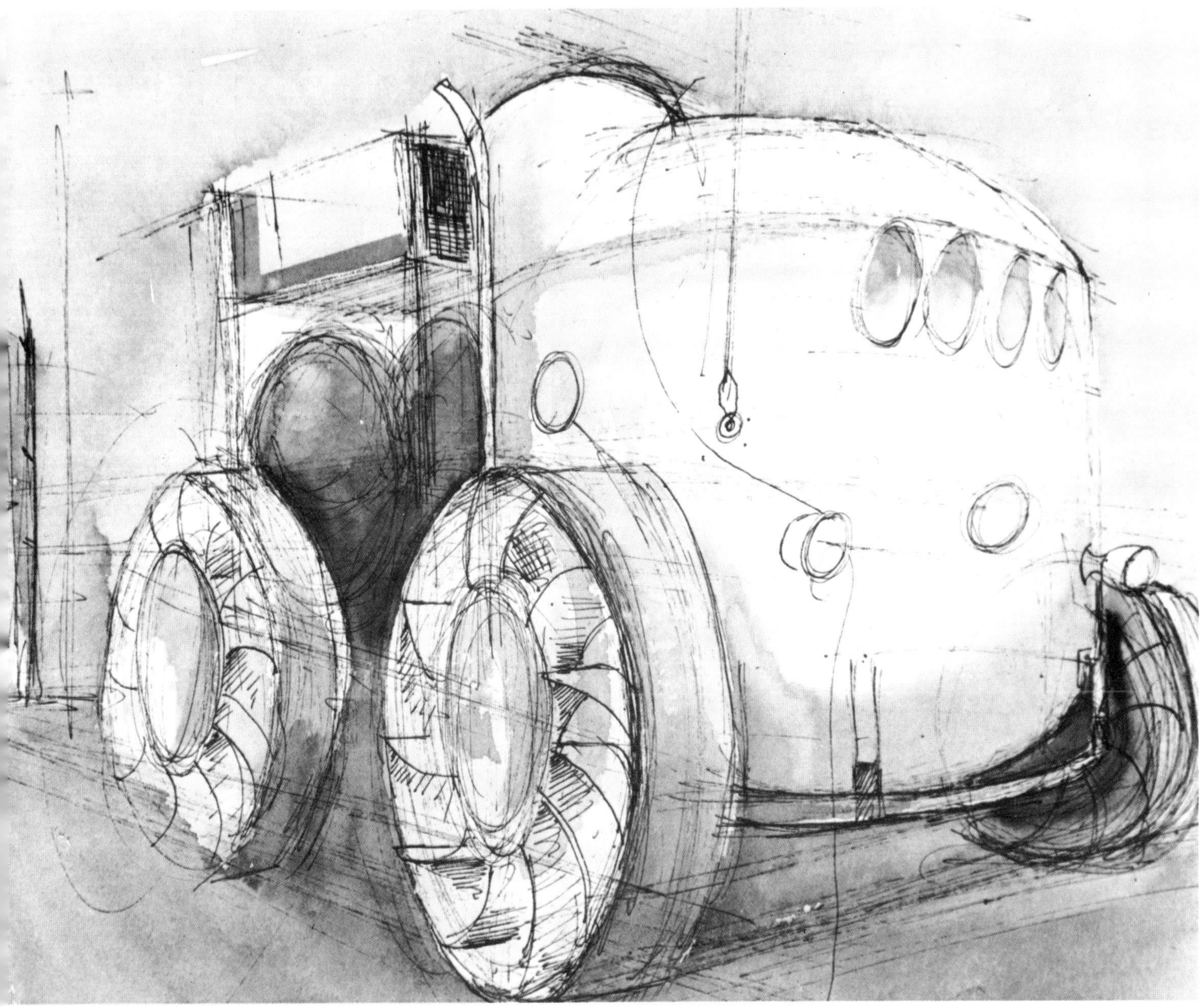

74. *Theodore Hancock.* MOON BUGGY. Ink and watercolor on paper, 9×12″

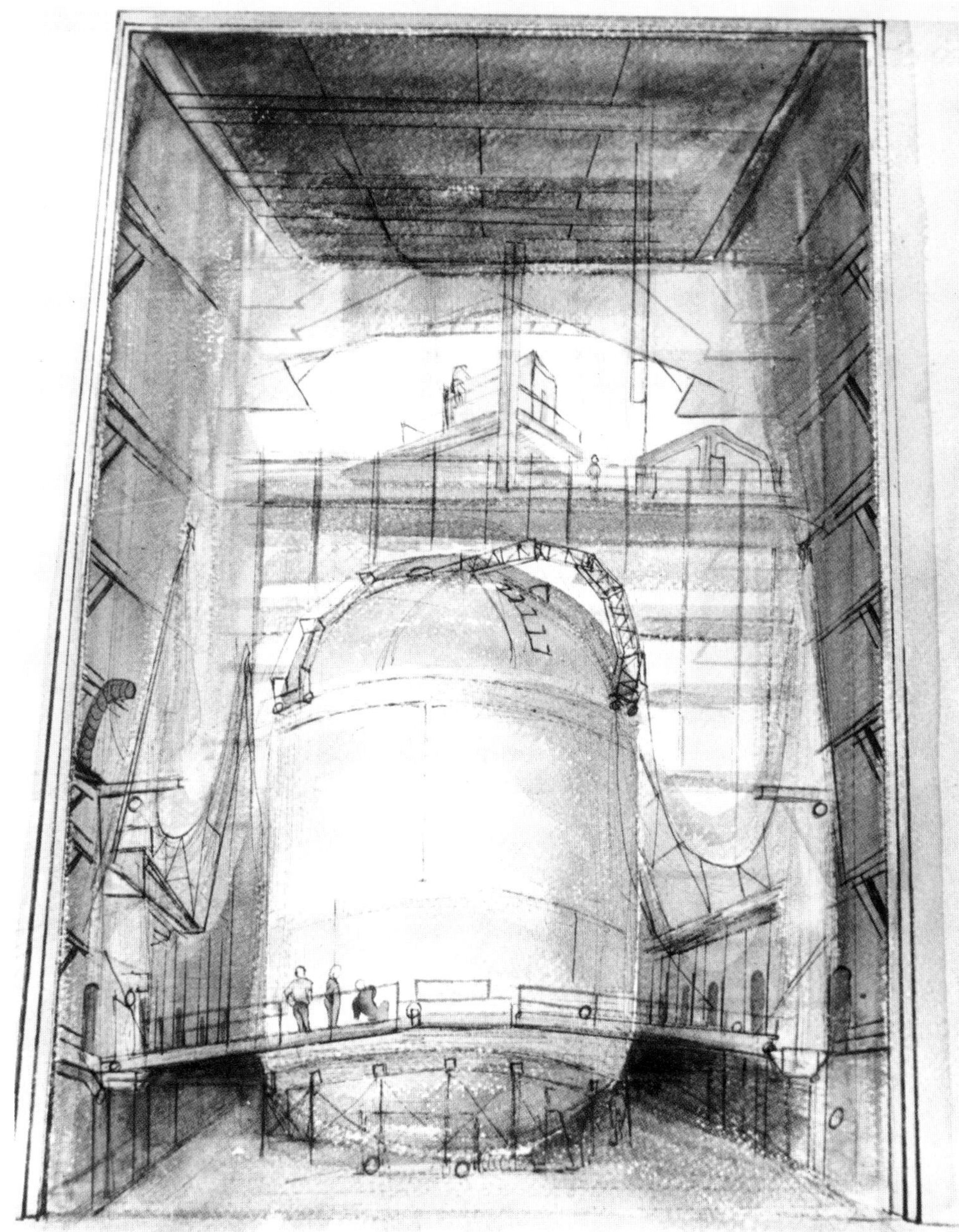

75. *Theodore Hancock.* SATURN COMPONENT
Ink and watercolor on paper, 22×15″

The fuel tank of a Saturn rocket is a paper-thin sack of light metal designed to withstand prodigious stresses and vast temperature changes. To keep these giant reservoirs from buckling, leaking, clogging, or collapsing like paper bags, and still not exceed the weight limits, was one of the Huntsville engineers' greatest triumphs.

77.
Paul Calle
KNIGHT'S ARMOR
Oil on canvas, 48×50″

The spacesuit, like the armor worn by a medieval knight, represents the most advanced technological research of the day. In effect it is a spacecraft shaped like a human being, designed to protect the tender creature within its confines against the fiercely hostile forces of space. It keeps temperature, humidity, pressure, and every other essential condition for life within very narrow limits, and so far its record has been an impressive tribute to the scientists who designed it.

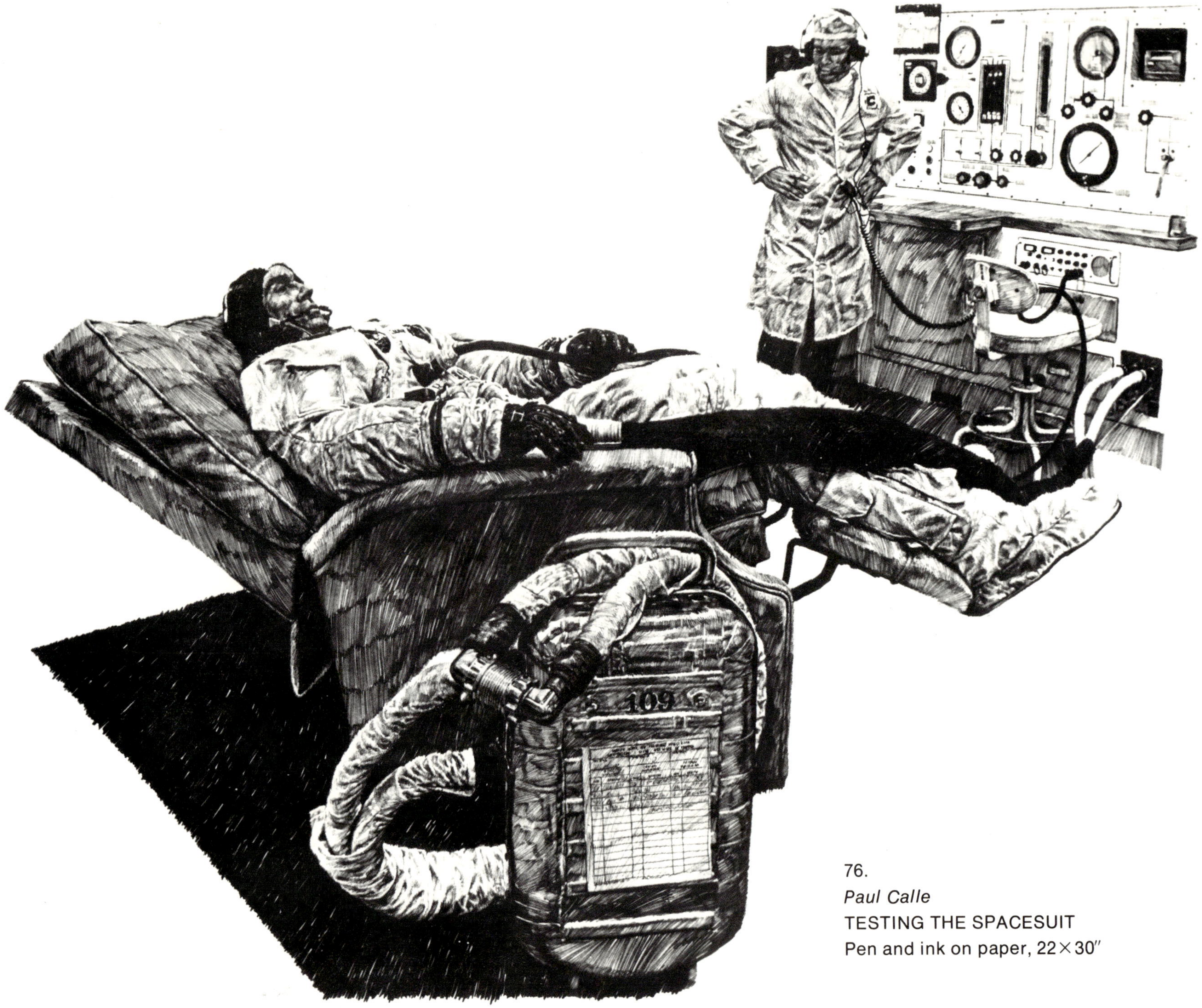

76.
Paul Calle
TESTING THE SPACESUIT
Pen and ink on paper, 22×30″

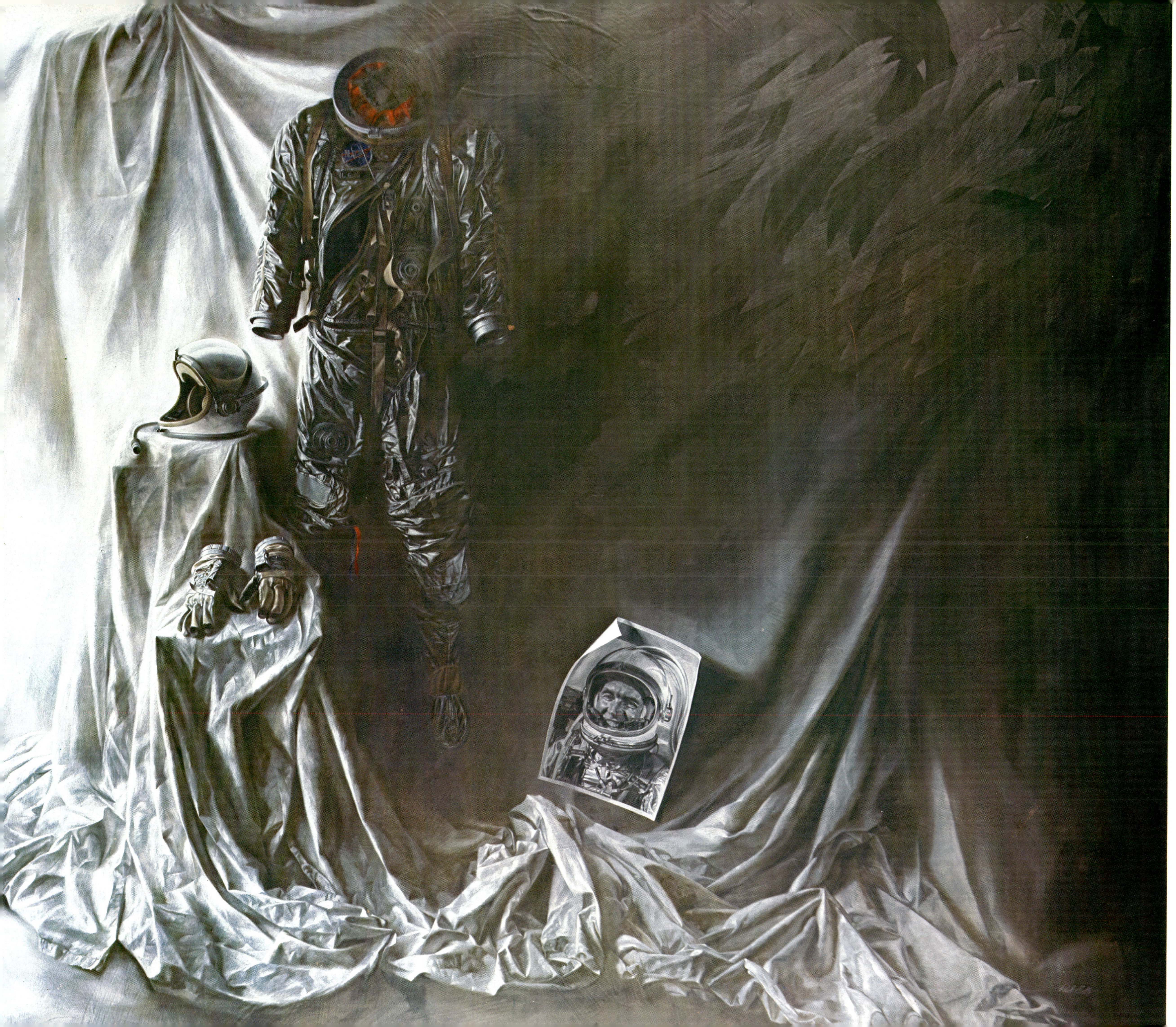

78.
Mitchell Jamieson
LADDERS
Acrylic on canvas, 71 × 45″

Jamieson often interprets the facts of vision in metaphorical terms. Thus the Saturn I launch pad is seen here as a fantasy of Oriental splendor, a source of wonder and beauty to the explorers at its base.

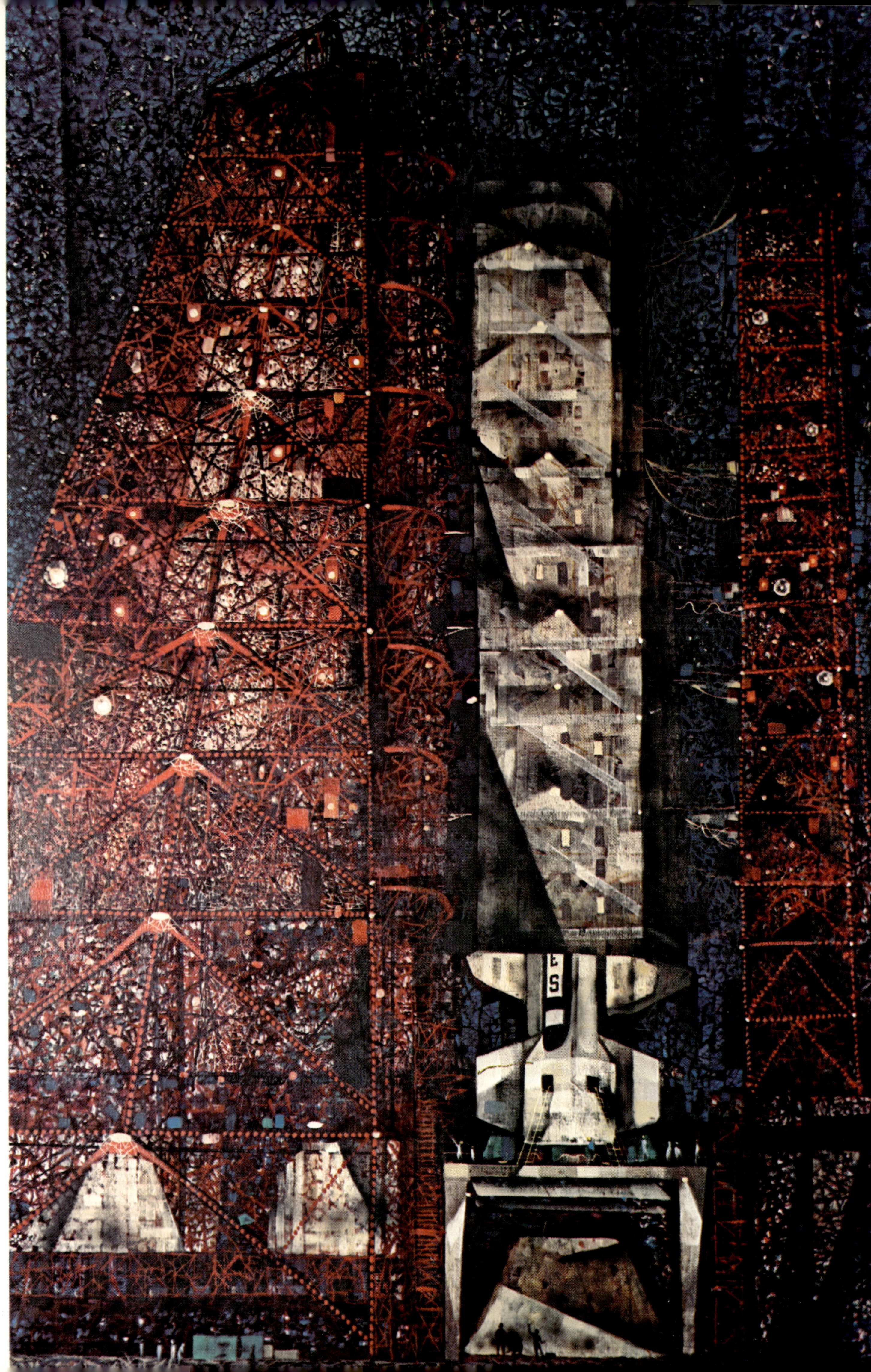

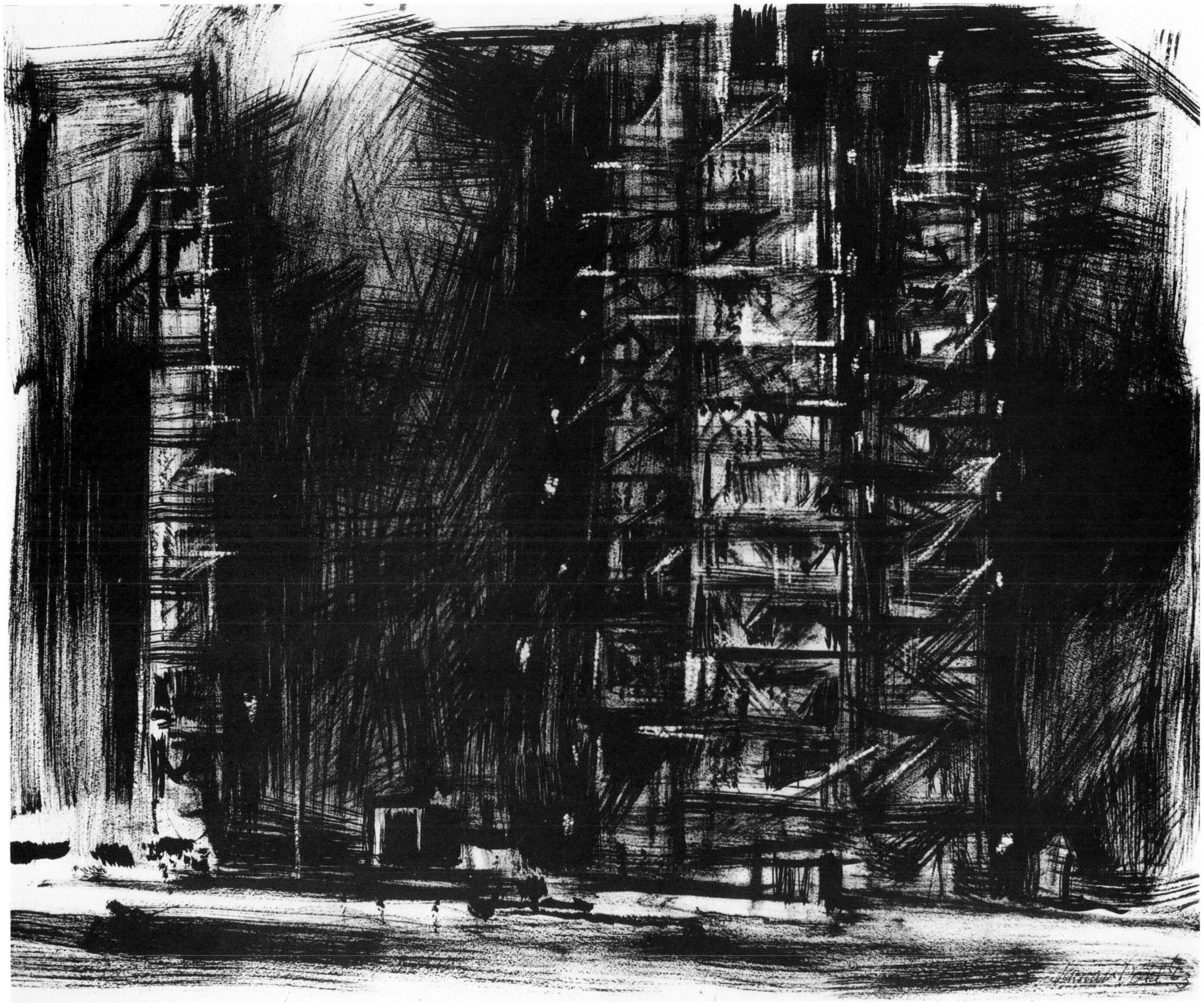

79. *Lamar Dodd.* SATURN GANTRY AT NIGHT. Ink on paper, 20×29″

81.
James Wyeth
WIDE LOAD
Watercolor on paper, 27×34″

To drive along roads on Cape Kennedy is to risk meeting vehicles that seem out of a nightmare fantasy. This one is a carrier, loaded with the first stage of an Atlas booster.

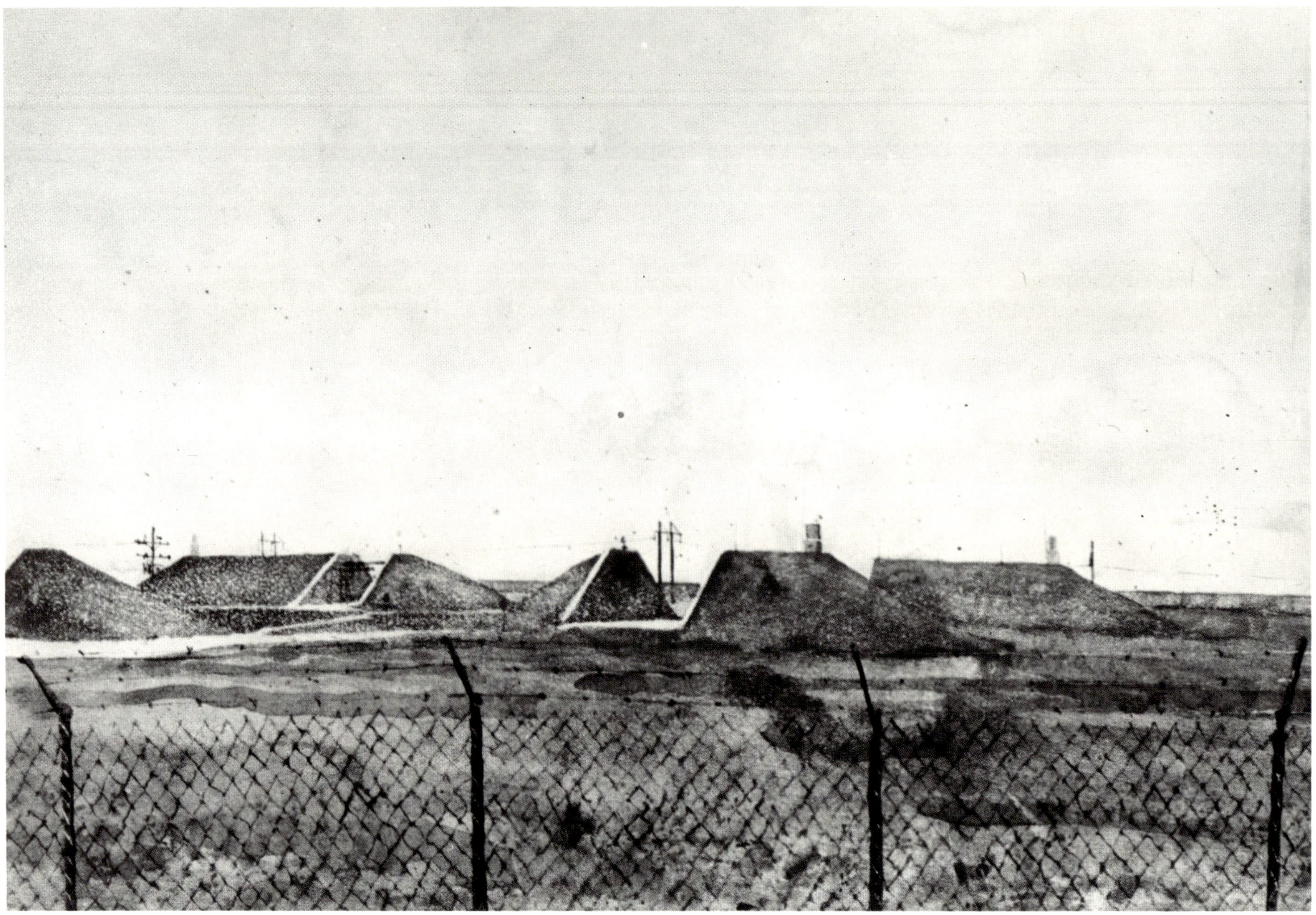

80.
James Wyeth
FUEL STORAGE
Watercolor on paper, 27×34″

Looking like the burial mounds of Viking warlords, the underground storage vaults for rocket fuels are strictly off limits. Many of the fuels used for rockets are dangerous and temperamental and fully merit such respectful isolation.

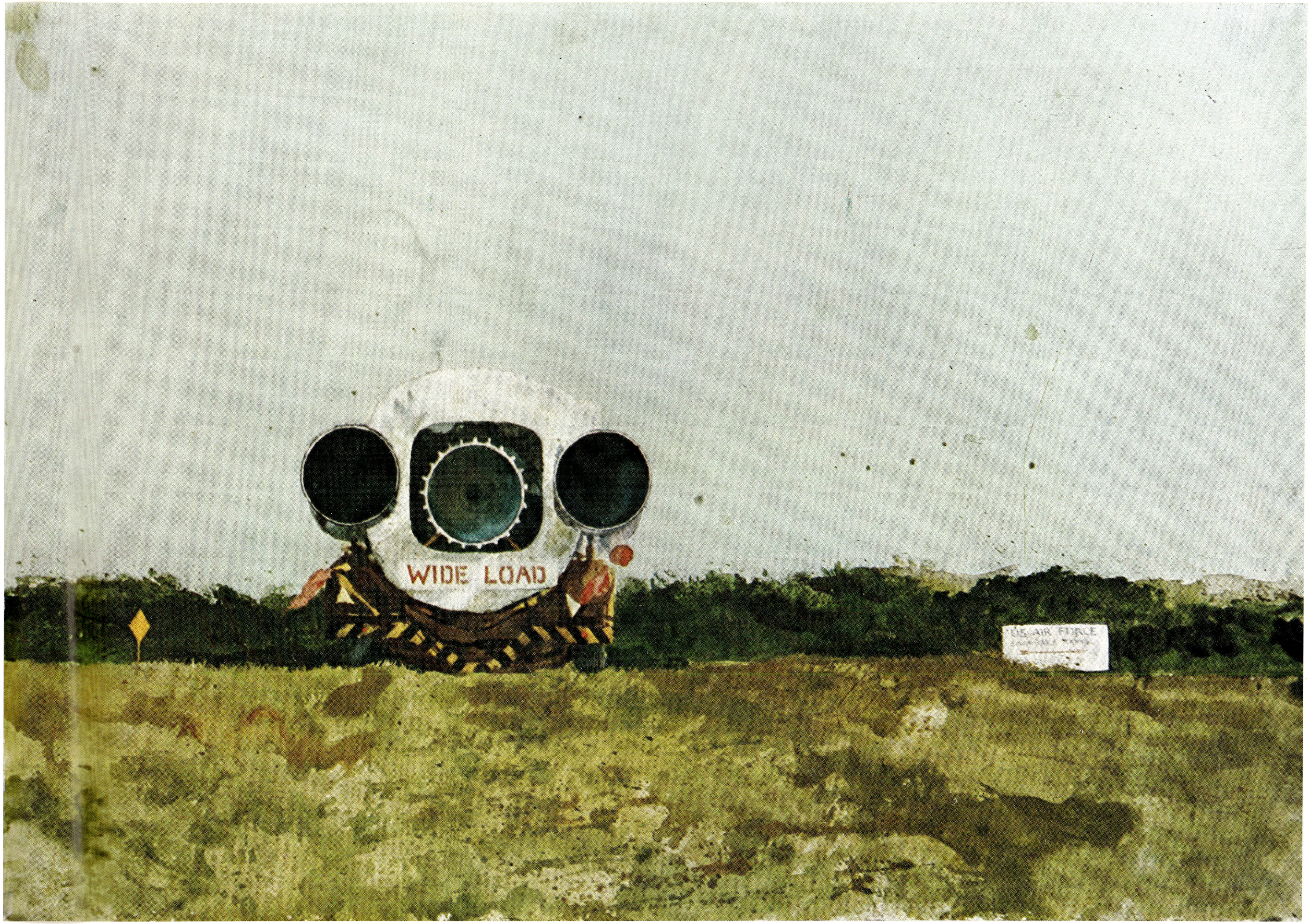
WIDE LOAD
US AIR FORCE

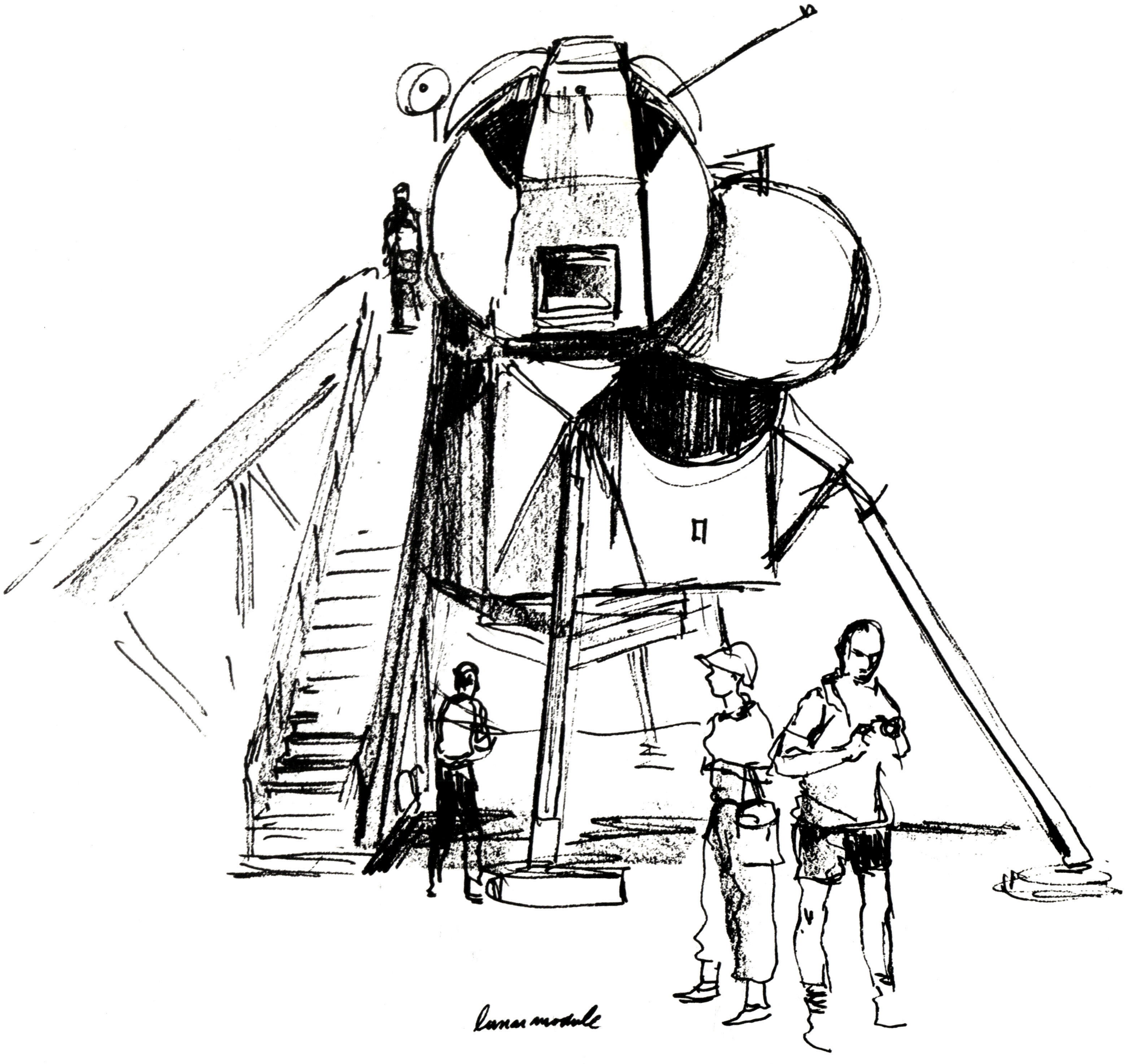

82.
Henry C. Pitz
SIGHTSEEING
Pencil on paper, 18 × 12″

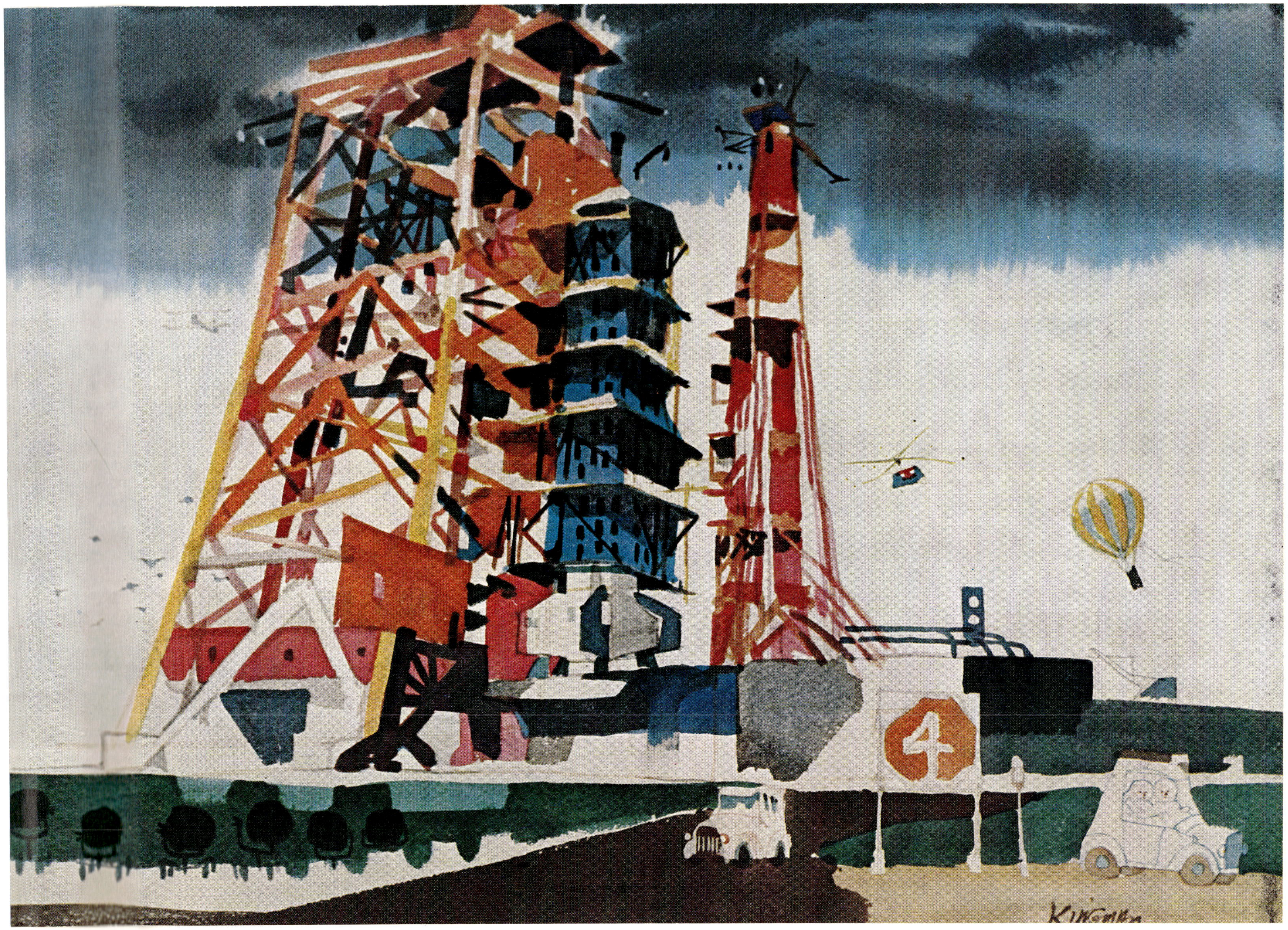

83.
Dong Kingman
HIGHER, FASTER, AND FARTHER
Watercolor on paper, 28×36″

Dong Kingman saw the Saturn I complex on Pad 34 in terms of a pagoda mysteriously transformed. Man, never content, must find new frontiers and new challenges to overcome. The bird, the balloon, and the early plane have something in common with the newest venture.

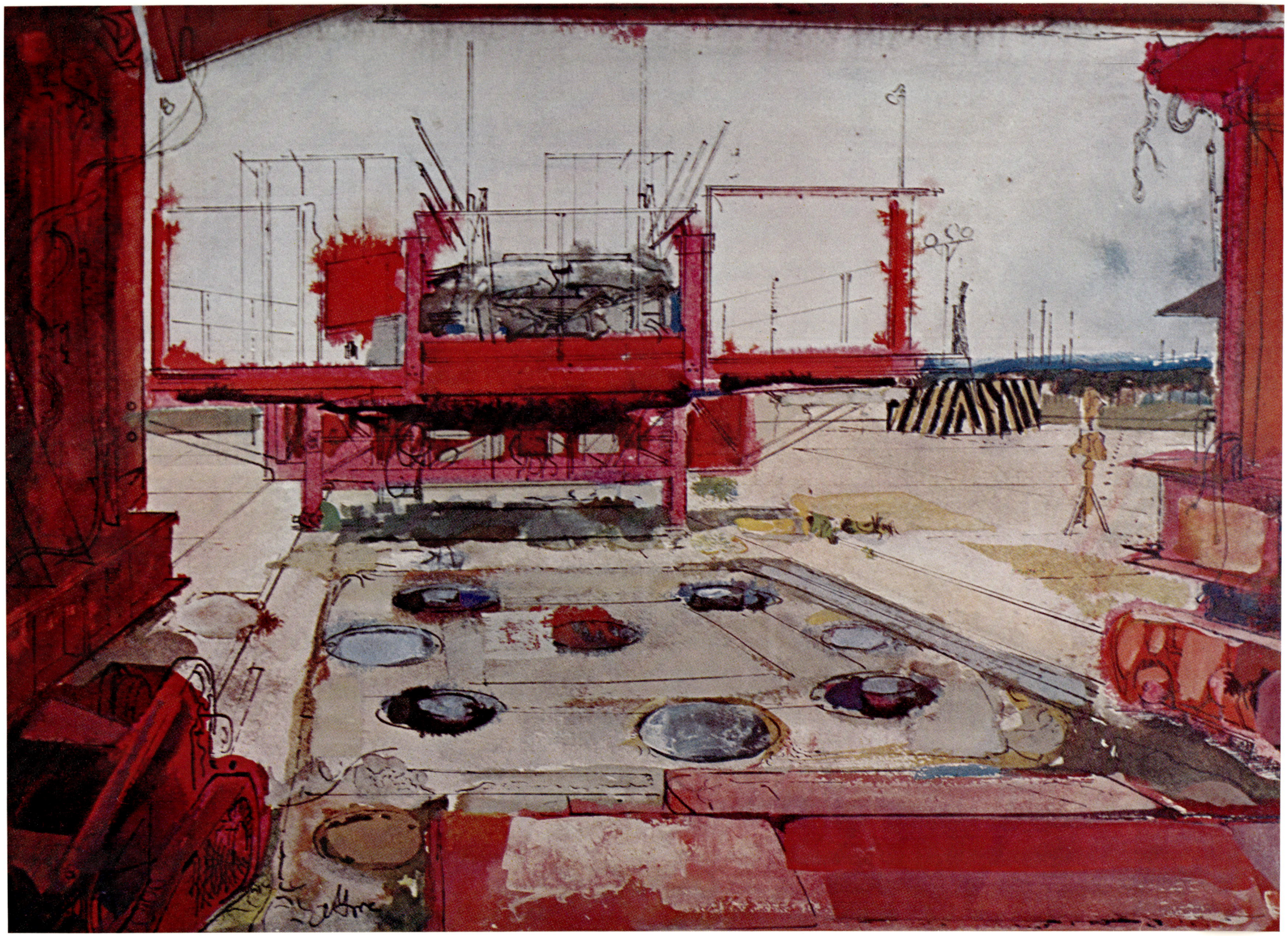

84.
Alfred McAdams
RECOLLECTION OF FIRE
Watercolor on paper, 21 × 27″

Part of the launch pad on an early rocket at Cape Kennedy. The Space Age has spawned an unbelievable assortment of strange machines and shapes. Plate 85 shows vacuum tanks at the Goddard Space Flight Center, near Washington.

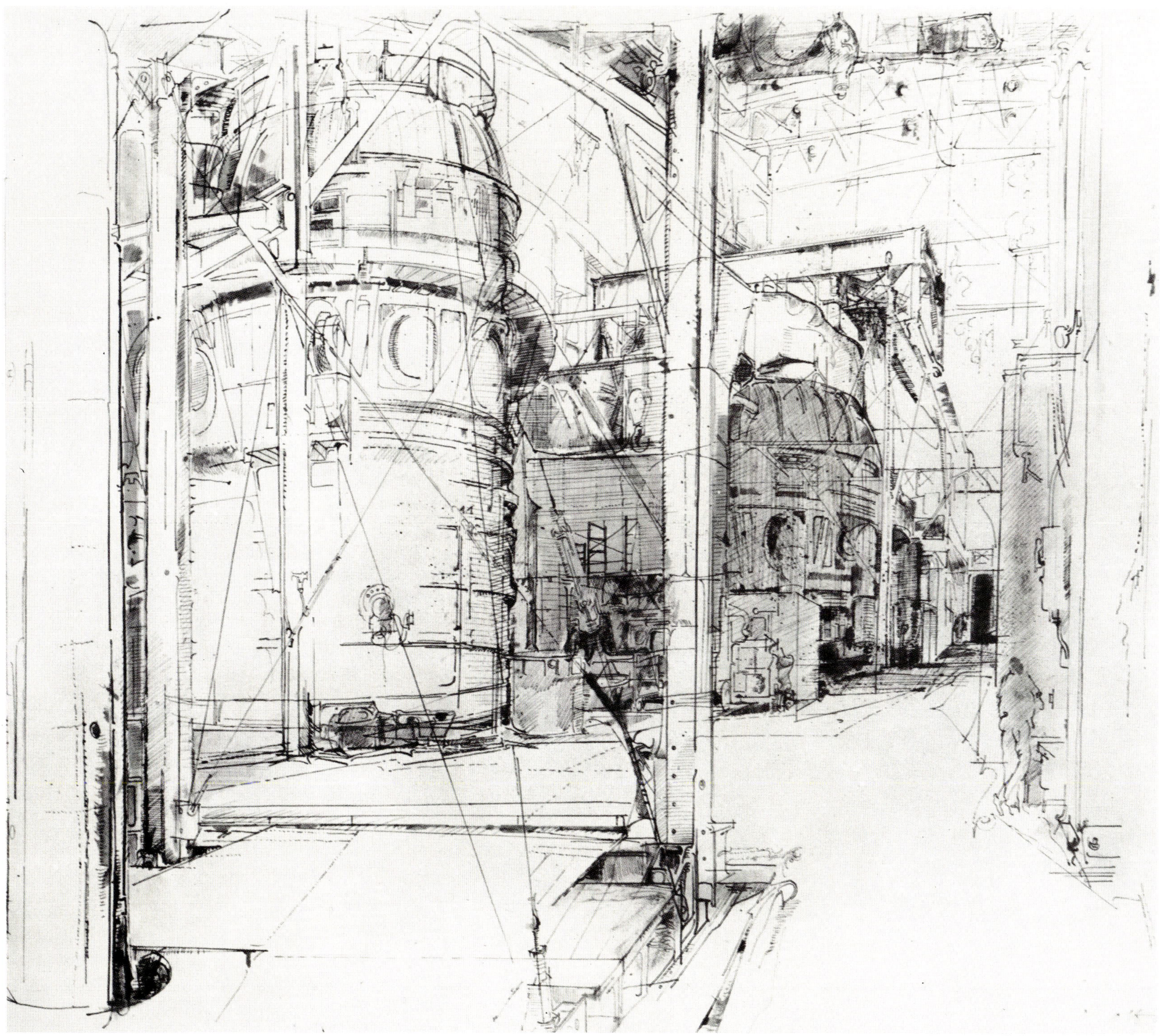

85. *Alfred McAdams.* VACUUM TANKS. Ink and wash on paper, 18×24″

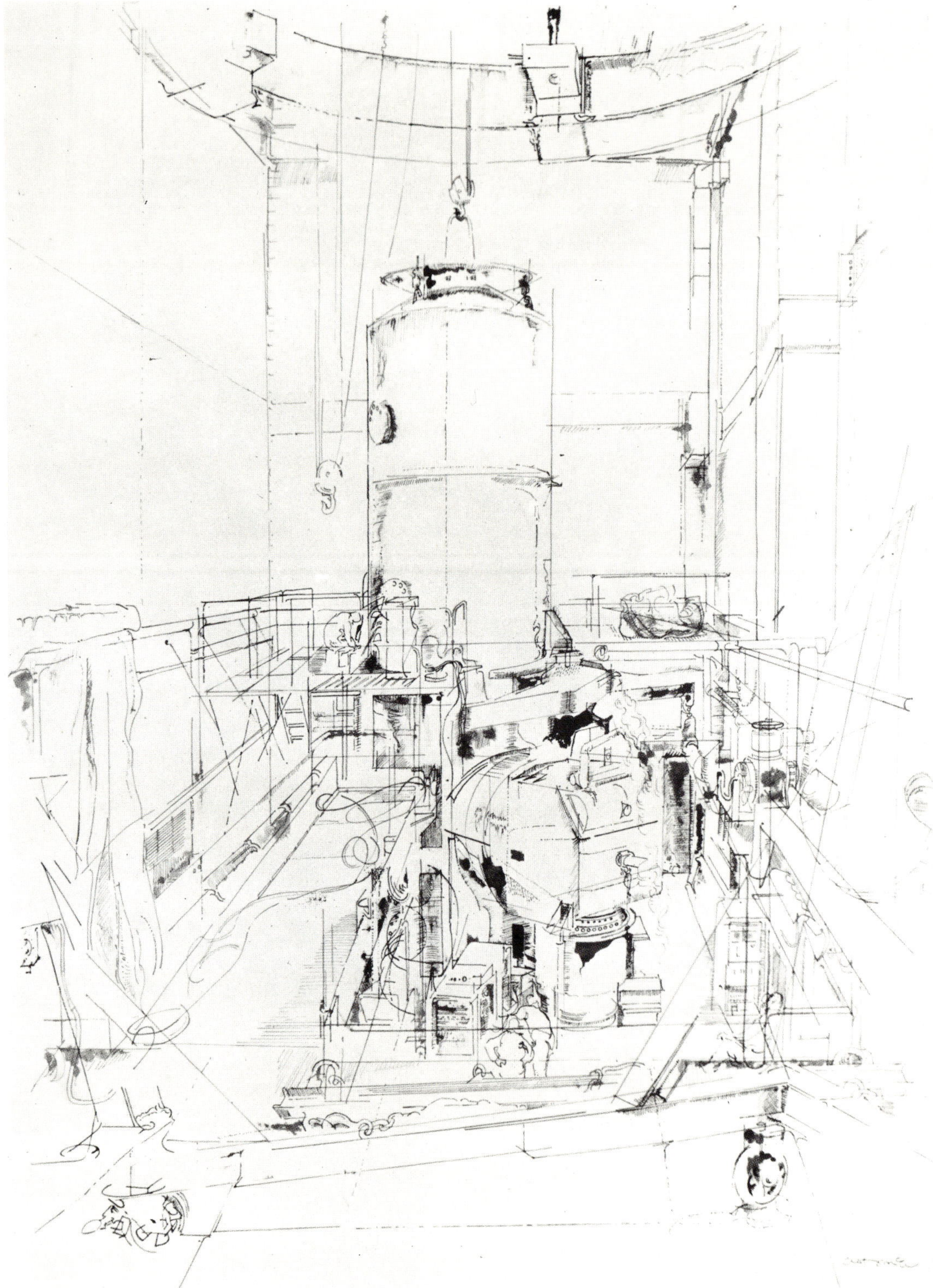

86.
Alfred McAdams
TESTING AT GODDARD
Ink and wash on paper, 24×18″

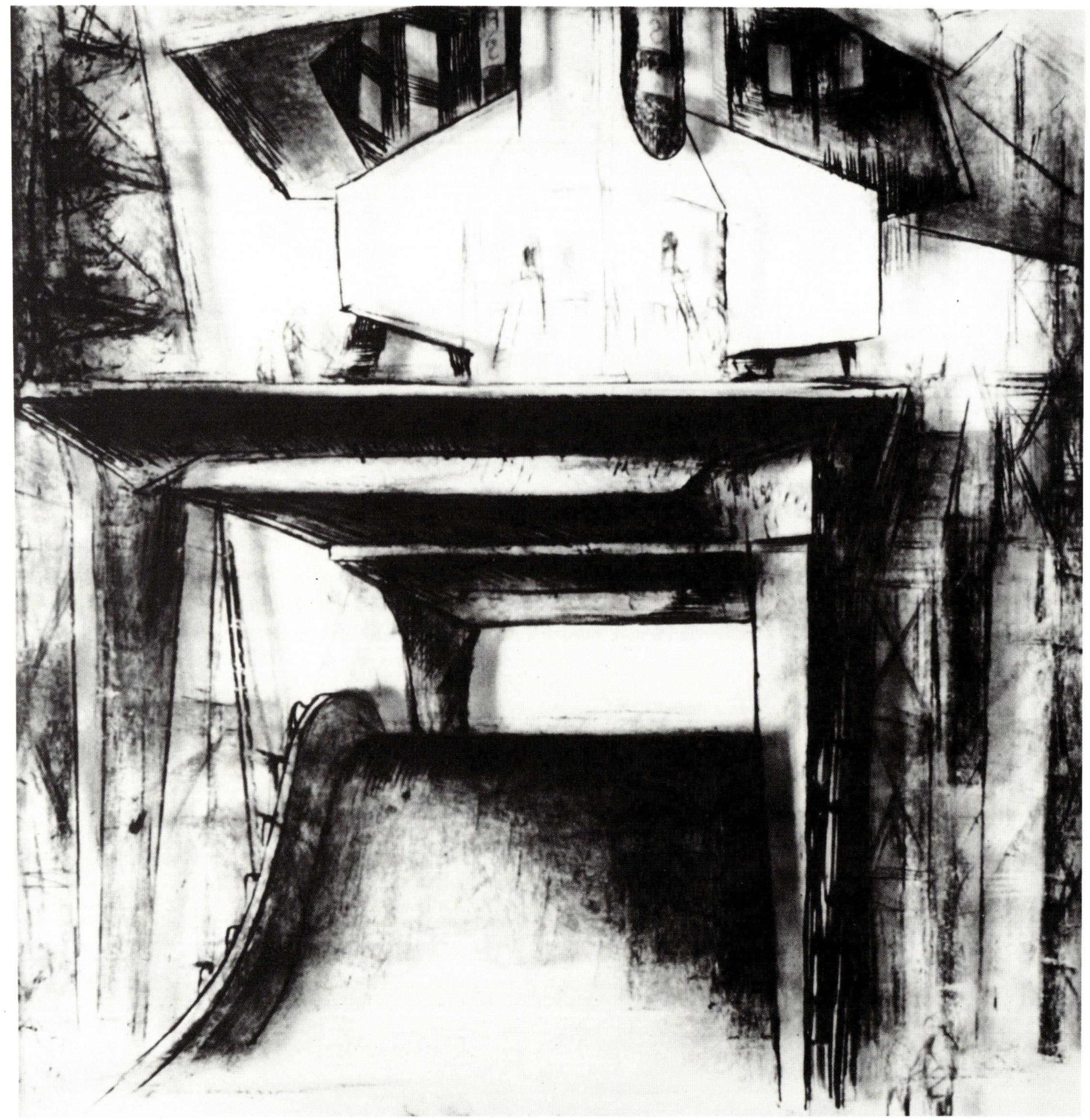

87. *Mitchell Jamieson.* FLAME DEFLECTOR. Ink on acetate, 7×7″

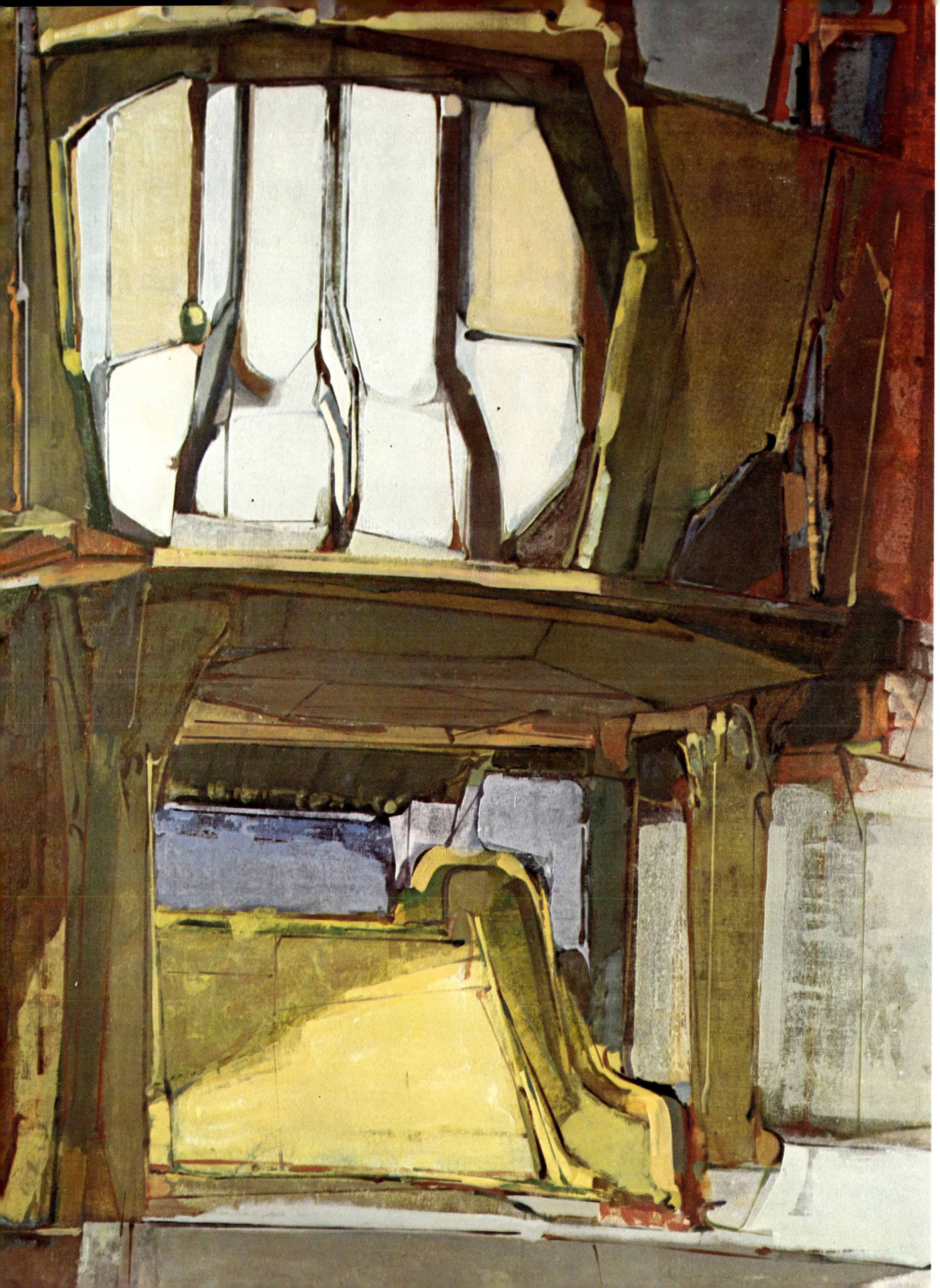

88.
Alfred McAdams
DEFLECTOR
Oil on canvas, 51 × 43″

Of all the strange and monstrous shapes of space technology, few can compare with the flame deflector. Placed under the rocket, it deflects the incredible blast from the engine. Painted a cheerful canary yellow and shaped vaguely like a hunched caterpillar, it does not seem to be destined for the inferno. Plate 86 shows rooms at the Goddard Space Flight Center where satellites are made and tested.

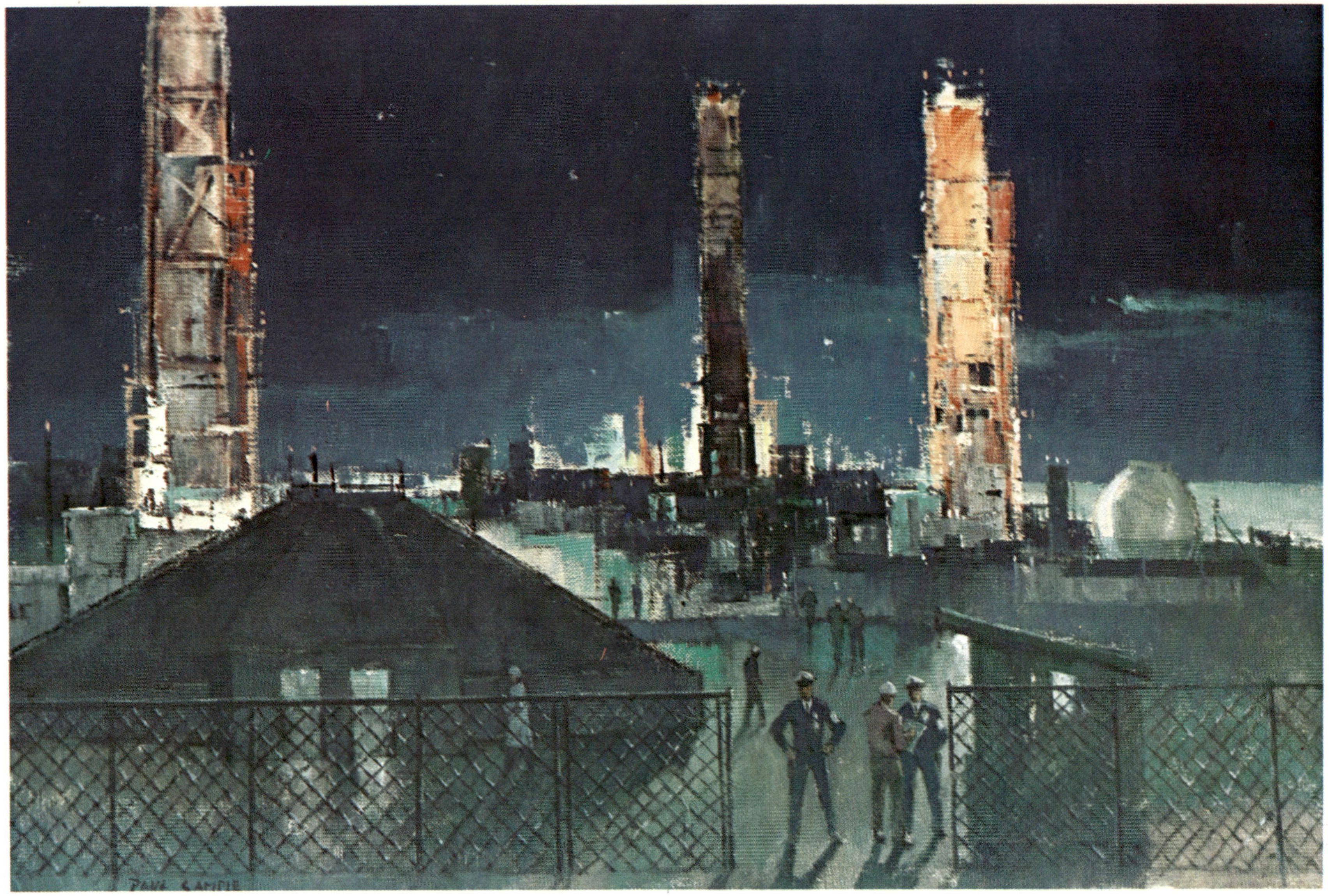

89.
Paul Sample
NIGHT CHECK
Acrylic on canvas, 24 × 36″

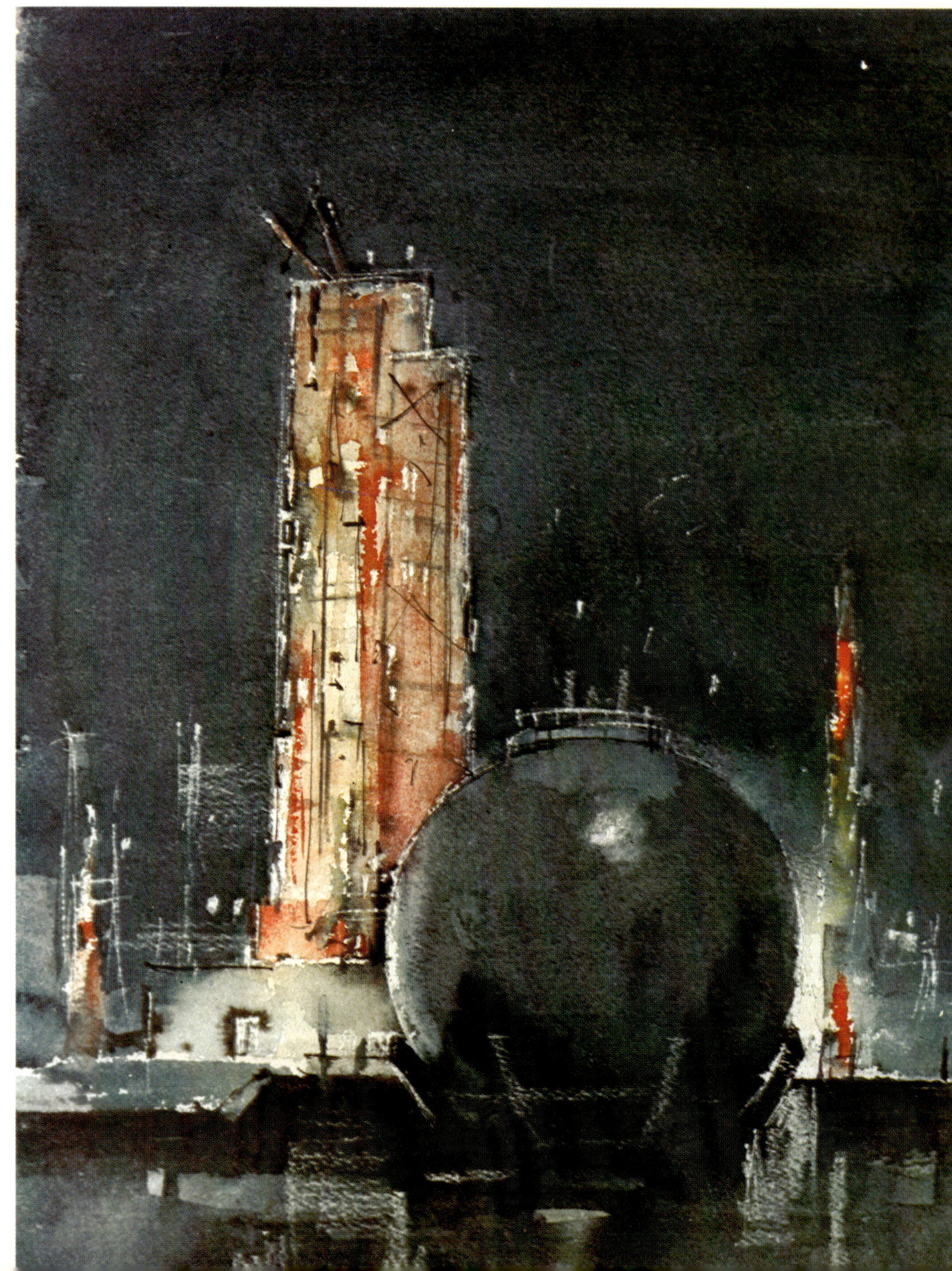

90.
Paul Sample
LIQUID FUEL
Watercolor on paper, 20 × 17″

In the days of the Mercury program in 1963, when a major launch was being readied, the Cape reminded one vaguely of a Luna Park or a theater strip. Flashing lights, crowds of people, traffic, and the hum of men at work. Each section was guarded by separate groups of security men, and only the right badge at the right place would gain admission. NASA was fortunate indeed to have Paul Sample among the first group of artists at the Cape. A series of brilliant watercolors resulted.

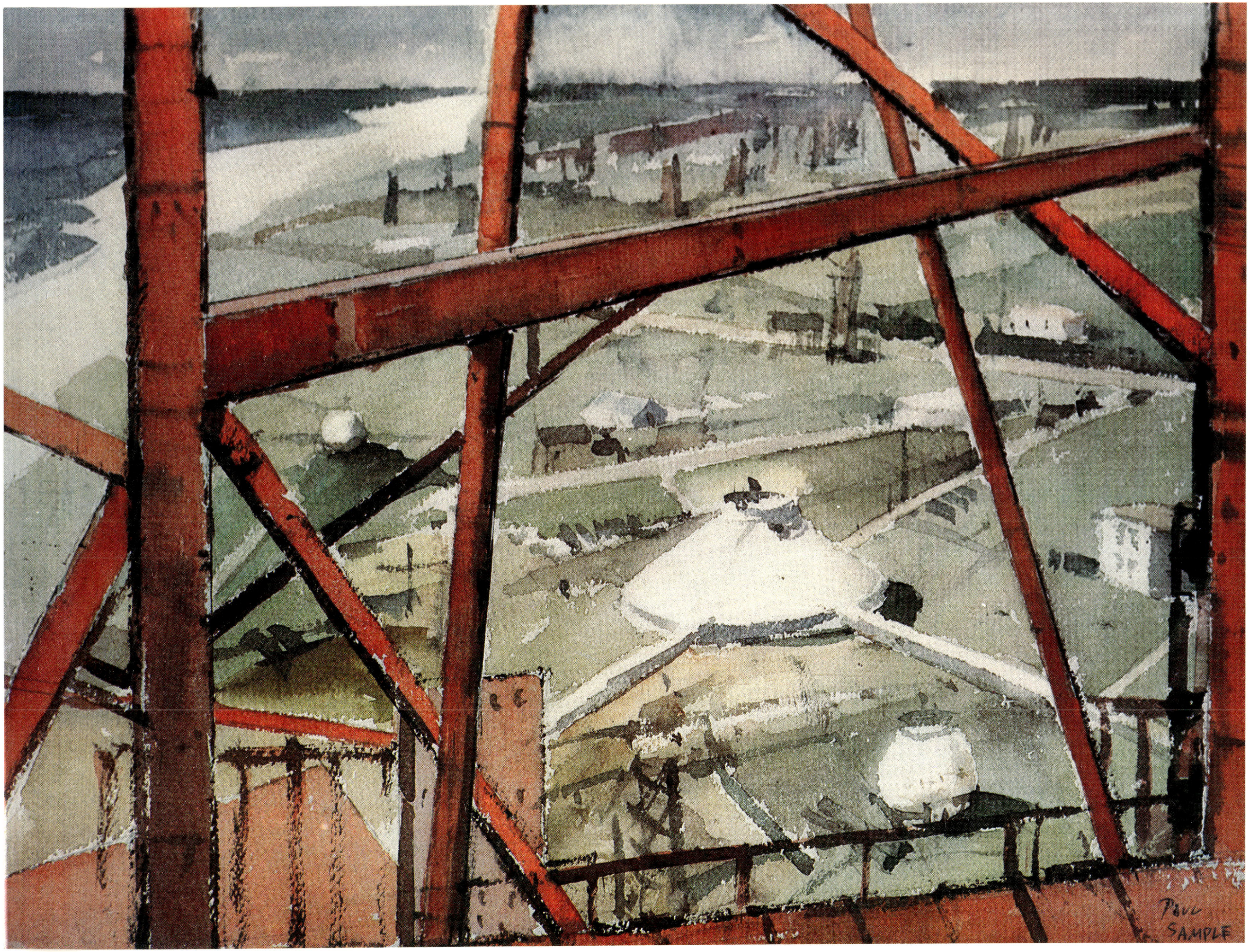

91. *Paul Sample*. ROW. Watercolor on paper, 17×20″

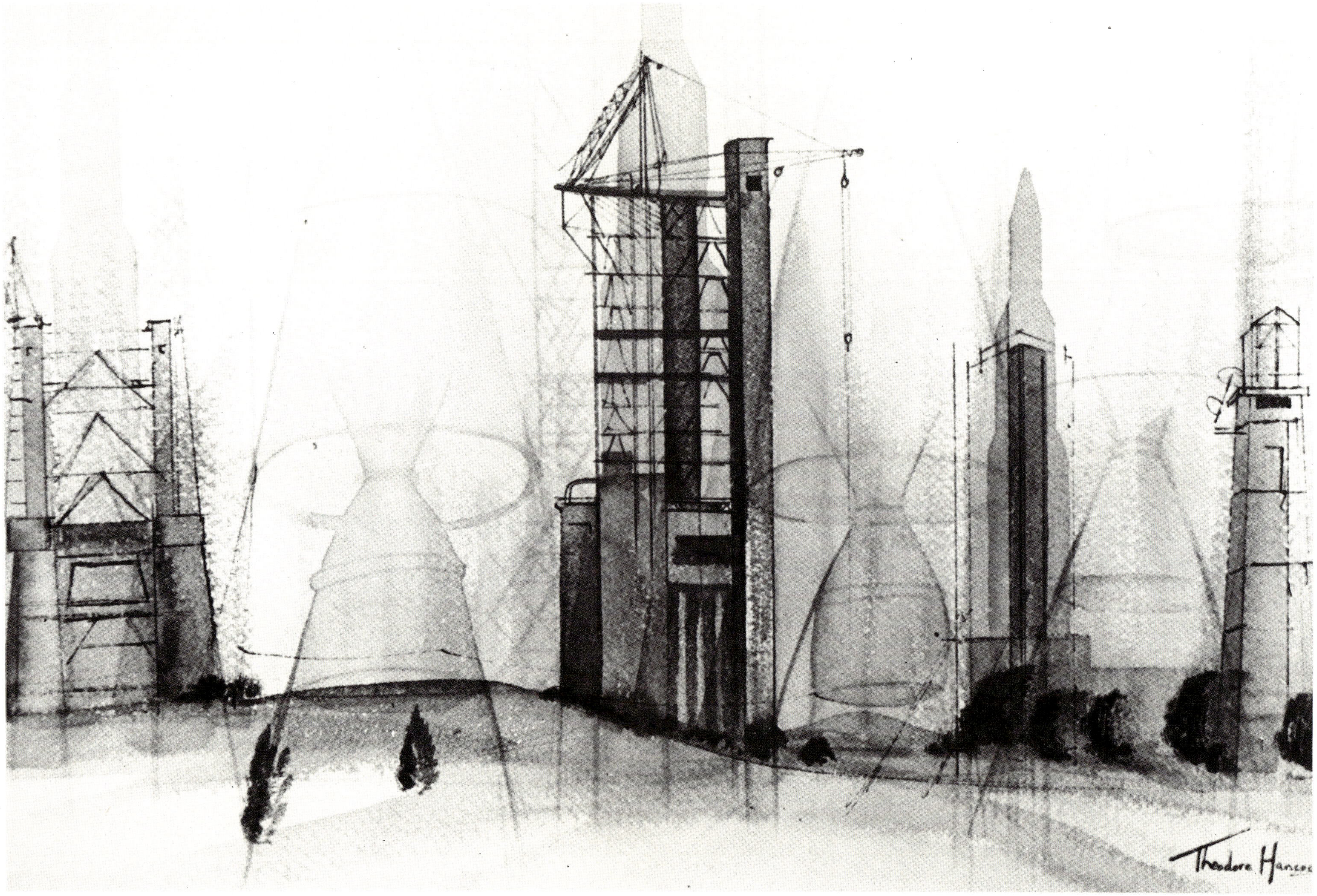

92. *Theodore Hancock.* THE SOUNDS OF ENGINES. Ink and watercolor on paper, 7 × 12″

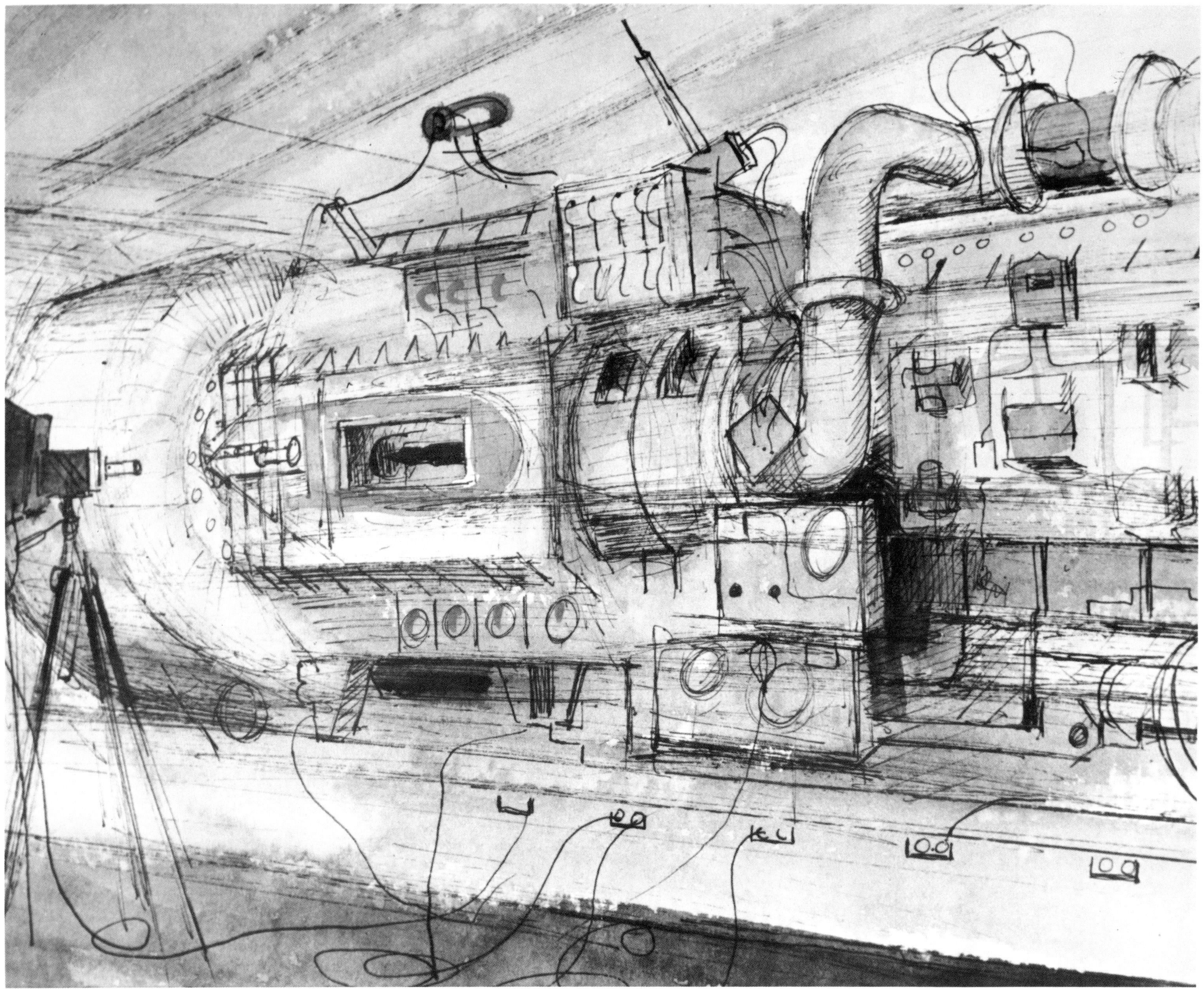

93.
Theodore Hancock
WIND TUNNEL
Ink and watercolor on paper, 9×12″

The wind tunnel of yesteryear has evolved during the Space Age into a metal tube shaped like a giant naval gun, where for brief seconds particles hurtle at velocities many times the speed of sound. Suspended in this cosmic cyclone are models of flying bodies which the nonspecialist finds hard to believe, but which are destined to pass through the earth's atmosphere, glowing white hot from the air friction. A camera records the action of friction and sound waves, and engineers can then deduce what shapes these vehicles of tomorrow may have.

95. *Theodore Hancock.* ALABAMA LANDSCAPE.
Ink and watercolor on paper, 7½ × 22″

Before the Space Age, Huntsville was a sleepy country town set among the red clay hills of Alabama and jealously guarding a glorious role in the Civil War. Here the Redstone arsenal was assigned to the Corps of Engineers and was used as a storage area. After World War II, many of the German scientists who had worked on the German rocket program were asked to continue their research on rocketry at this base. Here Wernher von Braun and his corps of highly trained specialists developed first the Redstone rocket, which put into orbit America's first satellite and first man, and later the family of Saturn rockets whose record of success has been little short of phenomenal. The English-born artist Theodore Hancock, perhaps unconsciously, saw the space city nestling among gentle hills as a medieval fortress town, with campaniles and strange engines of war rising from its cluster of barracks and workshops.

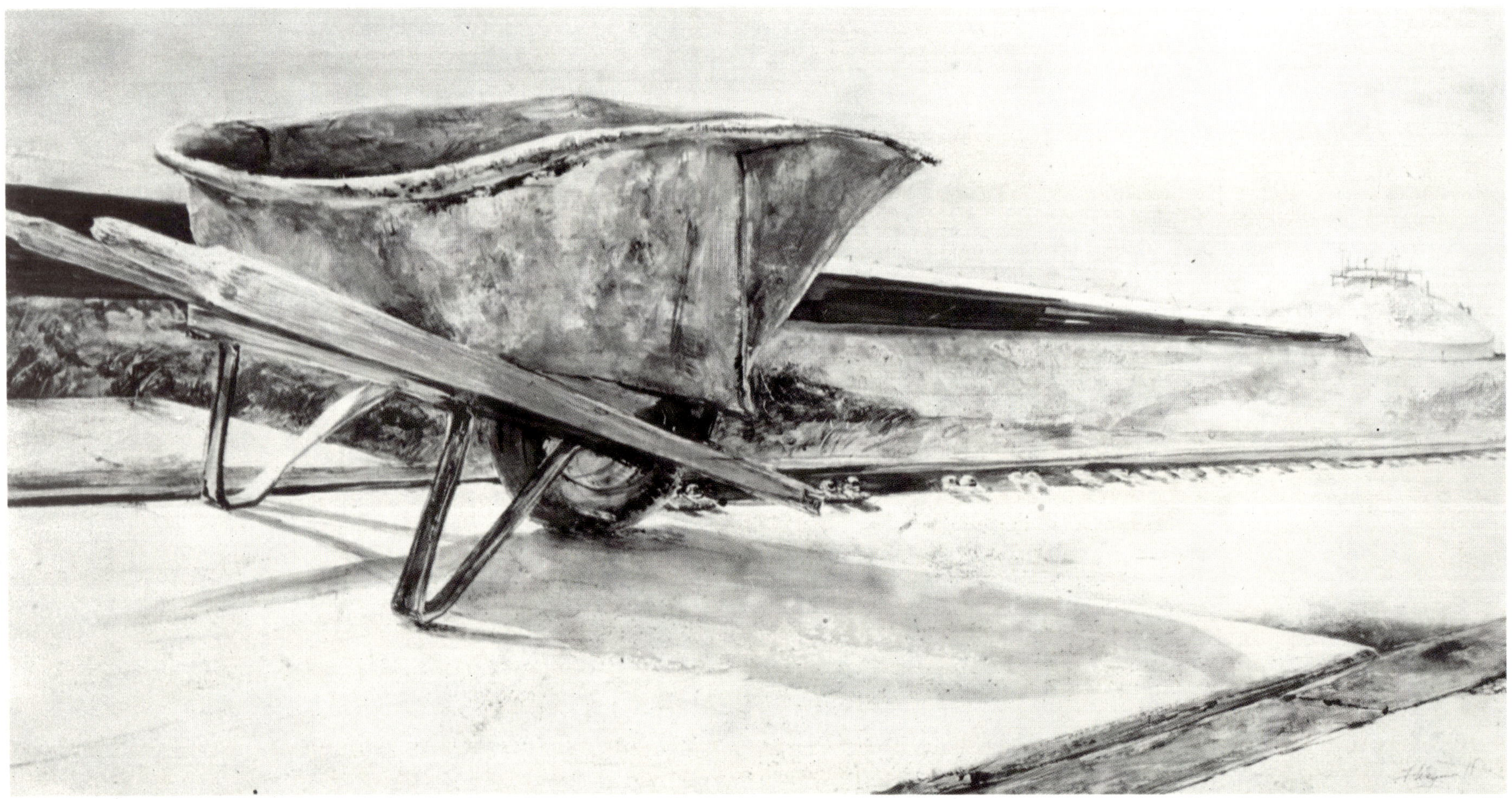

94.
George Weymouth
BARROW
Watercolor on paper, 24 × 37″

Cape Kennedy has many marvels of space technology, but weeds still grow, sand blows, rail lines run along the seaside, and wheelbarrows still serve their purpose.

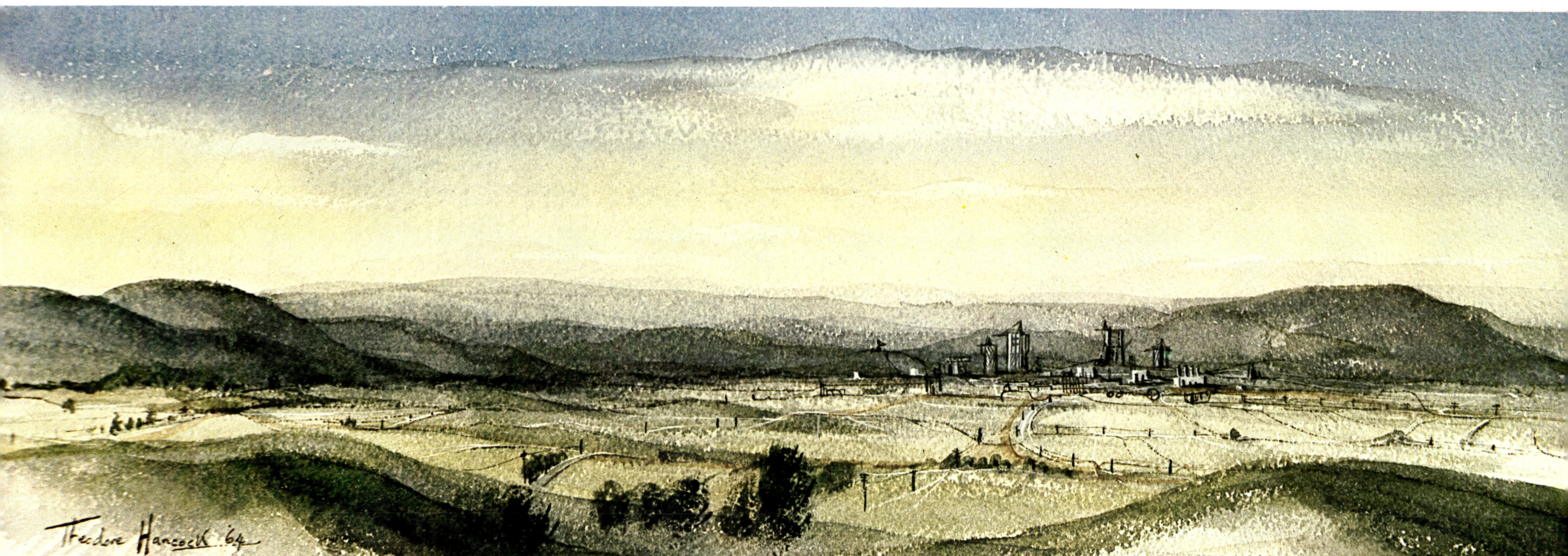
Theodore Hancock '64

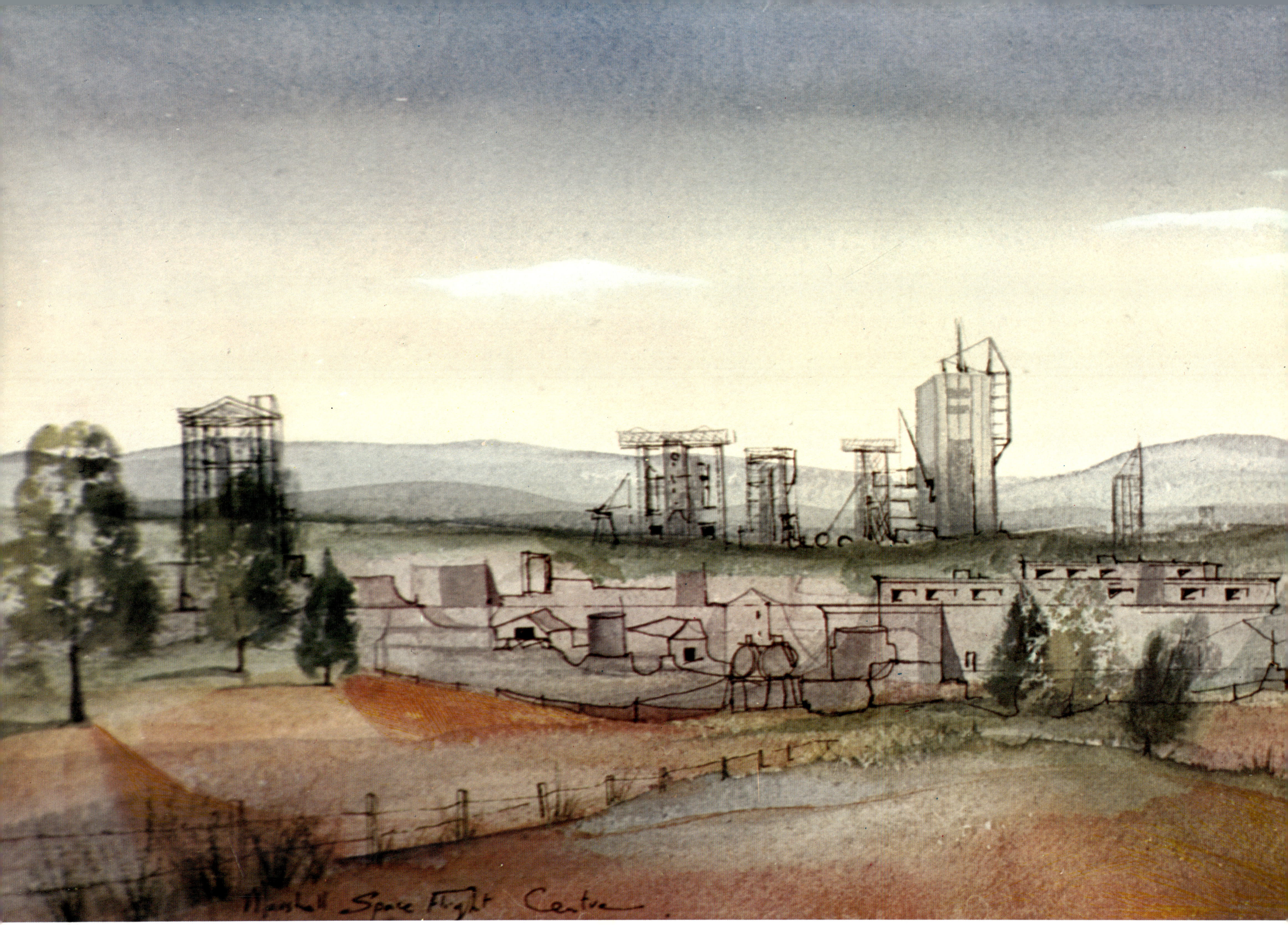
Marshall Space Flight Centre

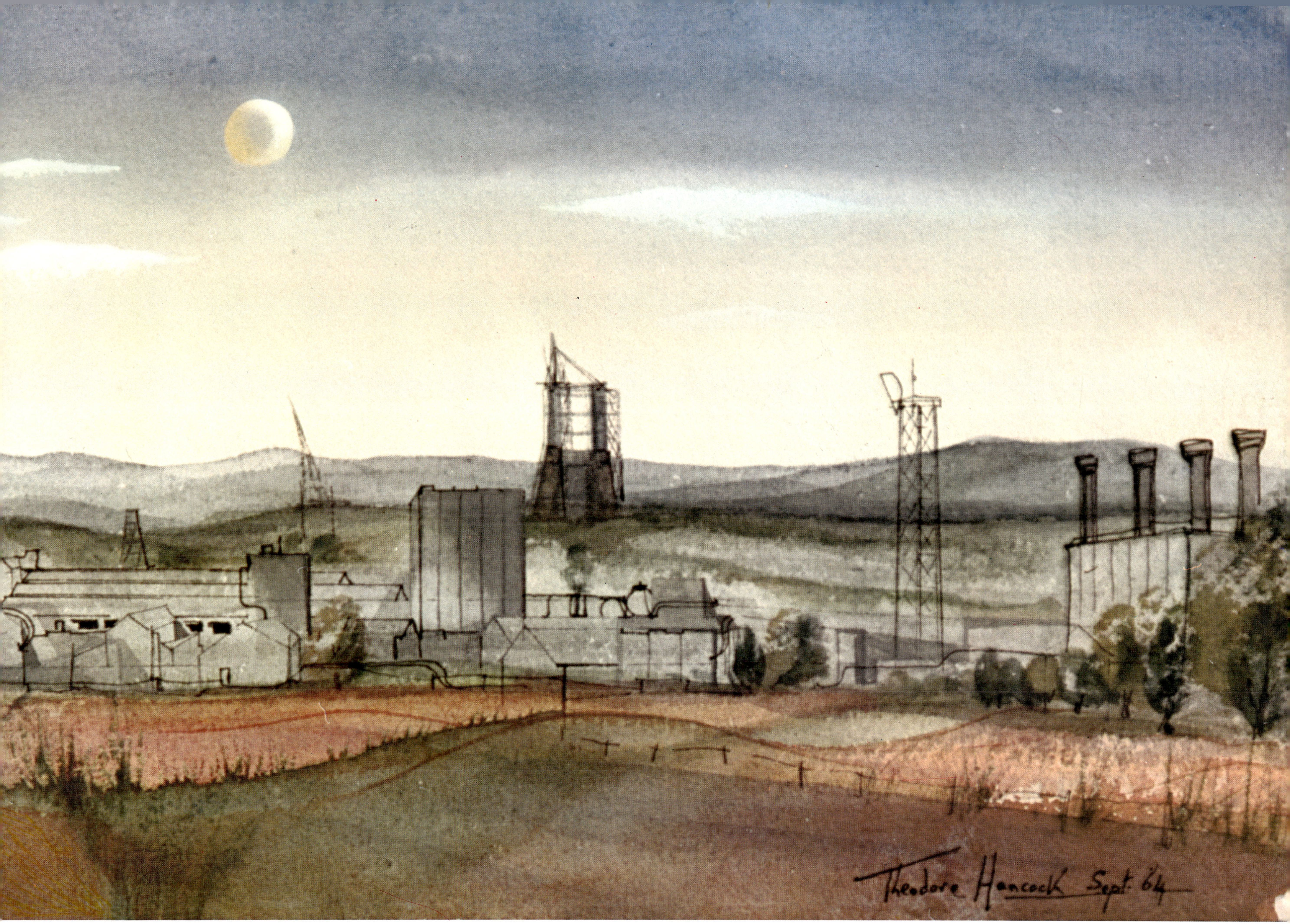

96. *Theodore Hancock.* MARSHALL. Ink and watercolor on paper, 7½ × 22″

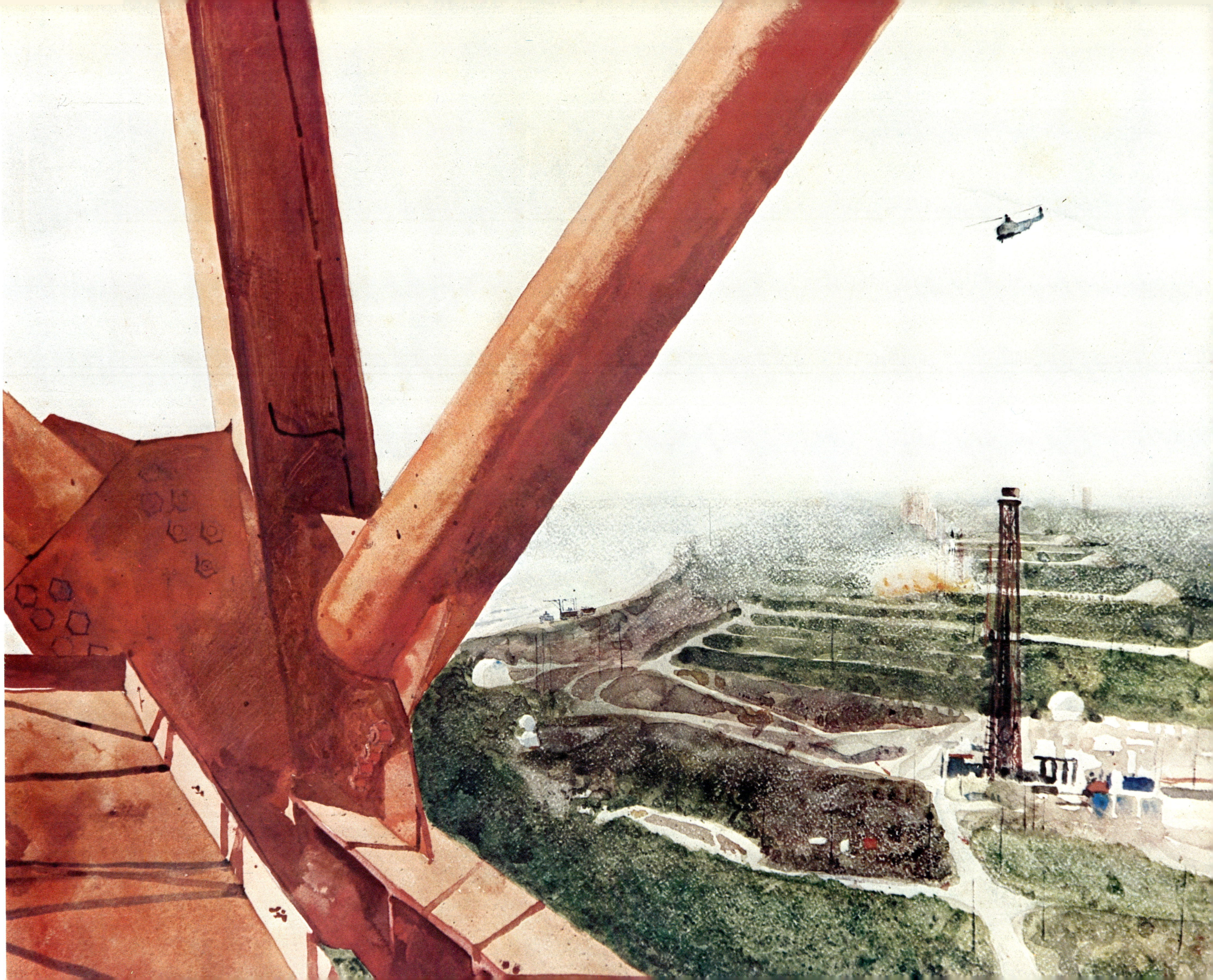

Eyewitness to Space

II

97.
James Wyeth
SUPPORT
Watercolor on paper, 21×24″

A view down gantry row at Cape Kennedy is like turning the pages of the history of America's space program. The earliest structures, now obsolete, rust and rot. Others are in mothballs, awaiting some future use.

98. Robert McCall *Rollout*
99. Paul Calle *Shirra and Stafford*
100. Robert Shore *Weightless*
101. Nicholas Solovioff *Obsolete Launch Pad*
102. Hugh Laidman *Old House on the Move*
103. William Thon *Space Age Landscape*
104. Nicholas Solovioff *Carrier Deck at Dawn*
105. Robert McCall *Copter Cockpit*
106. Franklin McMahon *Flag Plot Room*
107. Robert McCall *Carrier Deck at Night*
108. Robert McCall *Gemini Recovery*
109. Franklin McMahon *Red Carpet*
110. Paul Calle *GT-3 Recovered*
111. Fletcher Martin *Flame Pit*
112. Paul Calle *Gemini Astronaut McDivitt*
113. Paul Calle *Gemini Spacecraft and Gantry*
114. Paul Calle *Gemini Liftoff*
115. Paul Calle *Inside Gemini Spacecraft*
116. L. Katzen *Moon Markers*
117. Yeffe Kimball *Moonscape*
118. Fred Messersmith, A. W. S. *Moonscape*
119. James P. Hendricks *Moon Craters*
120. Paul Calle *Down to 14*
121. Lamar Dodd *Saturn Gantry*
122. Peter Hurd *Tracking*
123. Hugh Laidman *Wet Morning*
124. Mitchell Jamieson *Carrier Antennae*
125. Nicholas Solovioff *Carrier Flight Deck*
126. Paul Arlt *Big Dish Antennae, Tananarive*
127. Theodore Hancock *On the Barge*
128. Theodore Hancock *Down the Mississippi*
129. Theodore Hancock *MSFC*
130. Paul Calle *Centrifuge*
131. Paul Calle *Rendezvous*
132. Paul Calle *Gemini White Room*
133. Norman Rockwell *Grissom and Young*
134. Nicholas Solovioff *Tracking Ship*
135. Franklin McMahon *Canary Island Station*
136. Paul Calle *Grids*
137. Ludek Pesek *With Saturn's Rings*

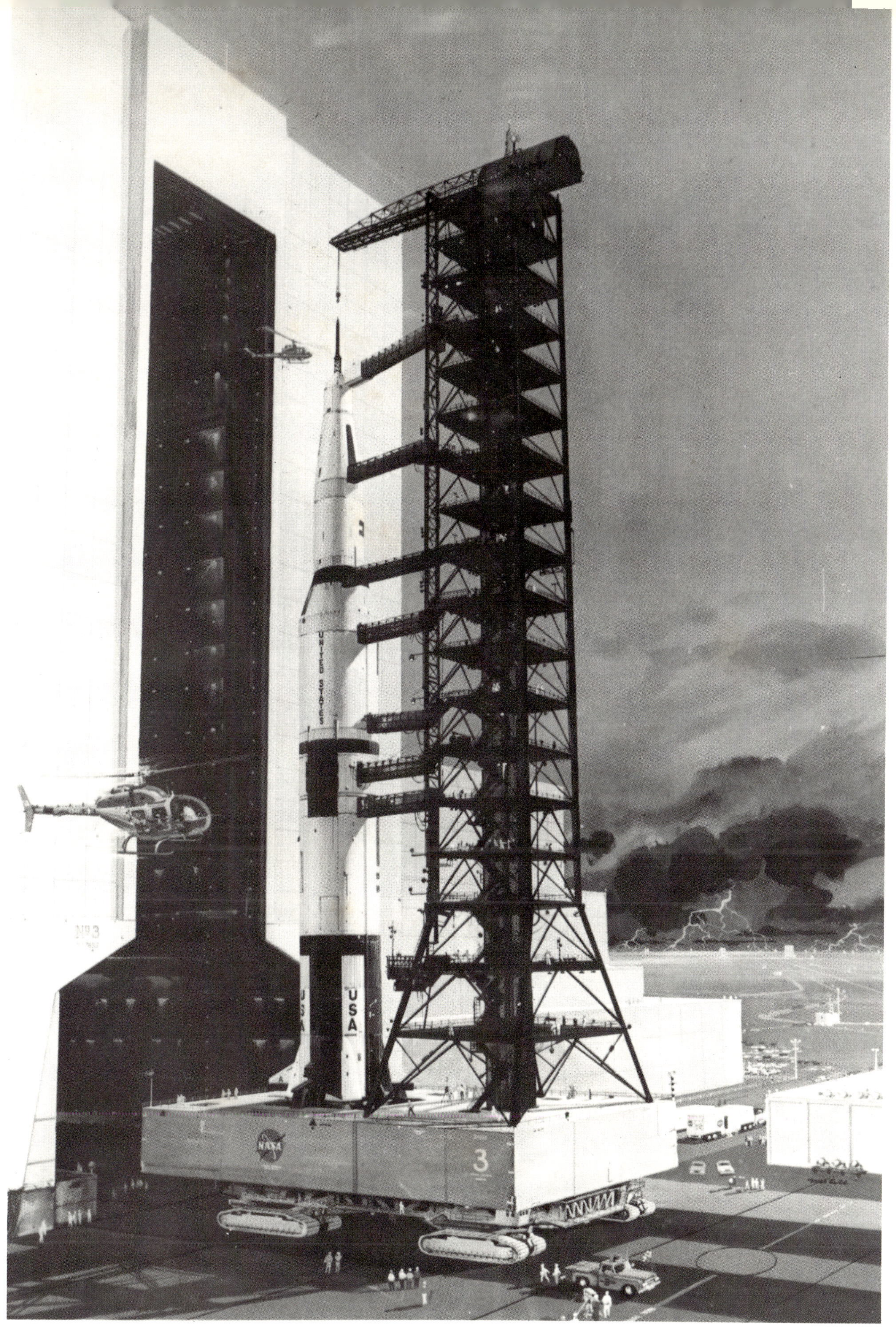

98. *Robert McCall.* ROLLOUT. Oil on canvas, 36×28". Reproduced courtesy of *Life* magazine

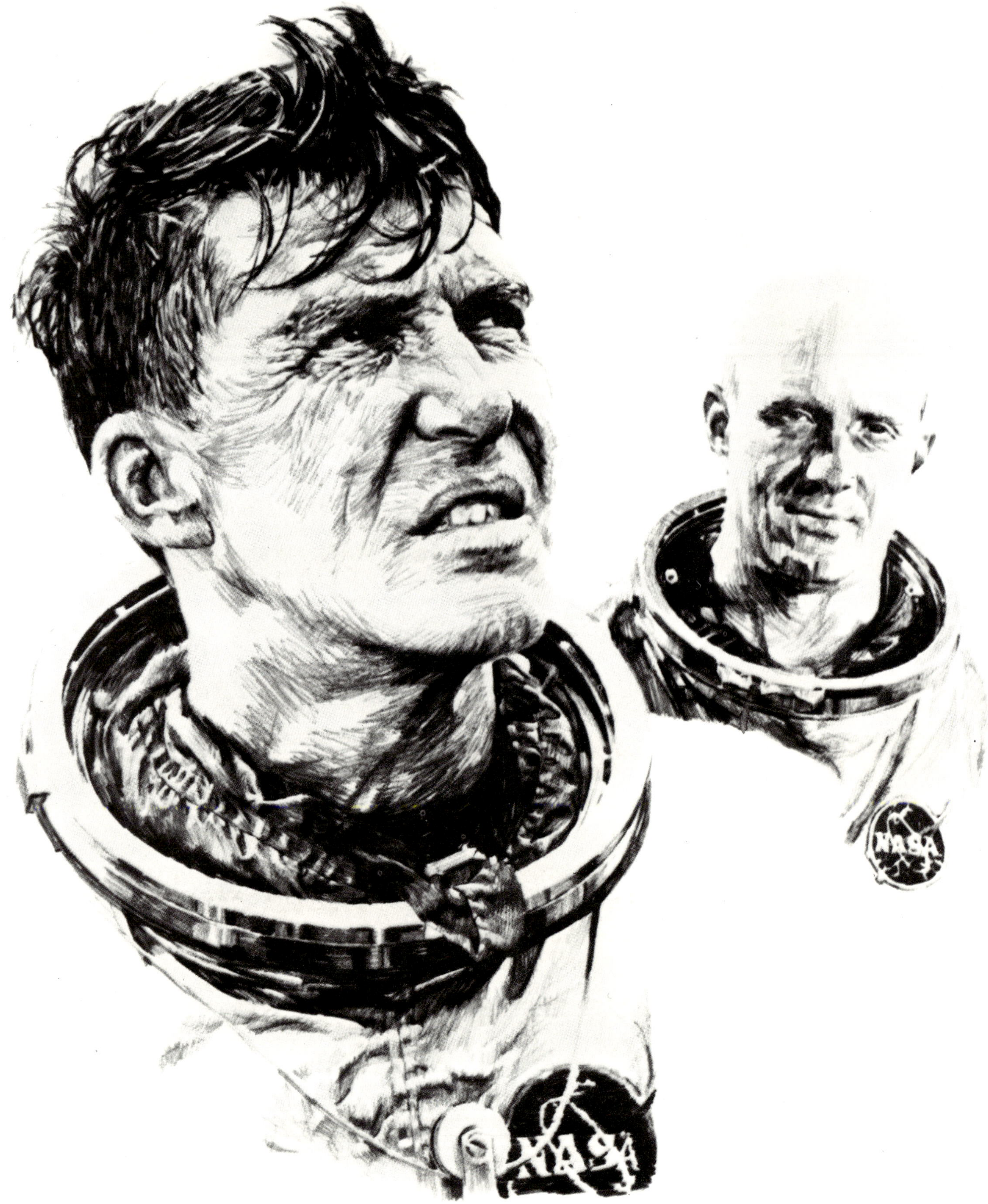

99.
Paul Calle
SHIRRA AND STAFFORD
Pencil on paper, 32×22″

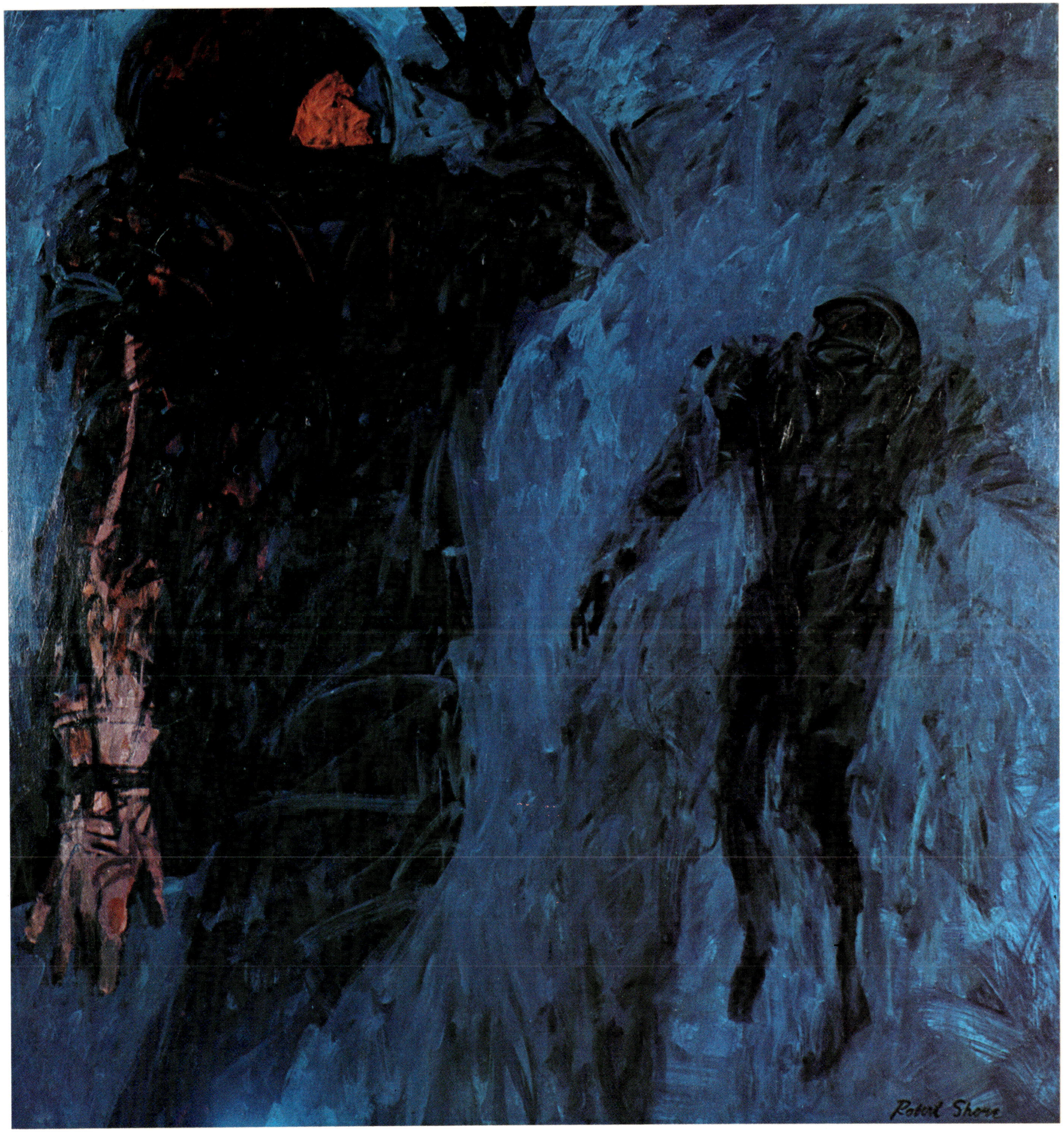

100.
Robert Shore
WEIGHTLESS
Oil on Masonite, 42×48″

To float, walk, and work in space is a challenge to the human machine that has no parallel in history. Astronauts were trained by scuba diving and in specially padded planes, where for brief periods zero G (weightlessness) could be attained. However, the early astronauts were unprepared for the exhilaration and the sense of unlimited freedom that space walking produces.

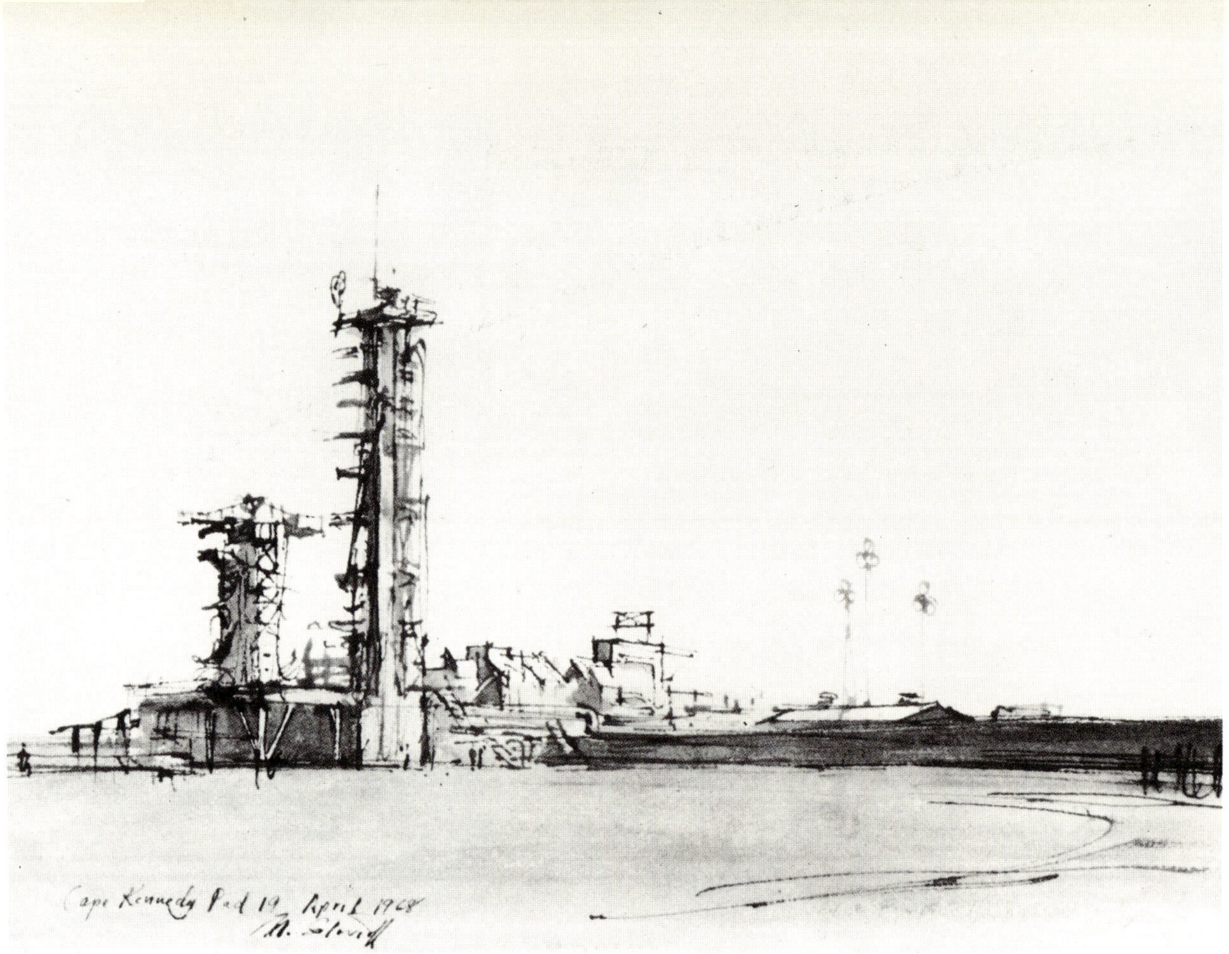

101.
Nicholas Solovioff
OBSOLETE LAUNCH PAD
Ink and wash on paper, 14×19″

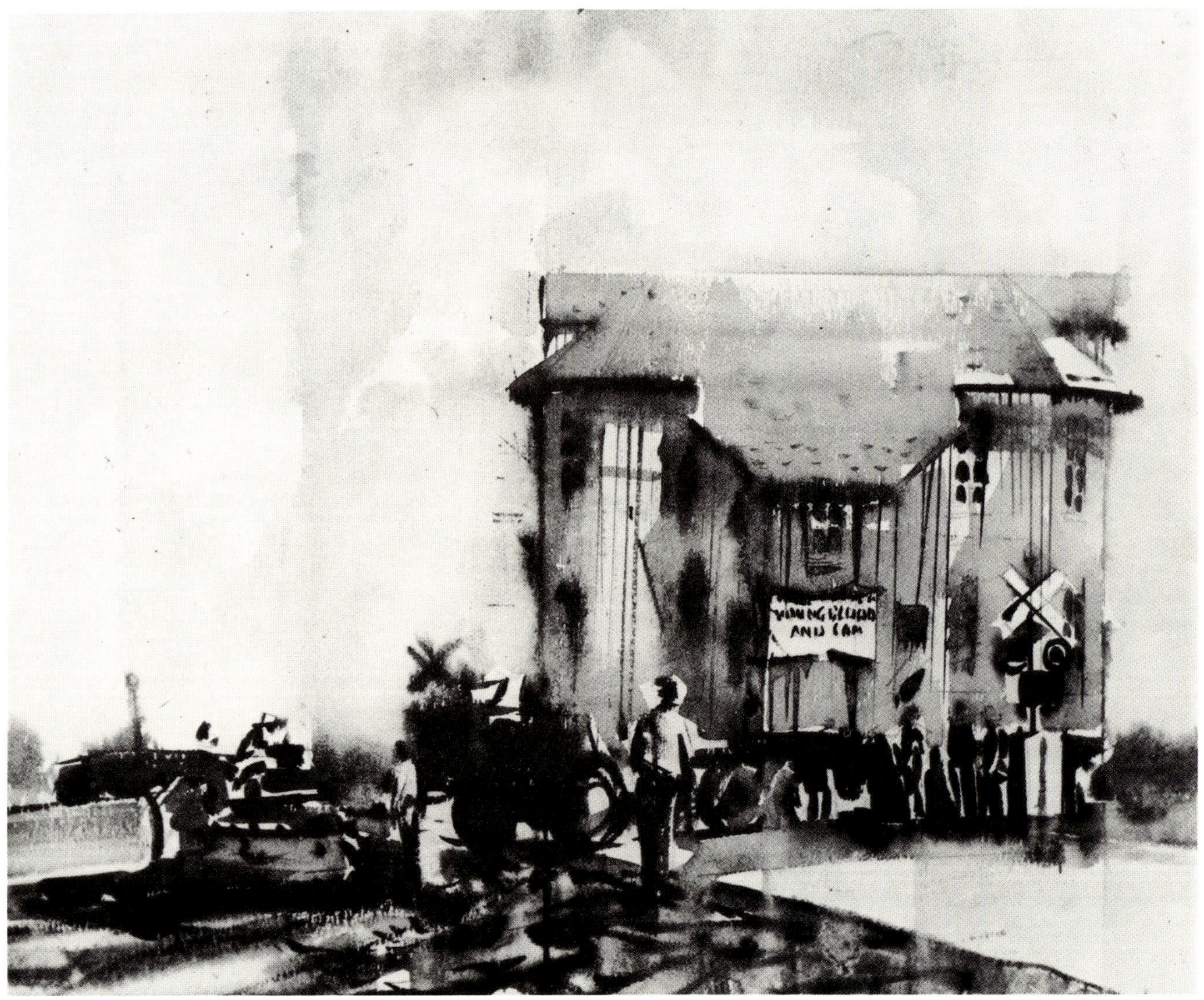

102.
Hugh Laidman
OLD HOUSE ON THE MOVE
Watercolor on paper, 28×36″

103.
William Thon
SPACE AGE LANDSCAPE
Watercolor on paper, 20×27″

The galloping technology of the Space Age makes equipment obsolete very soon after it has been used. Here a gantry of the Mercury era rusts in the subtropic sea air; most of its delicate equipment has been pirated by engineers working on later projects, and now it is used chiefly by sea birds who find the girders ideal nesting grounds. In the foreground, jungle growth moves in to reclaim its rights.

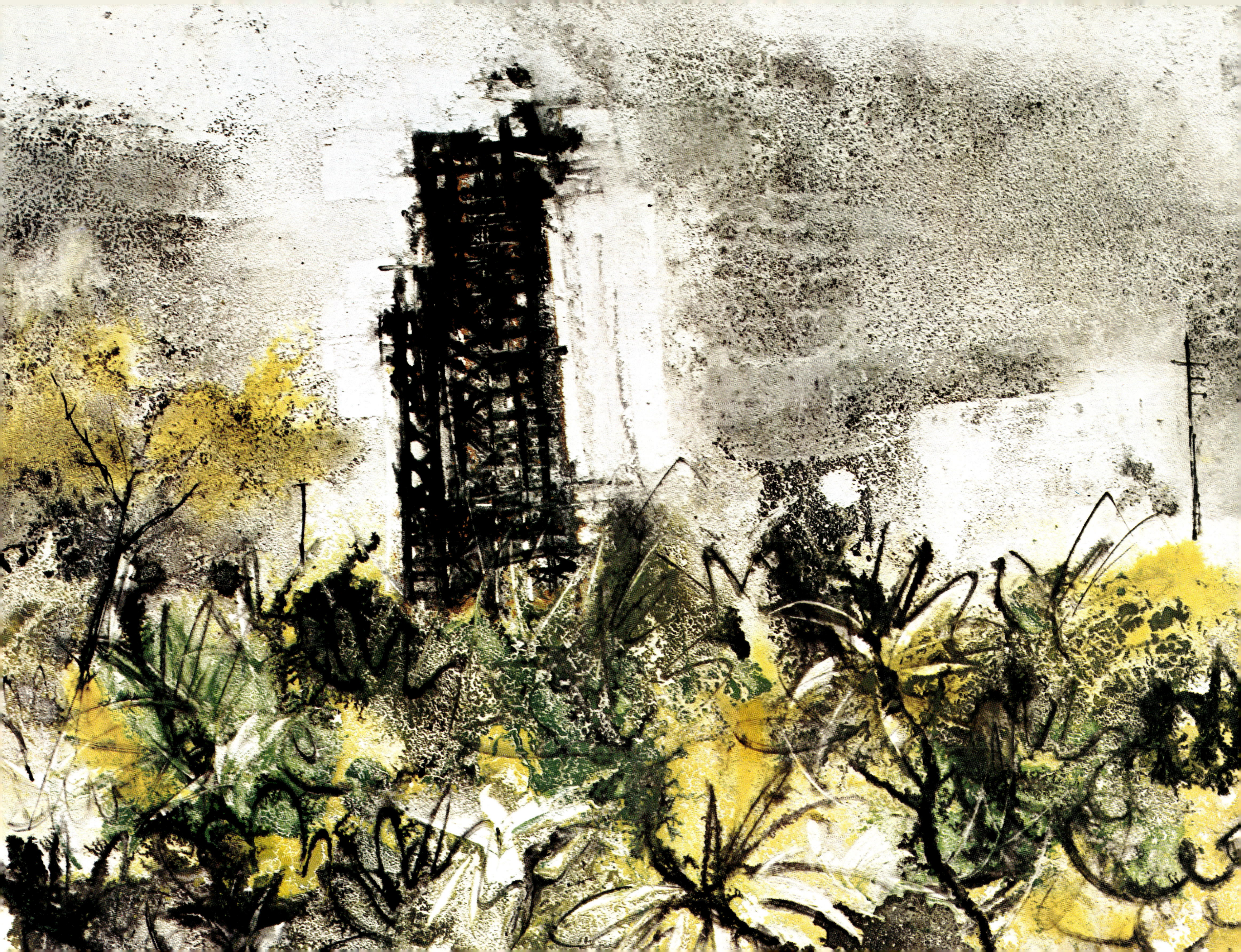

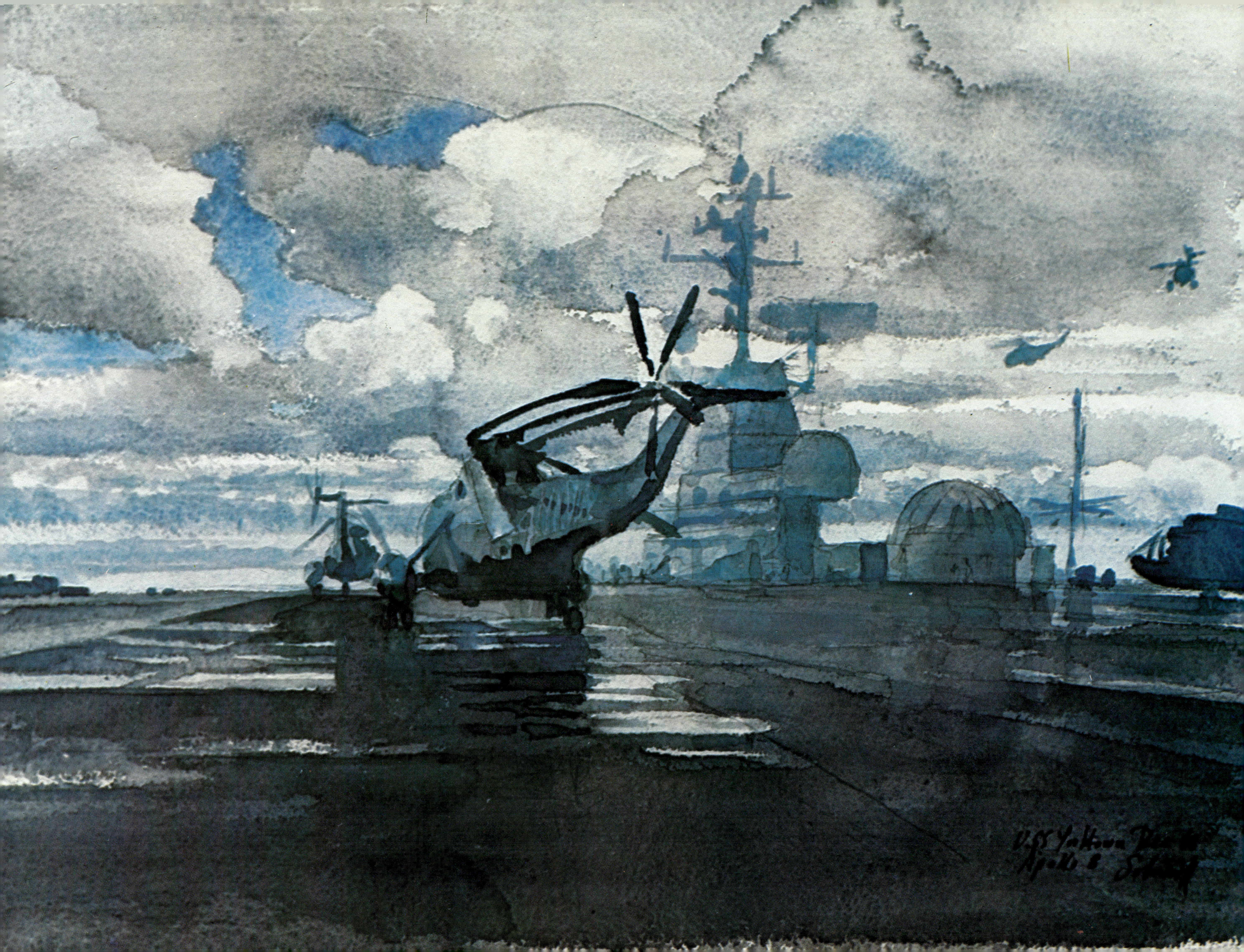
USS Yorktown
Apollo 8

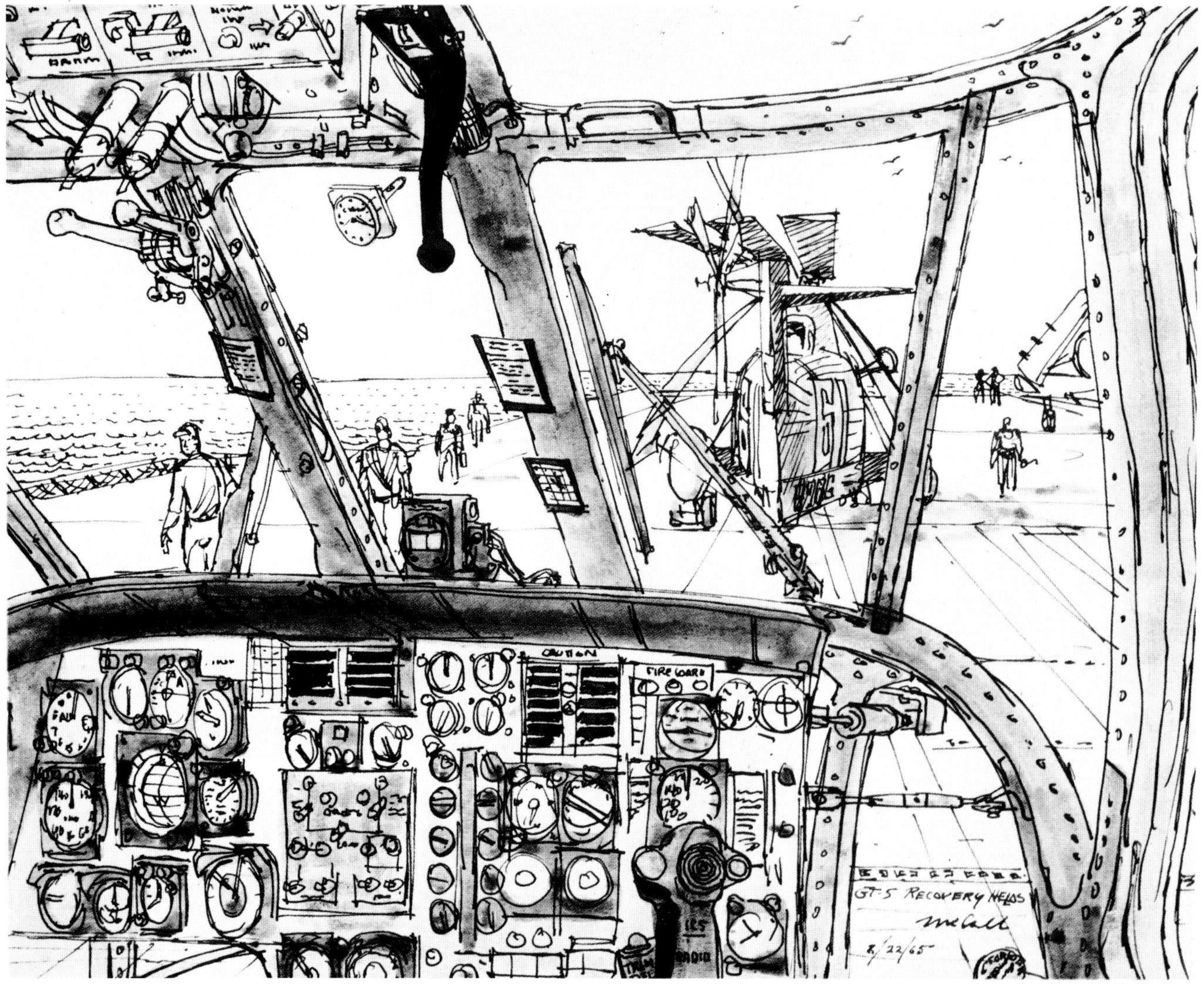

105.
Robert McCall
COPTER COCKPIT
Pen and ink on paper, 11×14″

104.
Nicholas Solovioff
CARRIER DECK AT DAWN
Watercolor on paper, 14×19″

The rain-drenched deck of the carrier, seeming like a vast and barren plain, supports helicopters with folded rotors and, in the distance, the circular tents put up by news media to shelter their broadcast antennae.

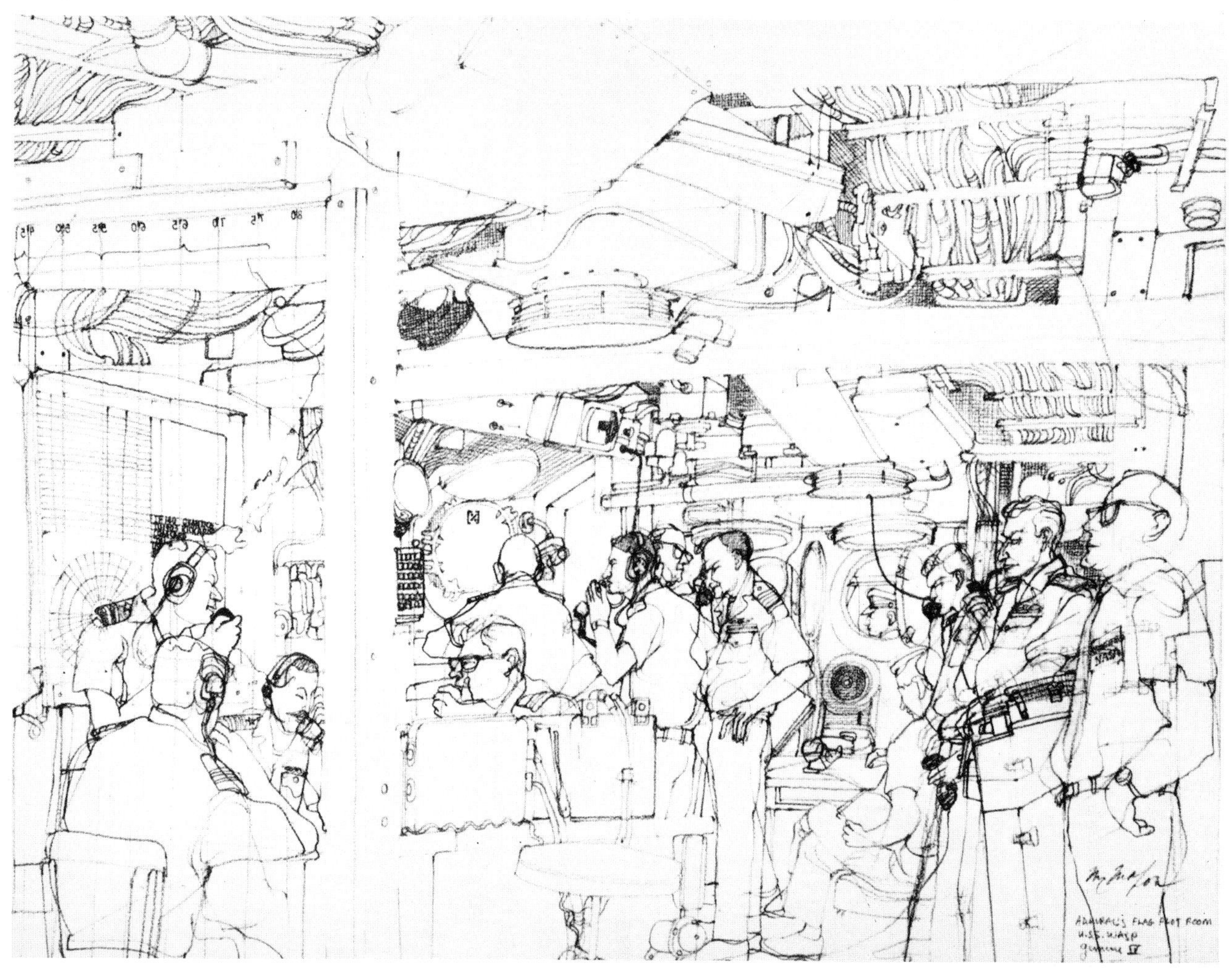

106.
Franklin McMahon
FLAG PLOT ROOM
Pencil on paper, 22×30″

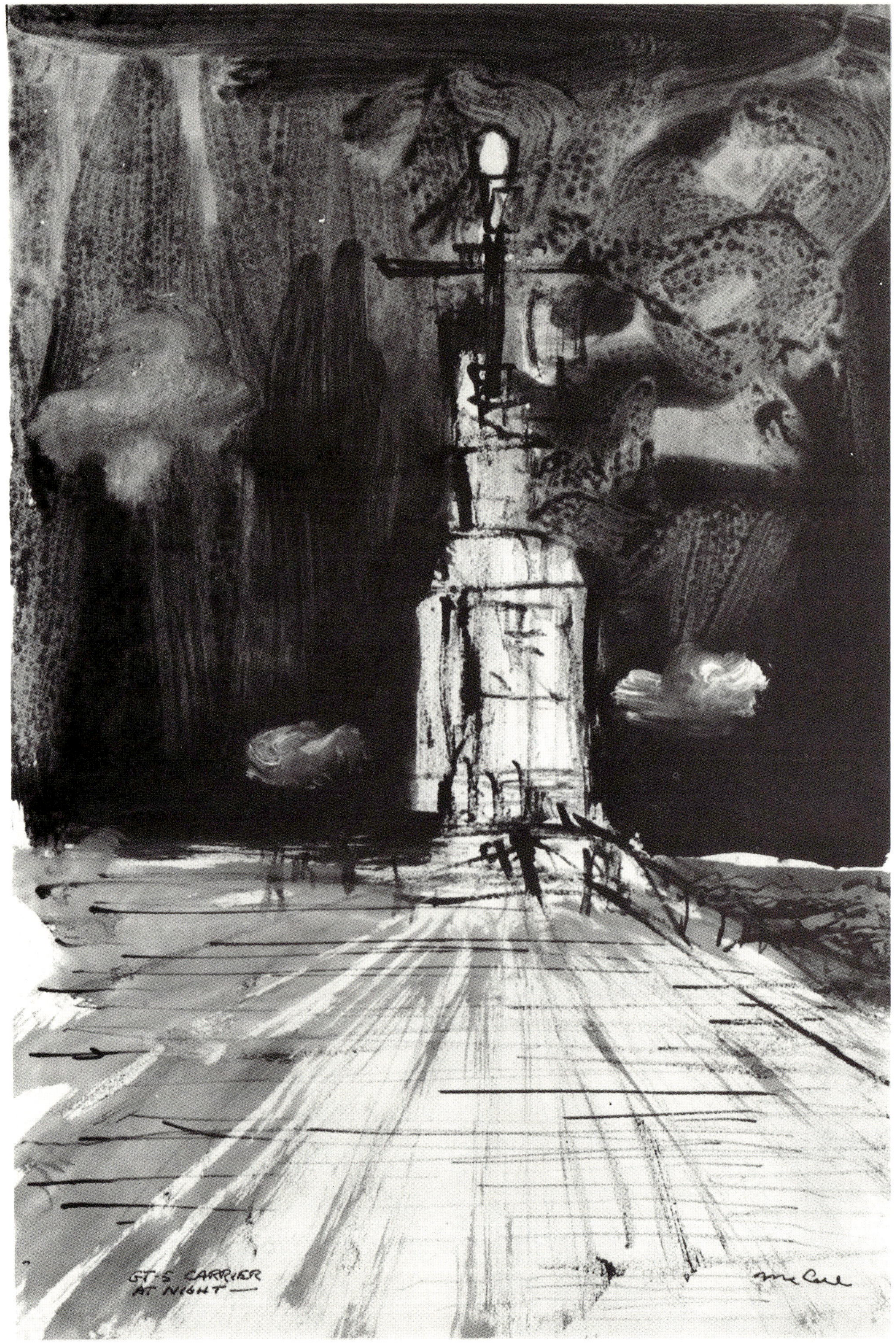

107.
Robert McCall
CARRIER DECK AT NIGHT
Watercolor on paper, 18×12″

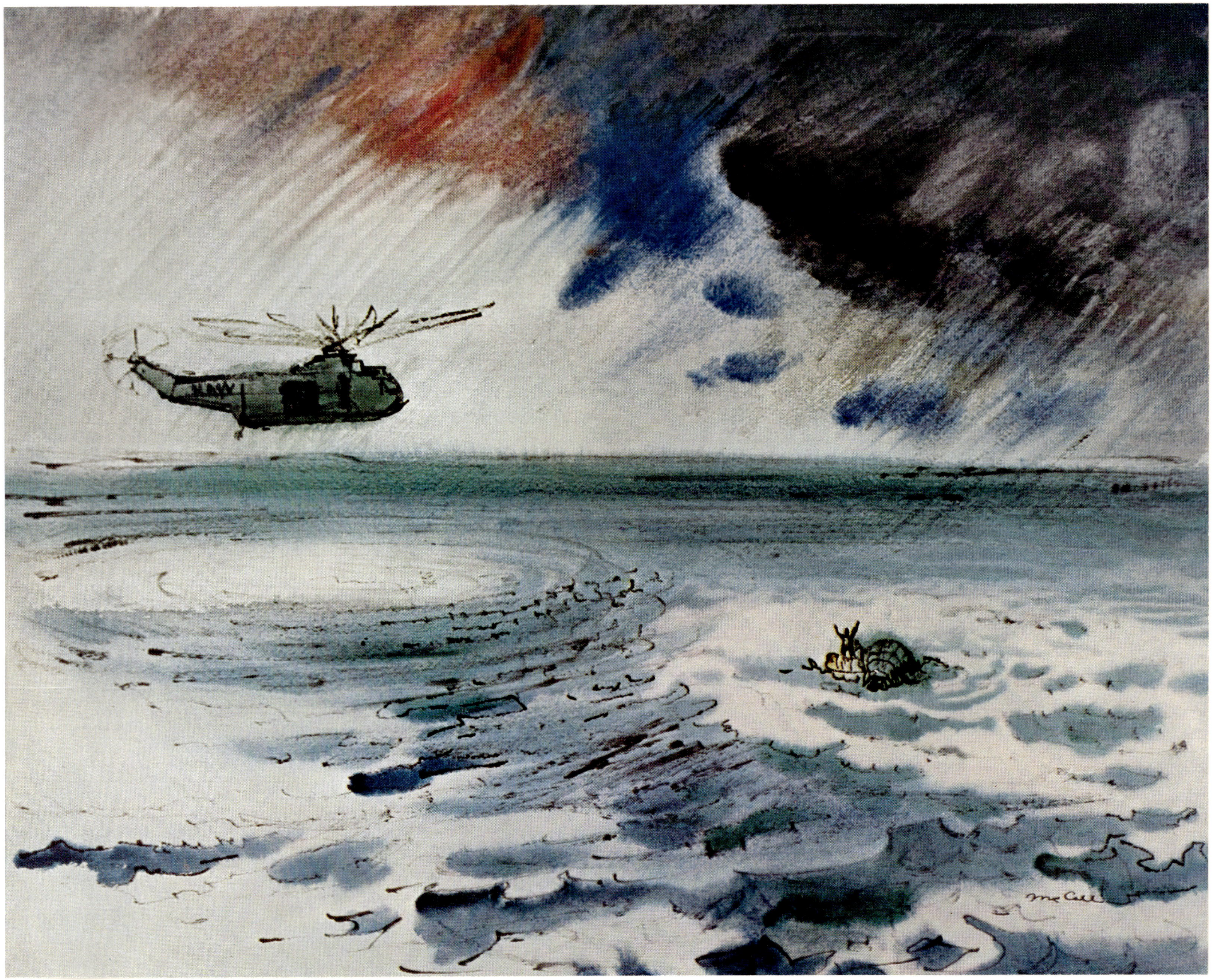

108.
Robert McCall
GEMINI RECOVERY
Watercolor on paper, 16×20″

The sound of a helicopter must have been music to the ears of the Gemini crew bobbing in their life rafts beside their downed craft. The occasion was the recovery of Gemini V, in 1965.

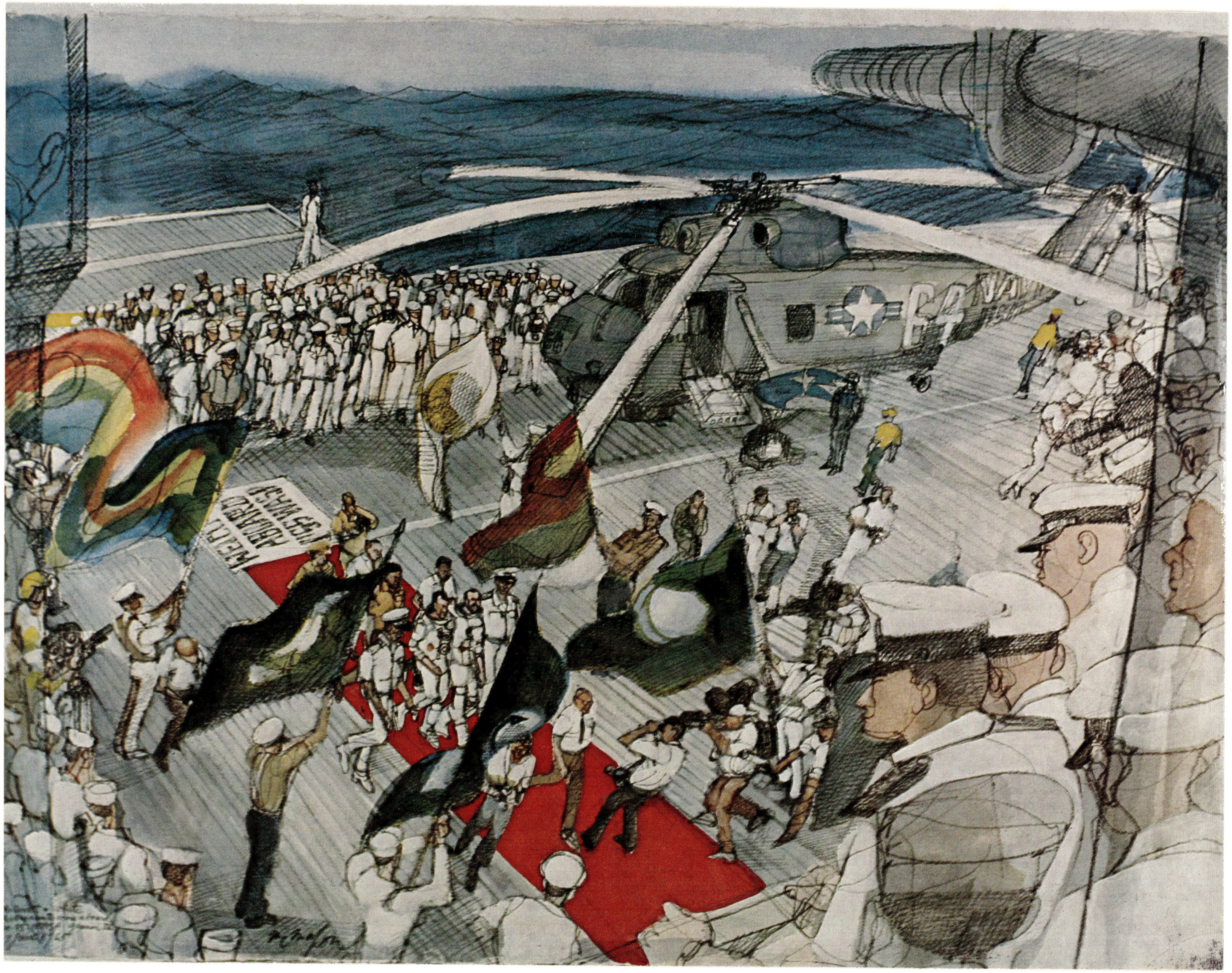

109.
Franklin McMahon
RED CARPET
Watercolor on paper, 22 × 30″

The band plays, the red carpet is rolled out, flags wave, the crowd cheers, photographers snap pictures, and triumphant men, home from the heavens, return to their own. McMahon has the ability to select only the significant facts from a scene such as this, and to leave a record that surrounds truth with a wiry line and a satirical sense of humor.

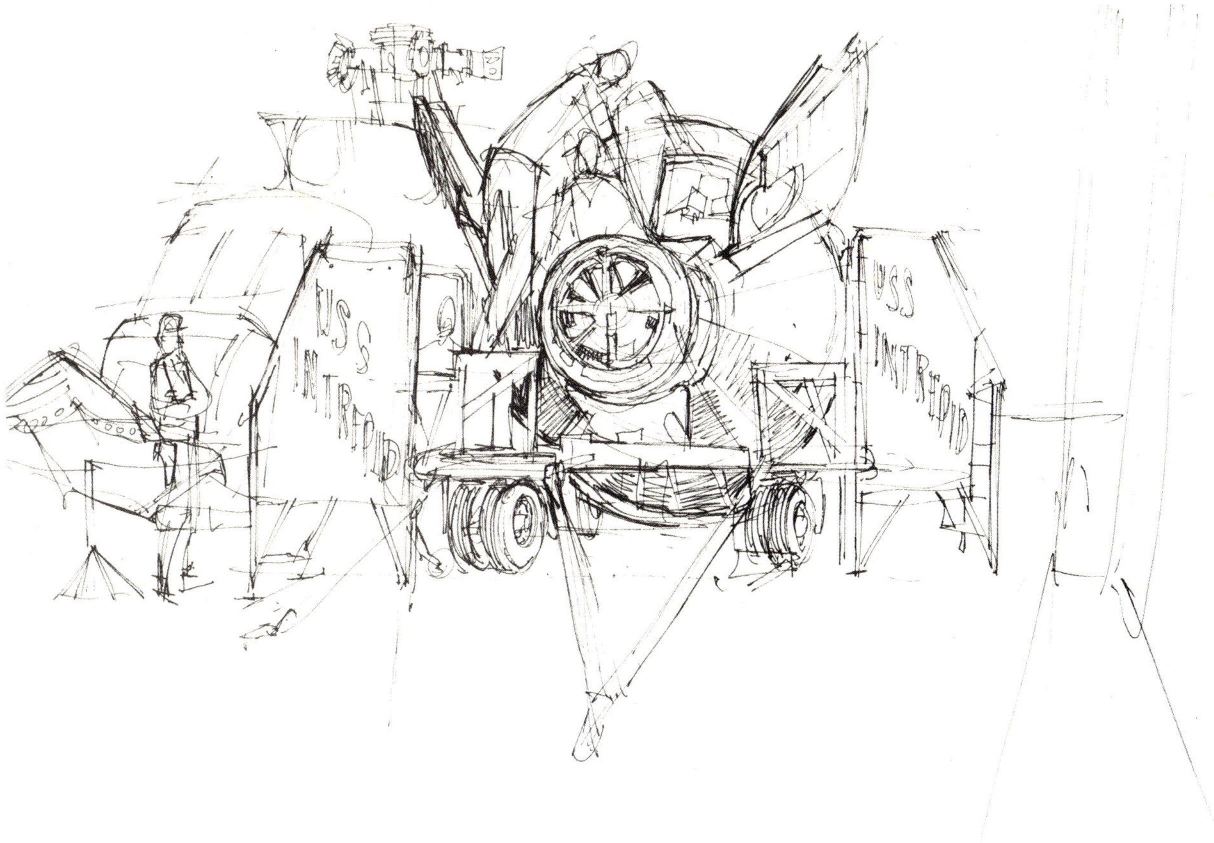

110. *Paul Calle.* GT-3 RECOVERED
Pen and ink on paper, 11×14″

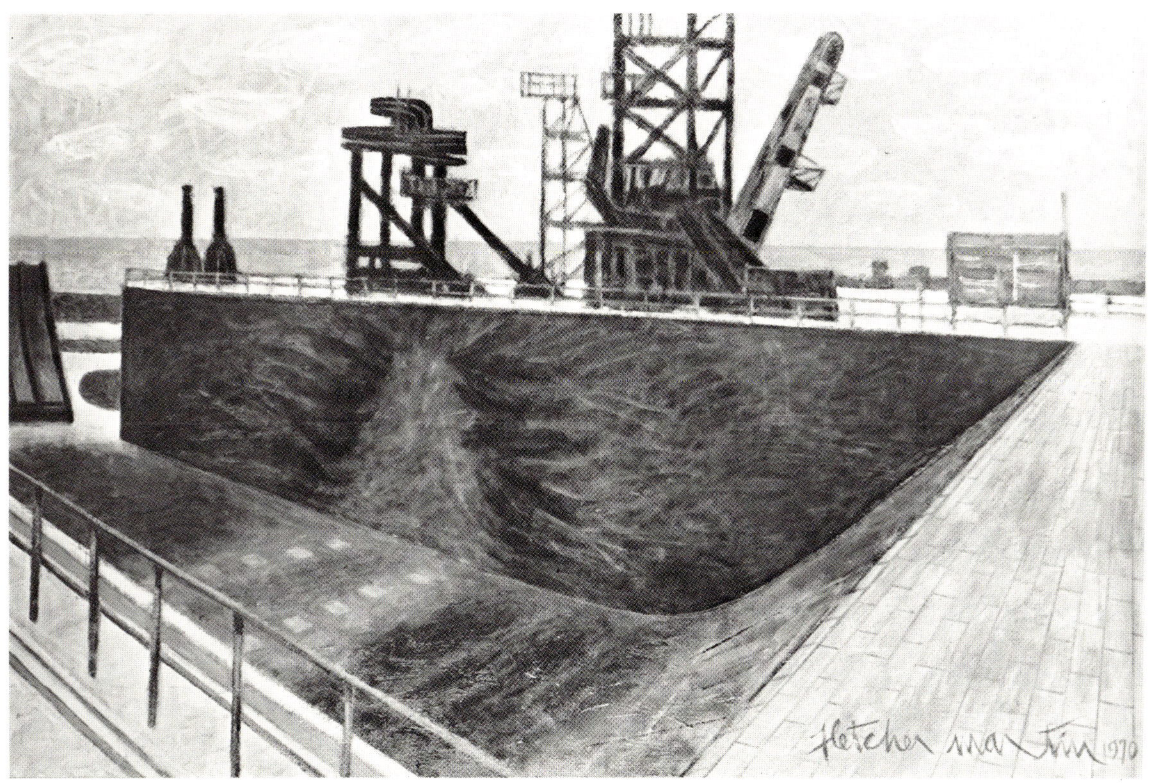

111.
Fletcher Martin
FLAME PIT
Oil on canvas, 24×36″

The fire pit where the Saturn V's exhaust flames are divided, diverted, and doused with millions of gallons of water looks like a disaster area after a launch. The blast lasts only about ten seconds, but in that time everything within range is scorched or incinerated. The claws and stanchions along the edge are used to hold the rocket in place.

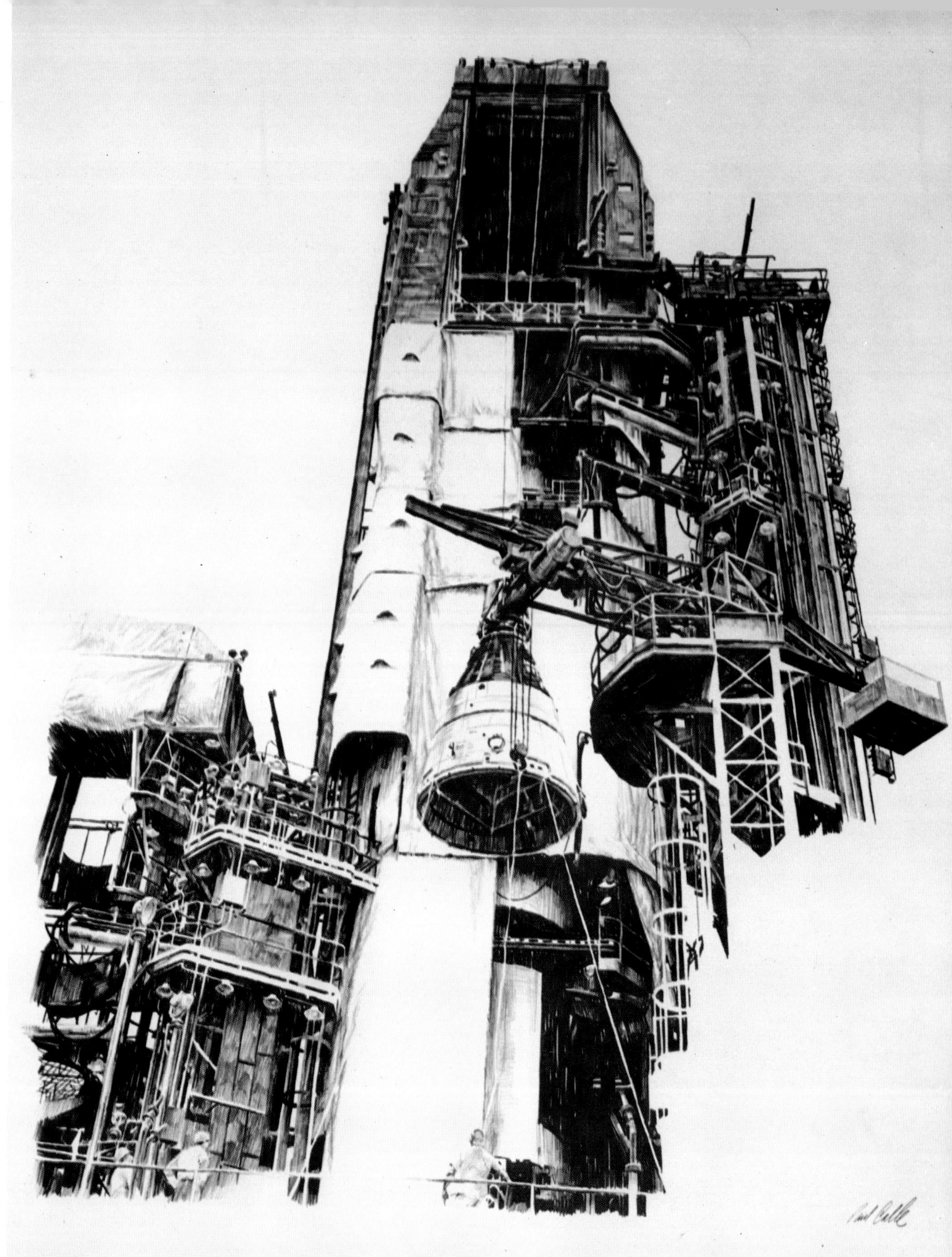

113.
Paul Calle
GEMINI SPACECRAFT AND GANTRY
Pencil on paper, 30×22″

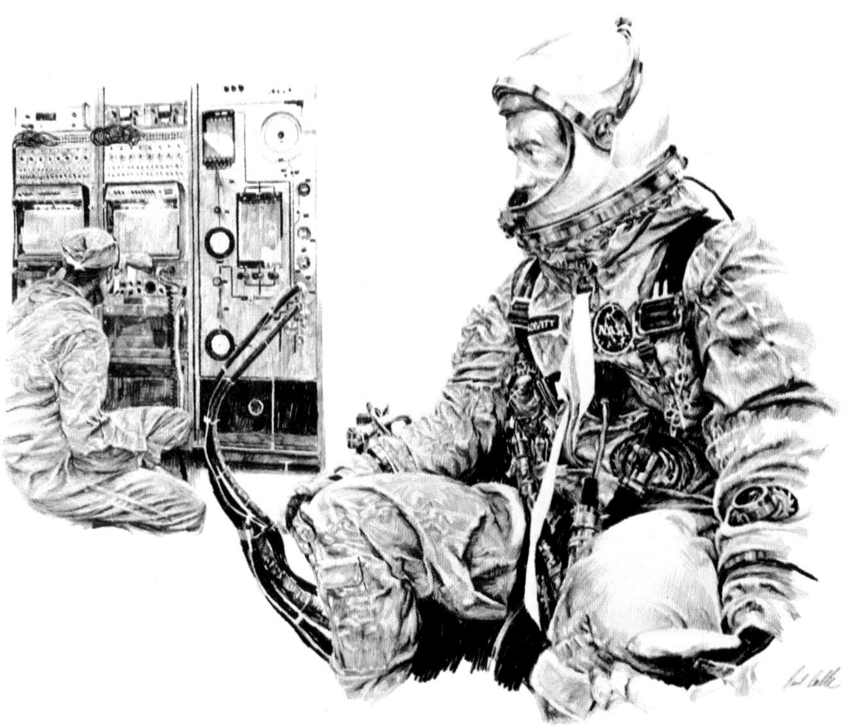

112. *Paul Calle*. GEMINI ASTRONAUT McDIVITT
Pencil on paper, 22×30″

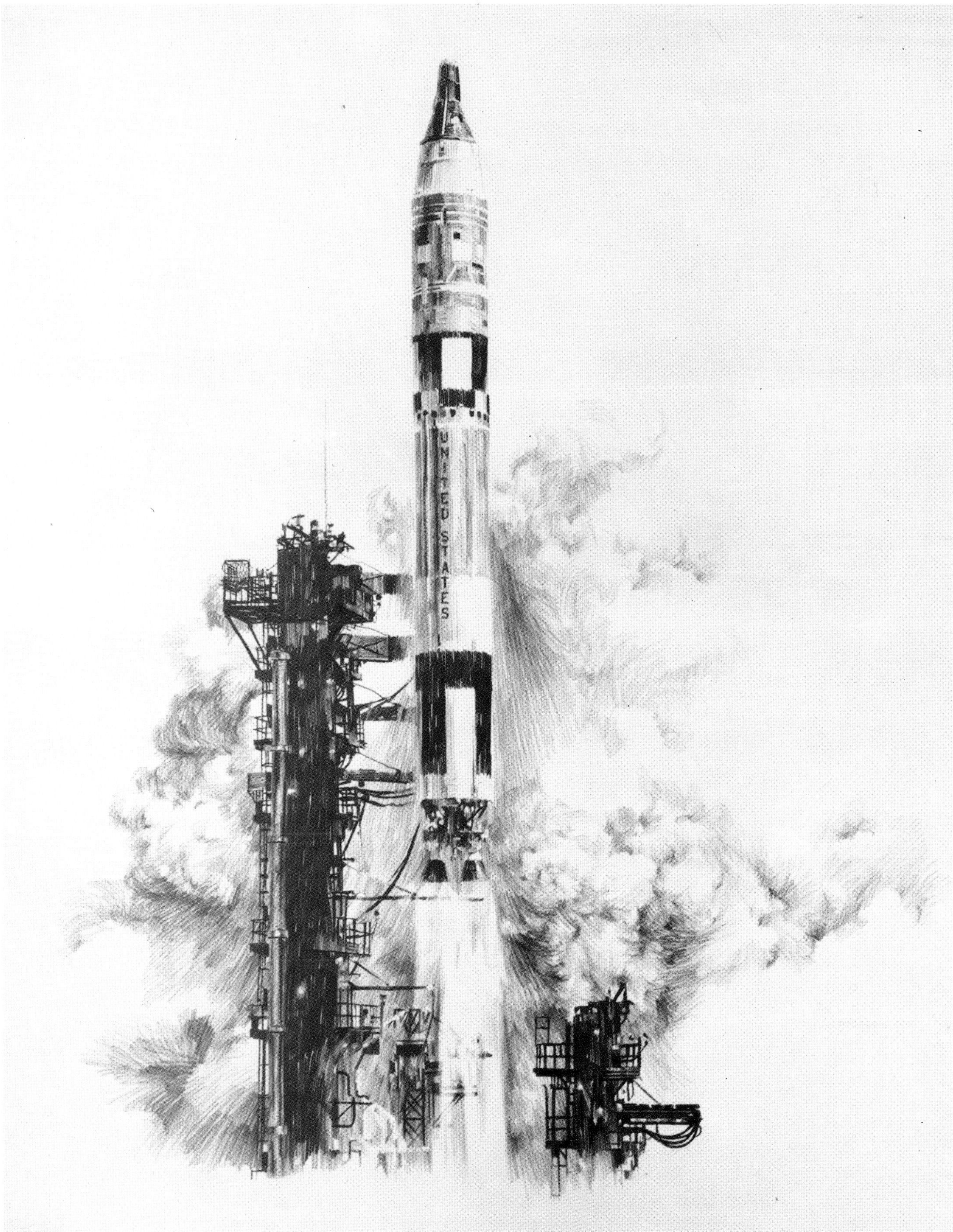

114.
Paul Calle
GEMINI LIFTOFF
Pencil on paper, 30×22″

The Gemini program was intended to test a number of vital functions in space: rendezvous and docking of two spacecraft, the ability of astronauts to leave the protection of their spacecraft and "walk" in space, and the ability of men and machines to endure for long periods the grueling conditions of space.

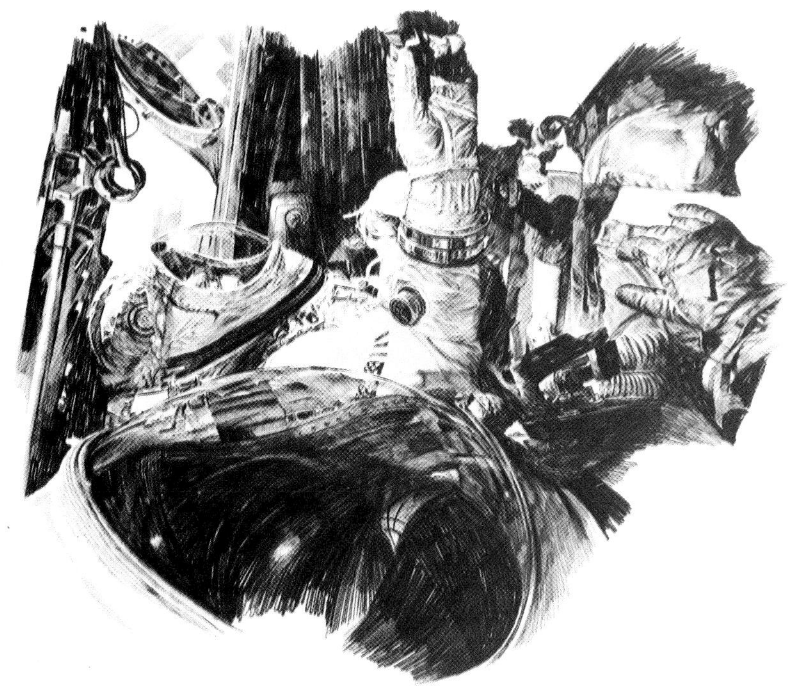

115. *Paul Calle*. INSIDE GEMINI SPACECRAFT
Pencil on paper, 22×30″

116.
L. Katzen
MOON MARKERS
Ink and plastic on paper, 20× 15″

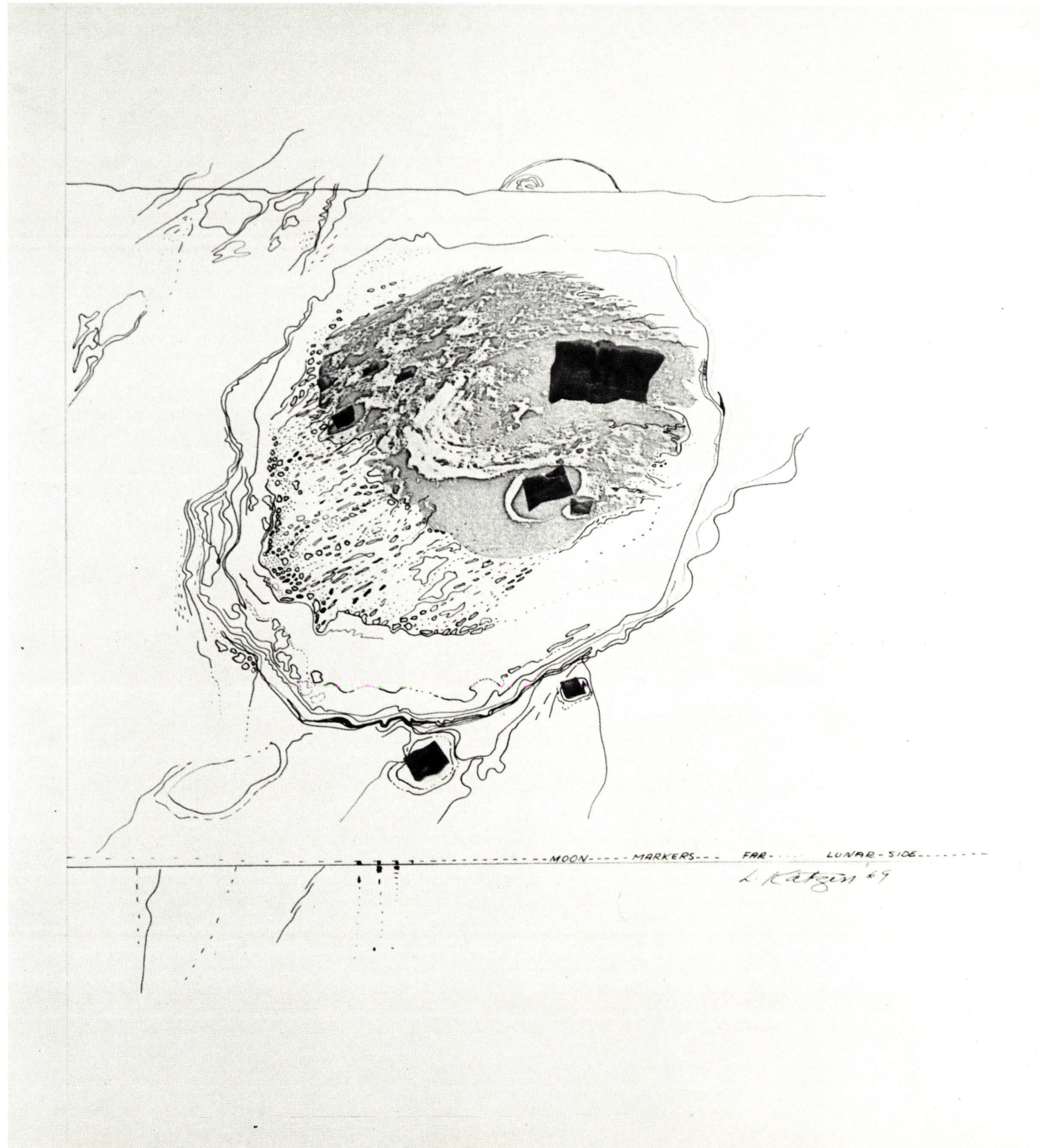

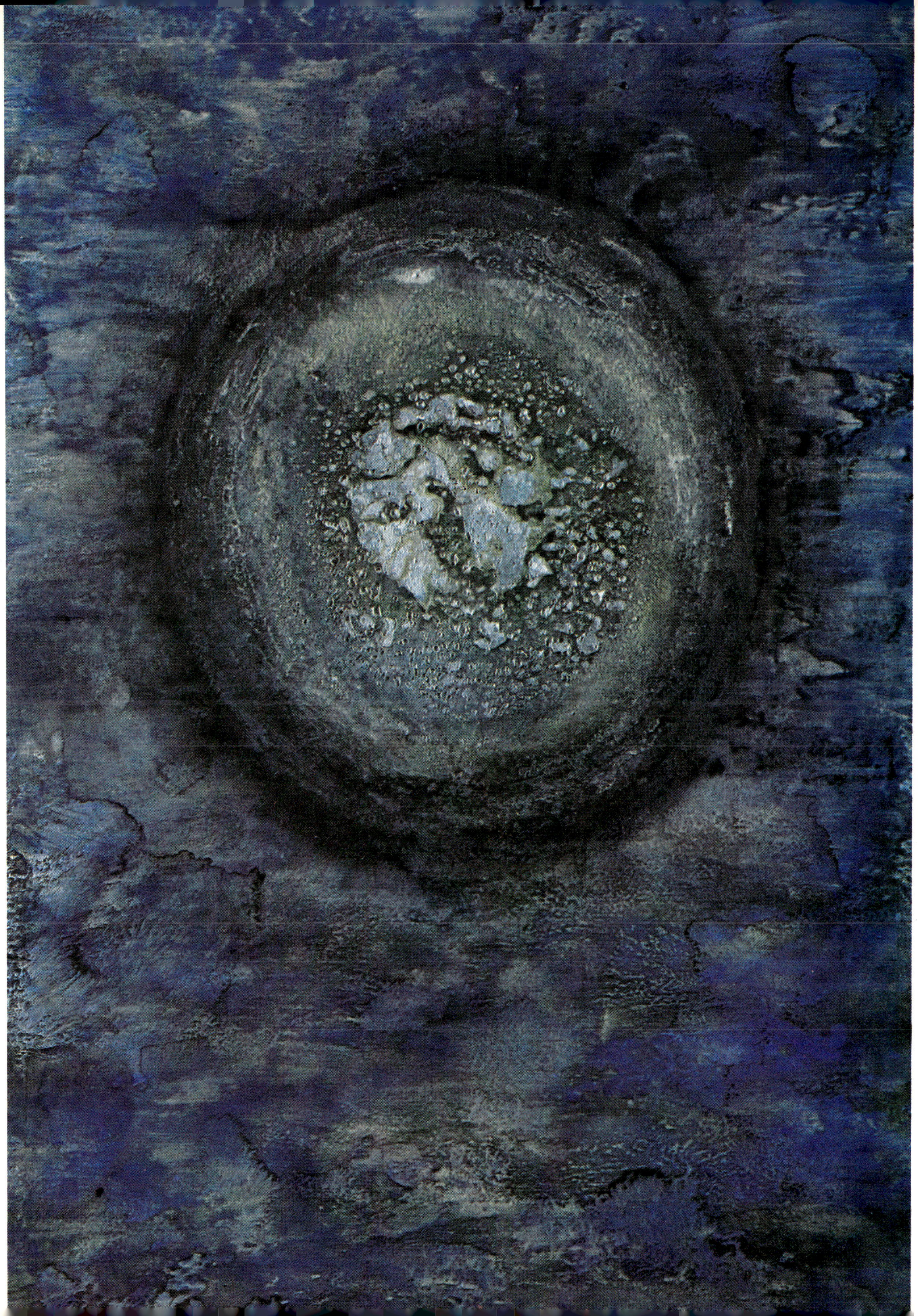

117.
Yeffe Kimball
MOONSCAPE
Acrylic on canvas, 6×4′

As the day when men would set foot on the moon drew nearer, artists began to speculate on the nature of our neighbor in the heavens. In fact, the surface turned out to be not unlike Yeffe Kimball's 1968 painting.

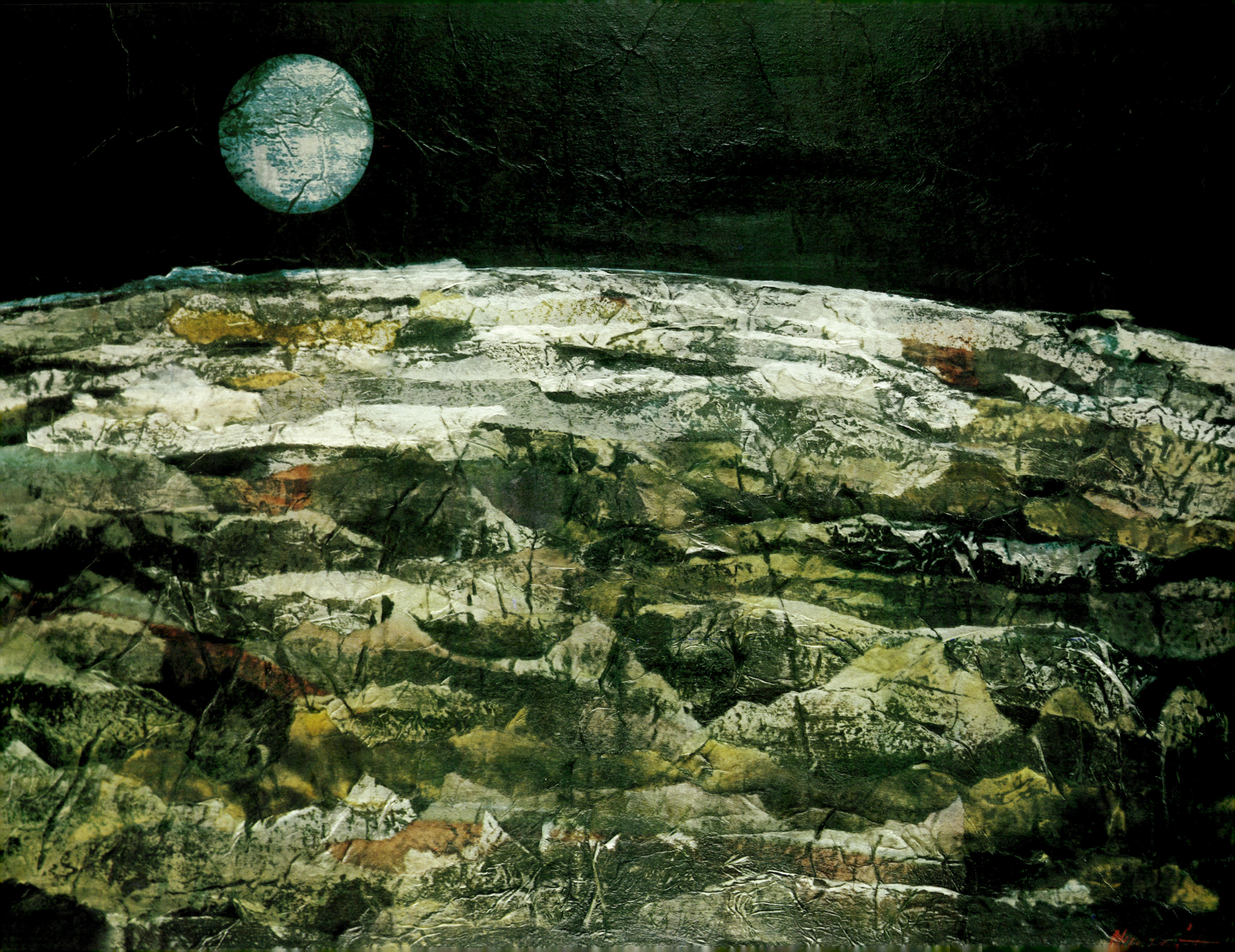

118.
Fred Messersmith, A.W.S.
MOONSCAPE
Acrylic collage on canvas, 34×48″
Reproduced courtesy of the artist

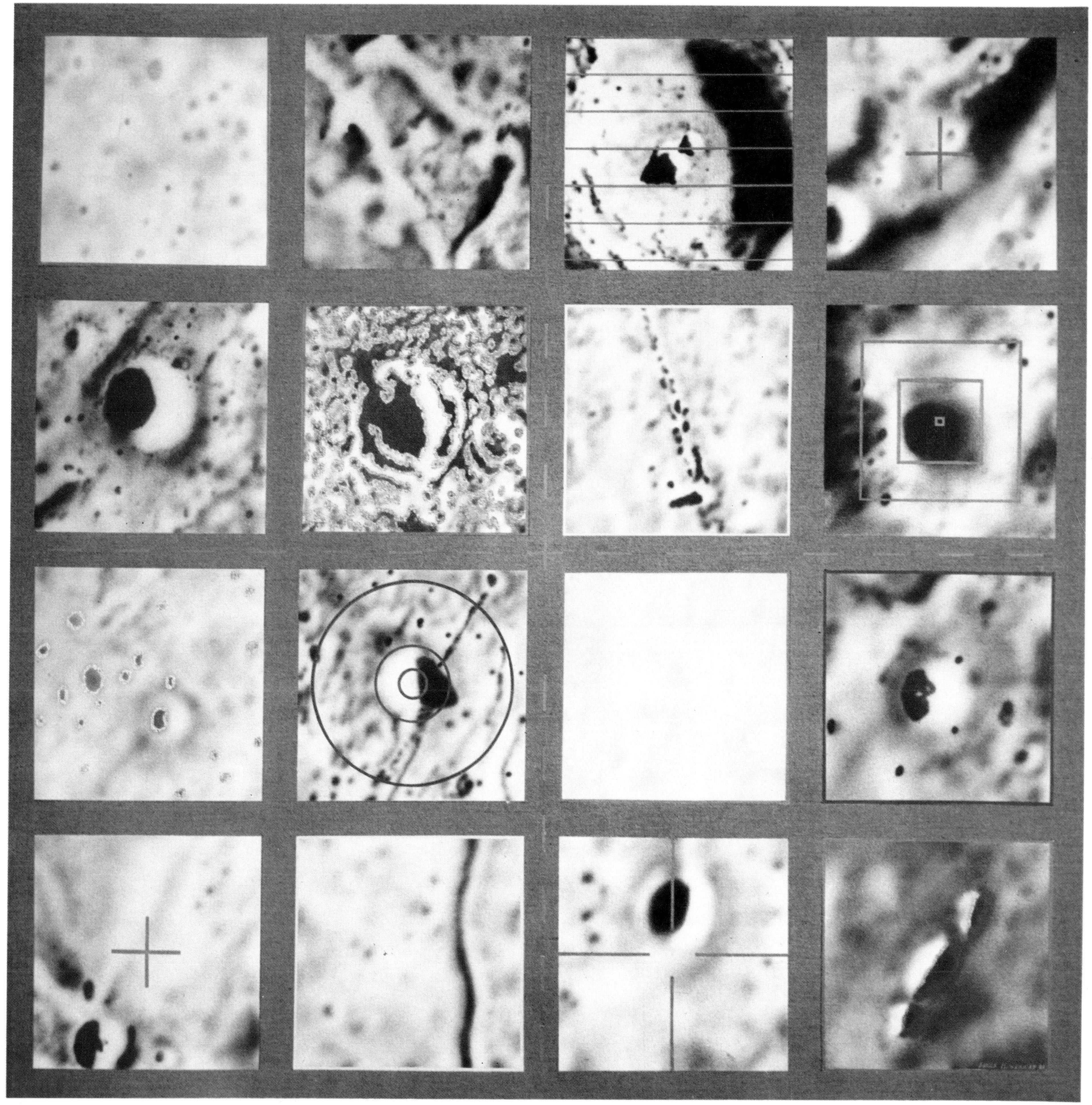

119. *James P. Hendricks*. MOON CRATERS. Oil on canvas, 4×4′
Reproduced courtesy of the artist and the Air and Space Museum, Smithsonian Institution

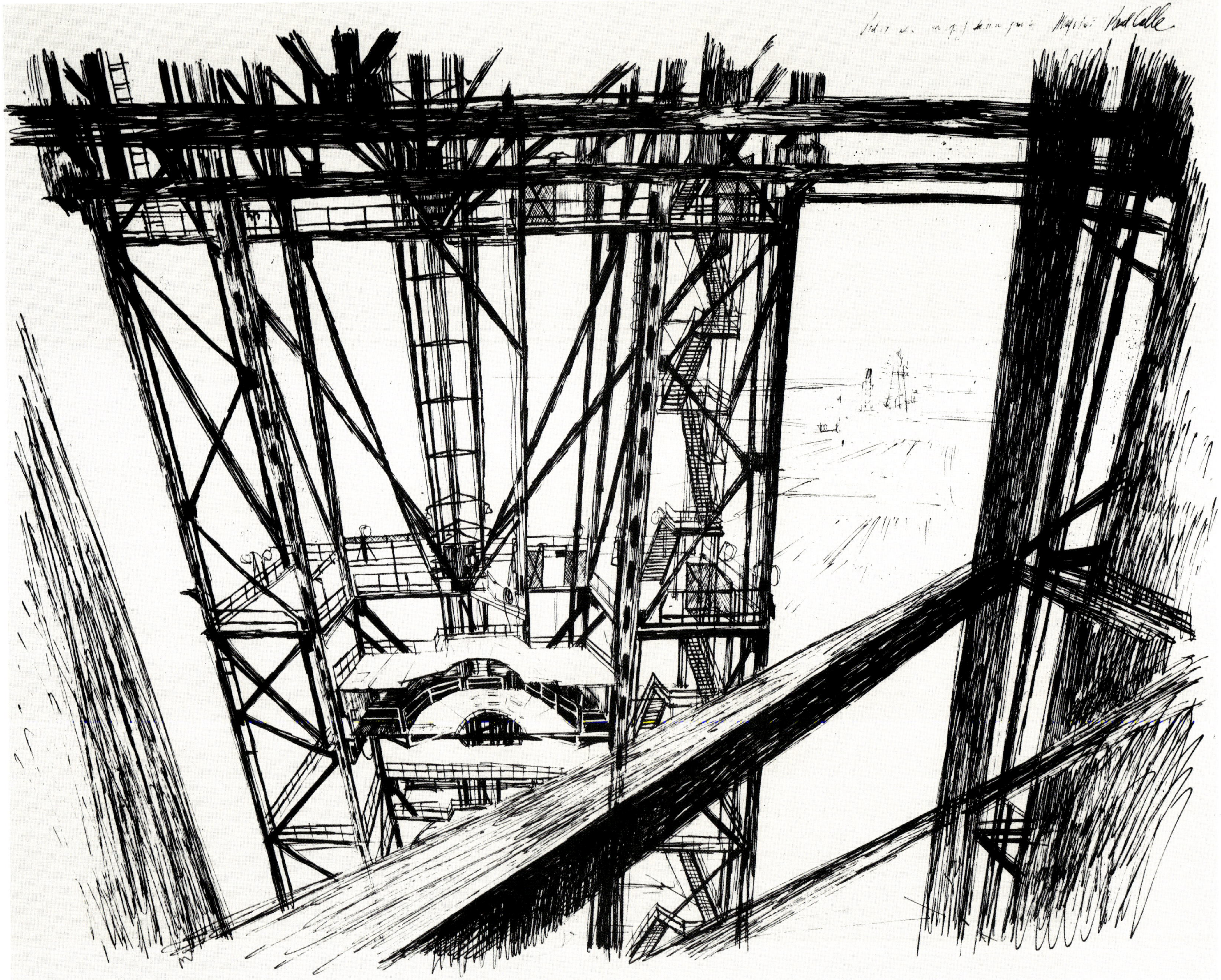

120. *Paul Calle*. DOWN TO 14. Pen and ink on paper, 28×35″

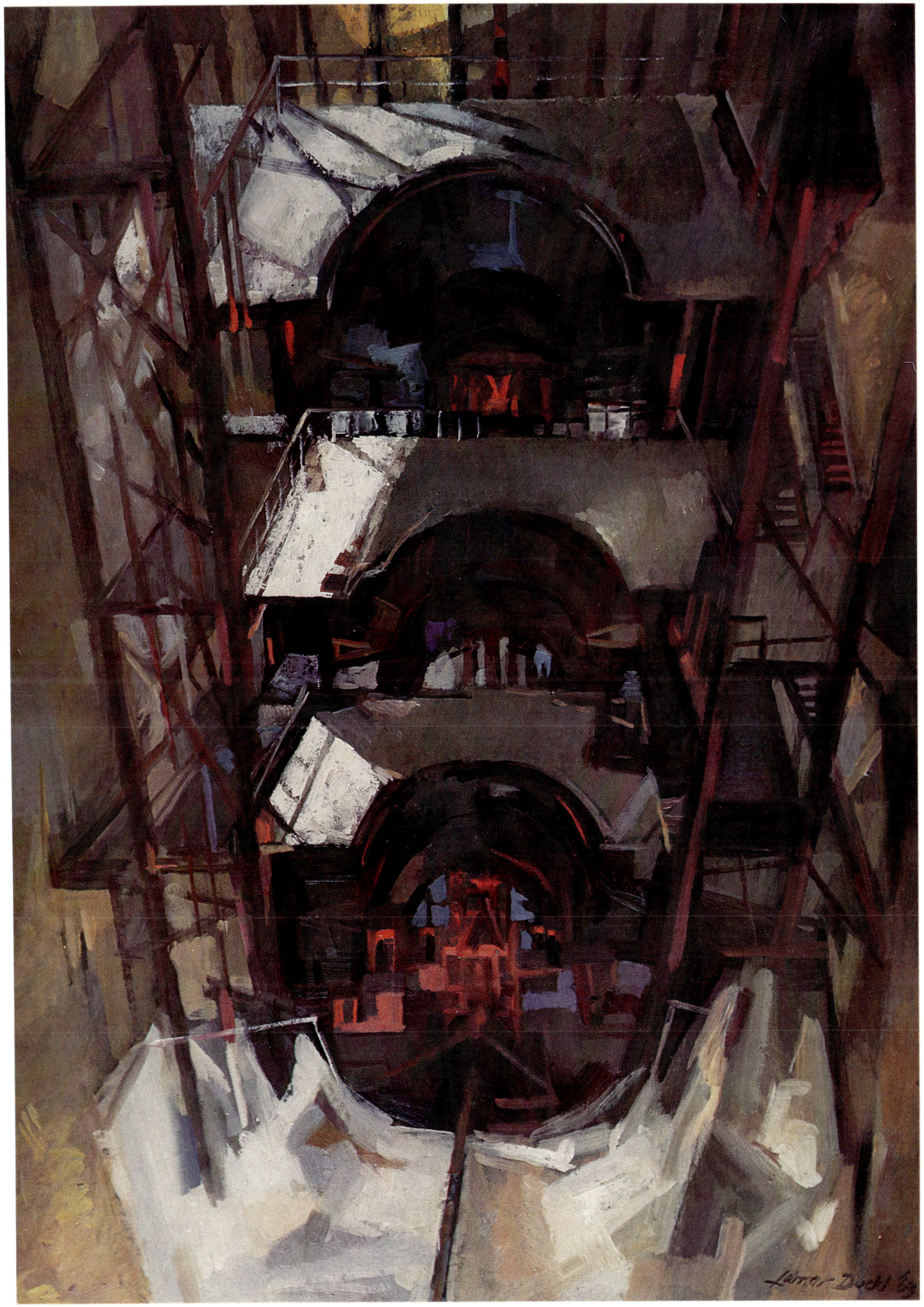

121.
Lamar Dodd
SATURN GANTRY
Oil on canvas, 47×35″

Views inside the service towers and gantries often combine strange beauty, plunging perspectives, and vertigo.

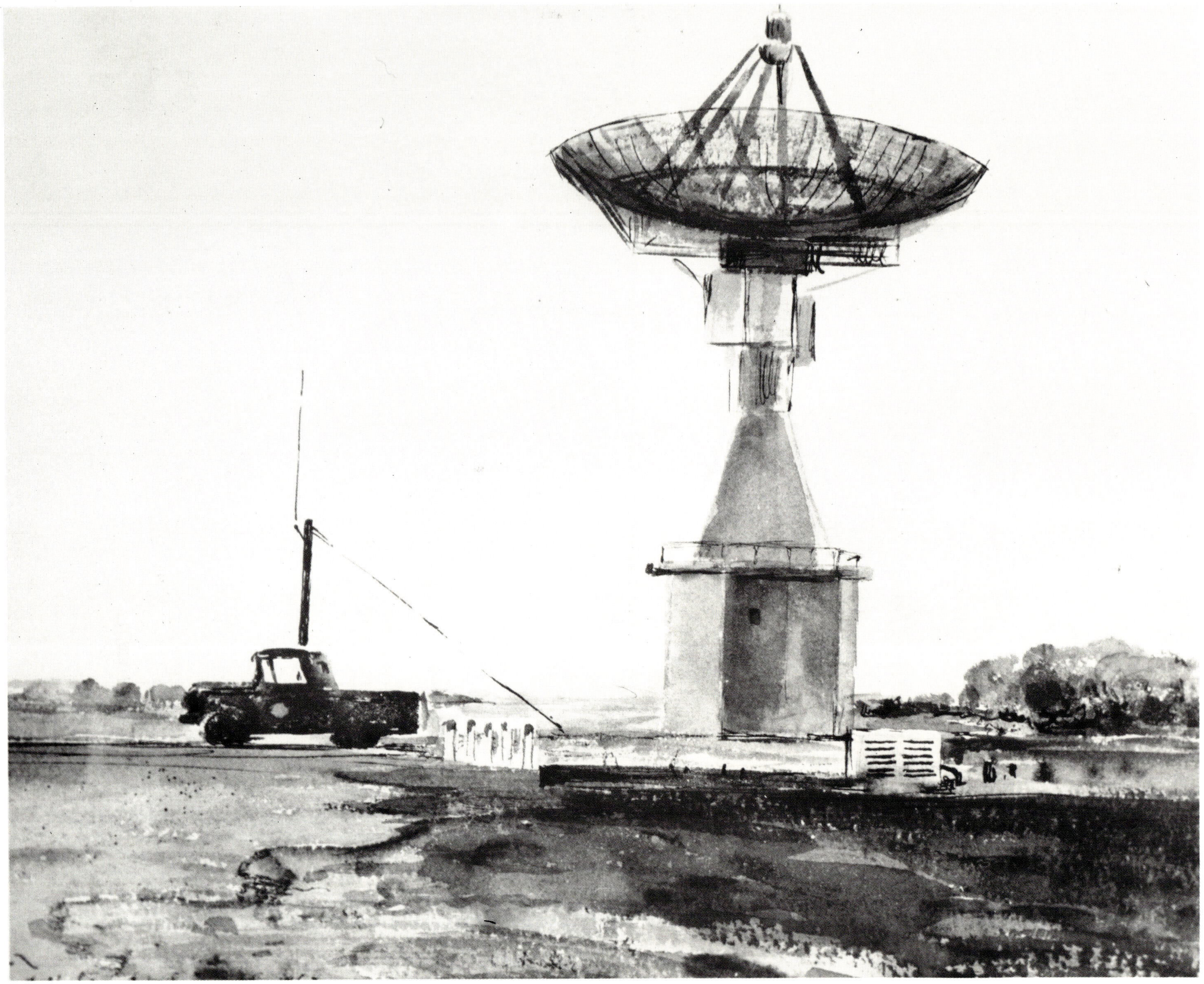

122. *Peter Hurd.* TRACKING. Watercolor on paper, 14×17″

123.
Hugh Laidman
WET MORNING
Watercolor on paper, 22×30″

One often thinks of Cape Kennedy as a sandy resort area with launch complexes along the seaside. In fact, much of the terrain is a subtropic jungle where wild boar, deer, rattlesnakes, and thousands of sea birds make their home. Here one of the many tracking antennae looks like the nest of a bird from another world.

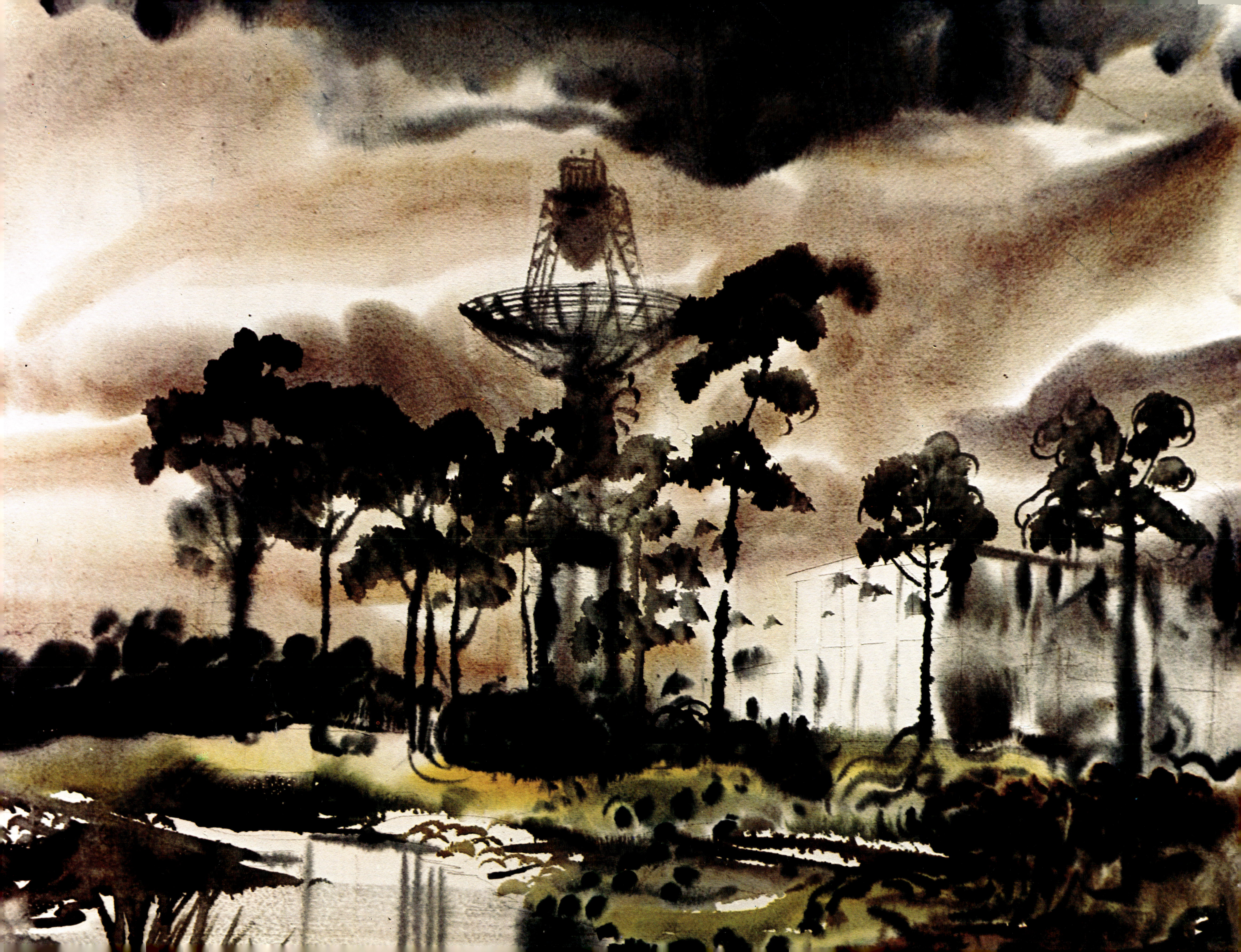

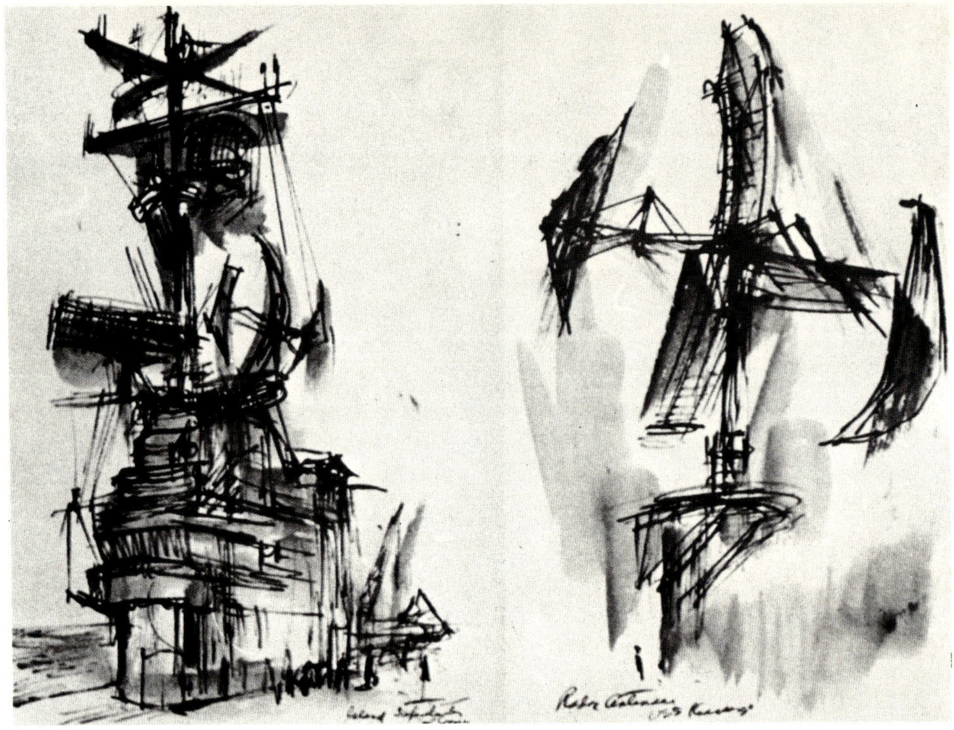

124.
Mitchell Jamieson
CARRIER ANTENNAE
Ink on paper, 9×12″

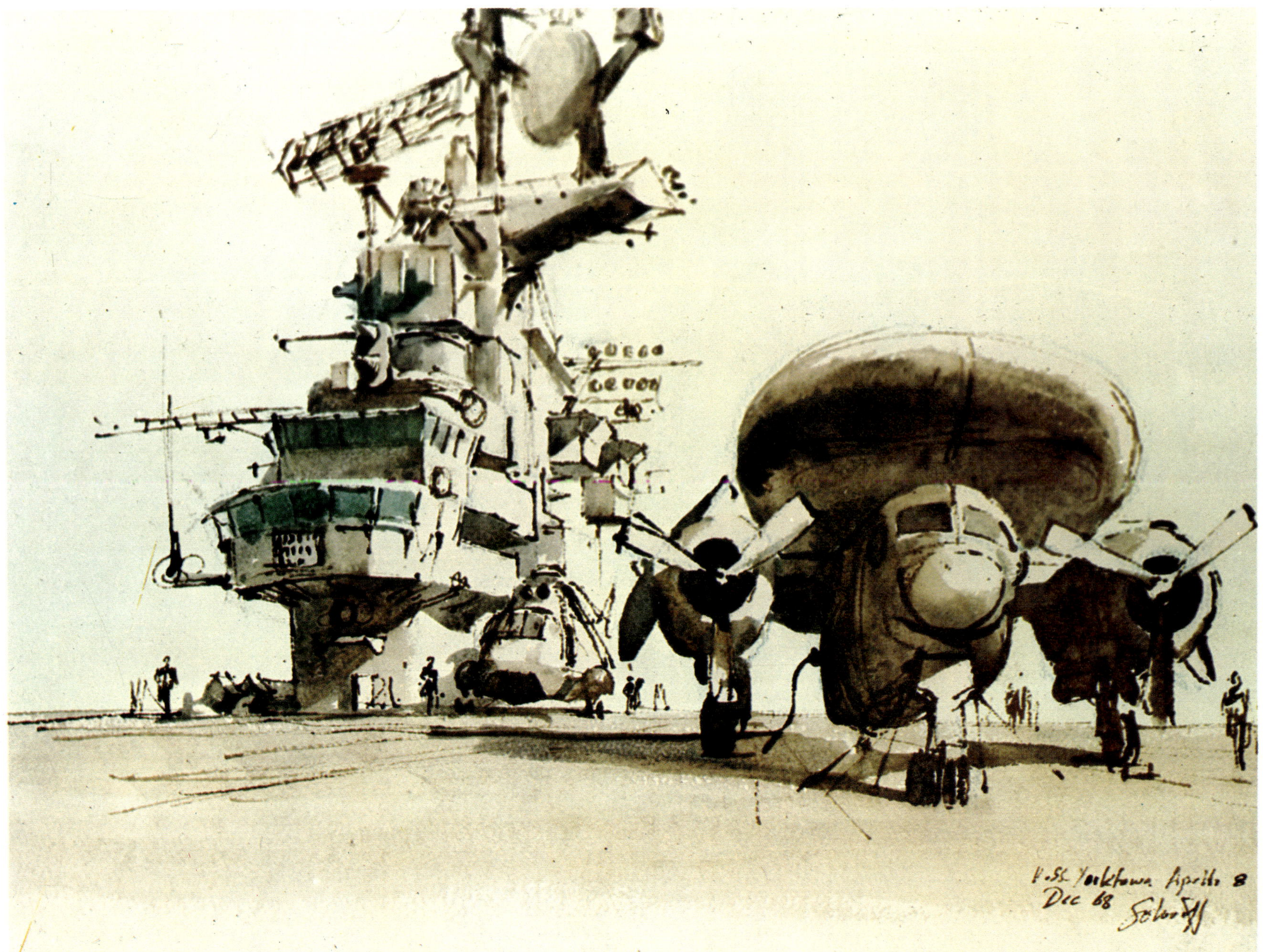

125.
Nicholas Solovioff
CARRIER FLIGHT DECK
Watercolor on paper, 14×19″

The demands of technology and the evolution of science are producing shapes that a few years ago even space science fiction could hardly have imagined. A search plane, its wings folded and its radar shrouded in a bulbous growth, waits on the deck of a carrier. On the "island" is arrayed a fantastic assortment of electronic equipment.

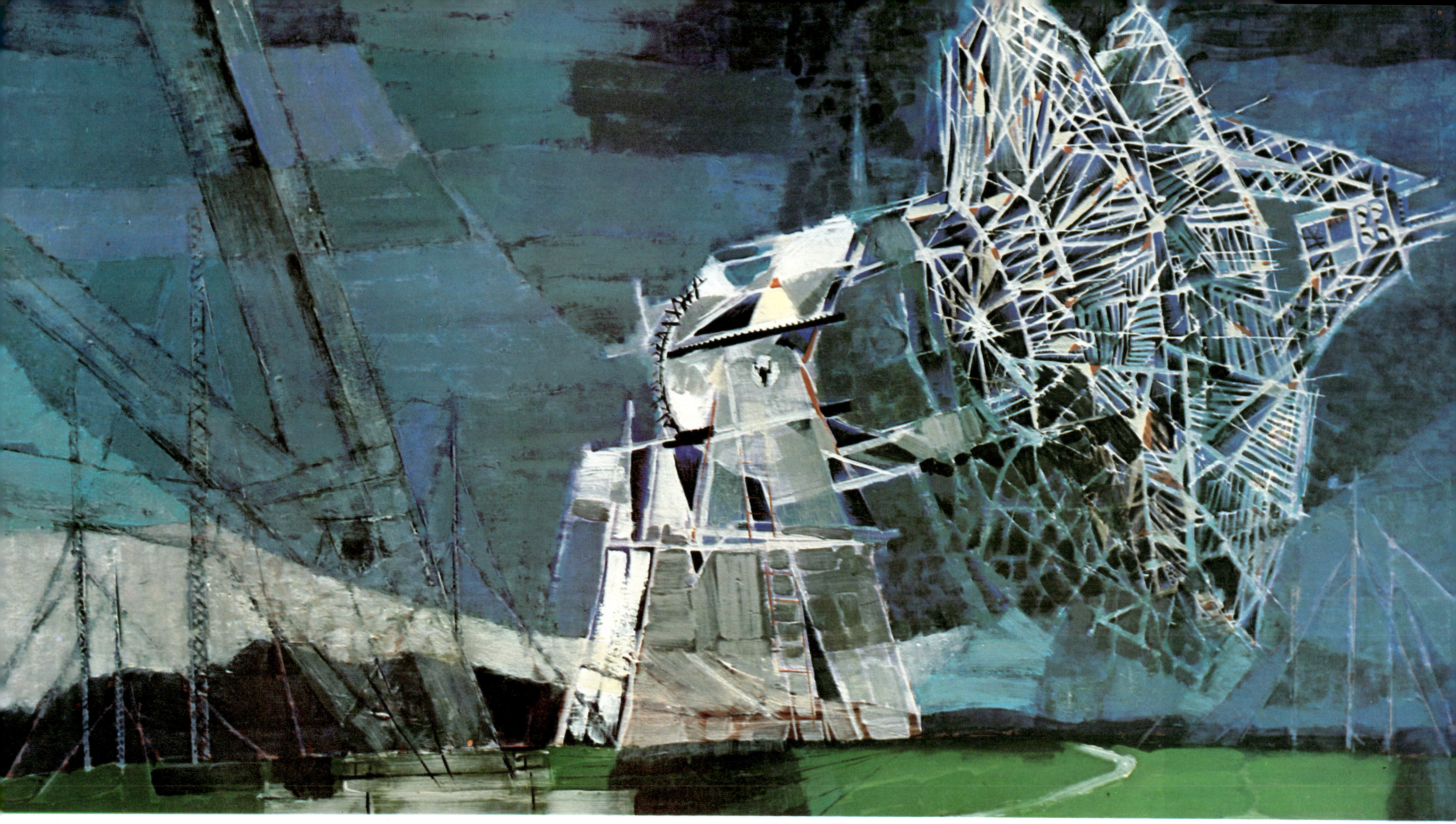

126.
Paul Arlt. BIG DISH ANTENNAE, TANANARIVE
Acrylic on canvas, 26×48″

Both geography and politics occasionally play a part in NASA's worldwide tracking system. Tananarive, off the east coast of Africa, is the site of a large antenna that probes space over the Indian Ocean.

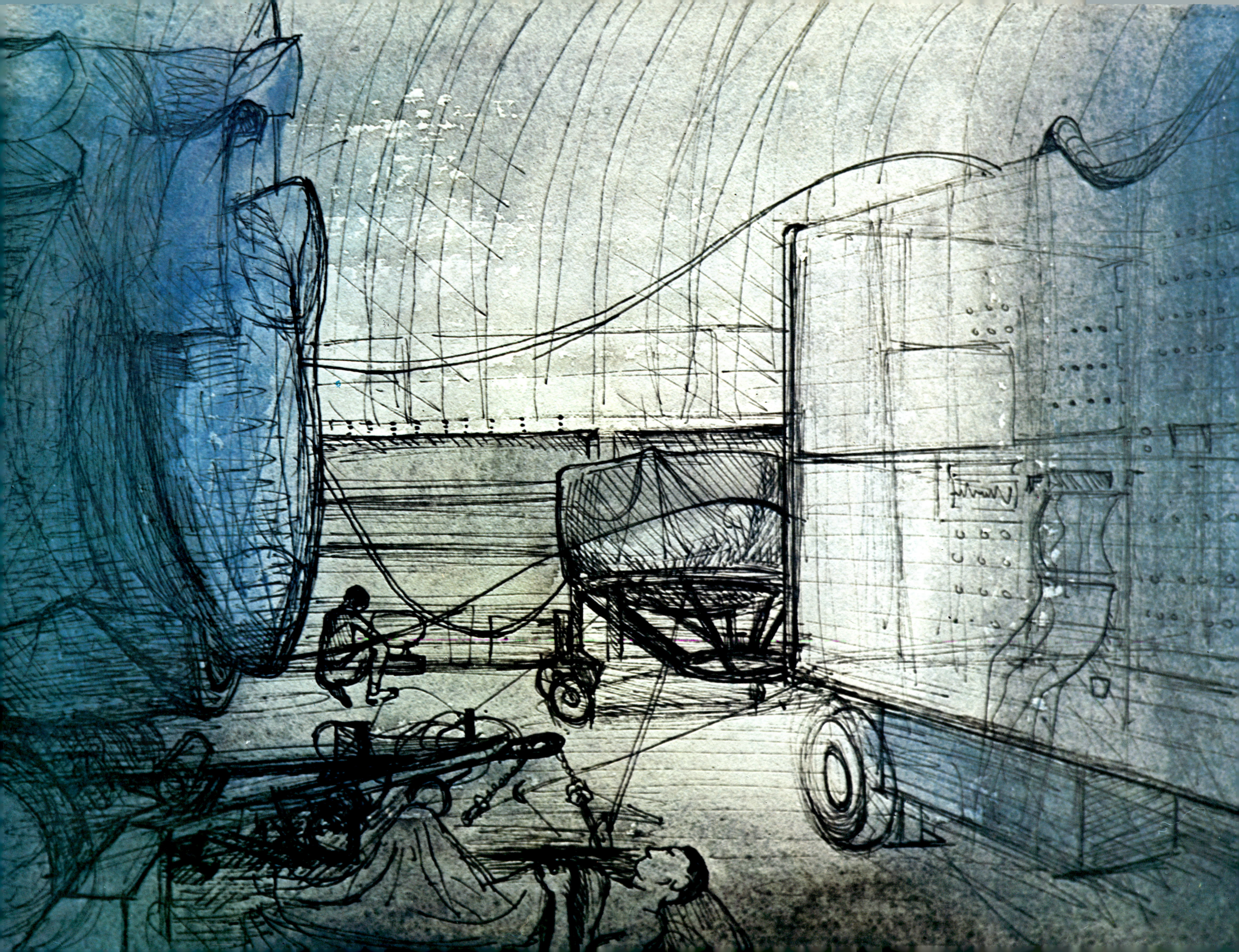

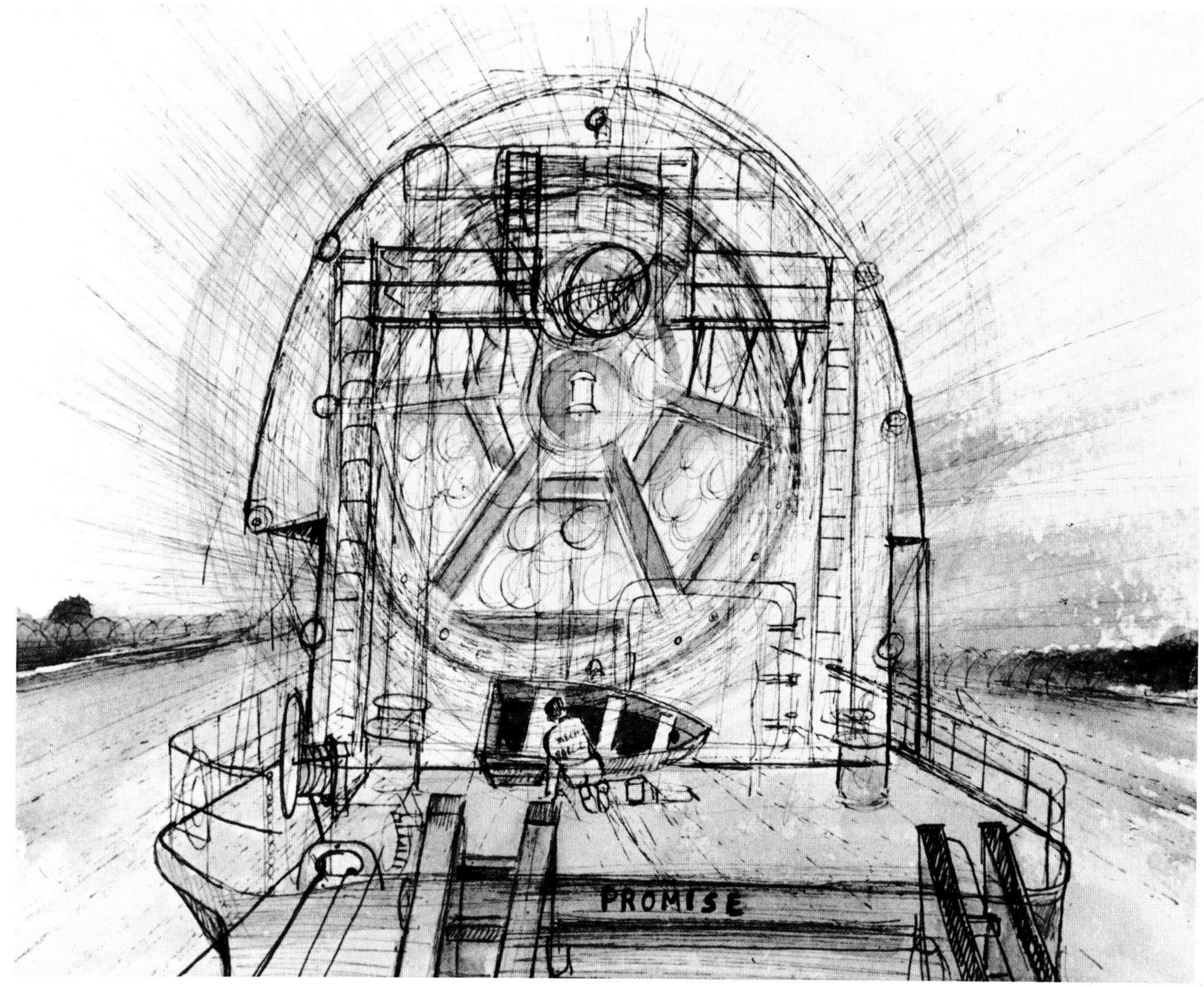

128.
Theodore Hancock
DOWN THE MISSISSIPPI
Ink and watercolor on paper, 9×12″

127.
Theodore Hancock
ON THE BARGE
Ink and watercolor on paper, 9×12″

Theodore Hancock made a trip from Huntsville, Alabama, to Cape Kennedy on board the Promise, a strange craft, half barge, half hangar, in which the first stage and engines of the Saturn V rocket traveled to the launch site. Inside the cavernous shell, supports for the fuel tanks stand like flying buttresses. On the deck, NASA technicians sleep beside their precious charge.

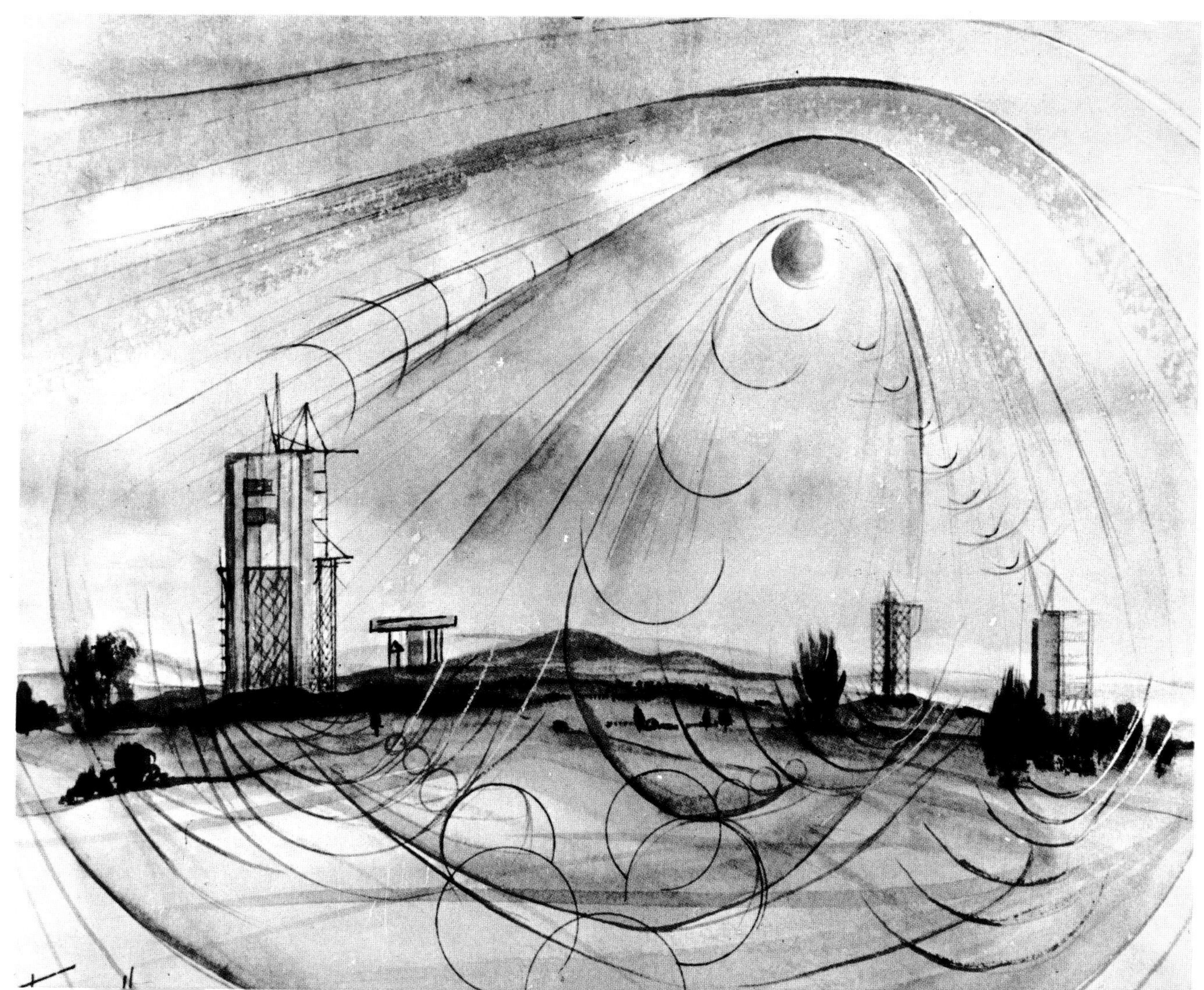

129.
Theodore Hancock
MSFC
Ink and watercolor on paper, 9×12″

The harvest moon seems to cast a spell over the Huntsville landscape with its Space Age structures.

130.
Paul Calle
CENTRIFUGE
Pencil on paper, 22×30″

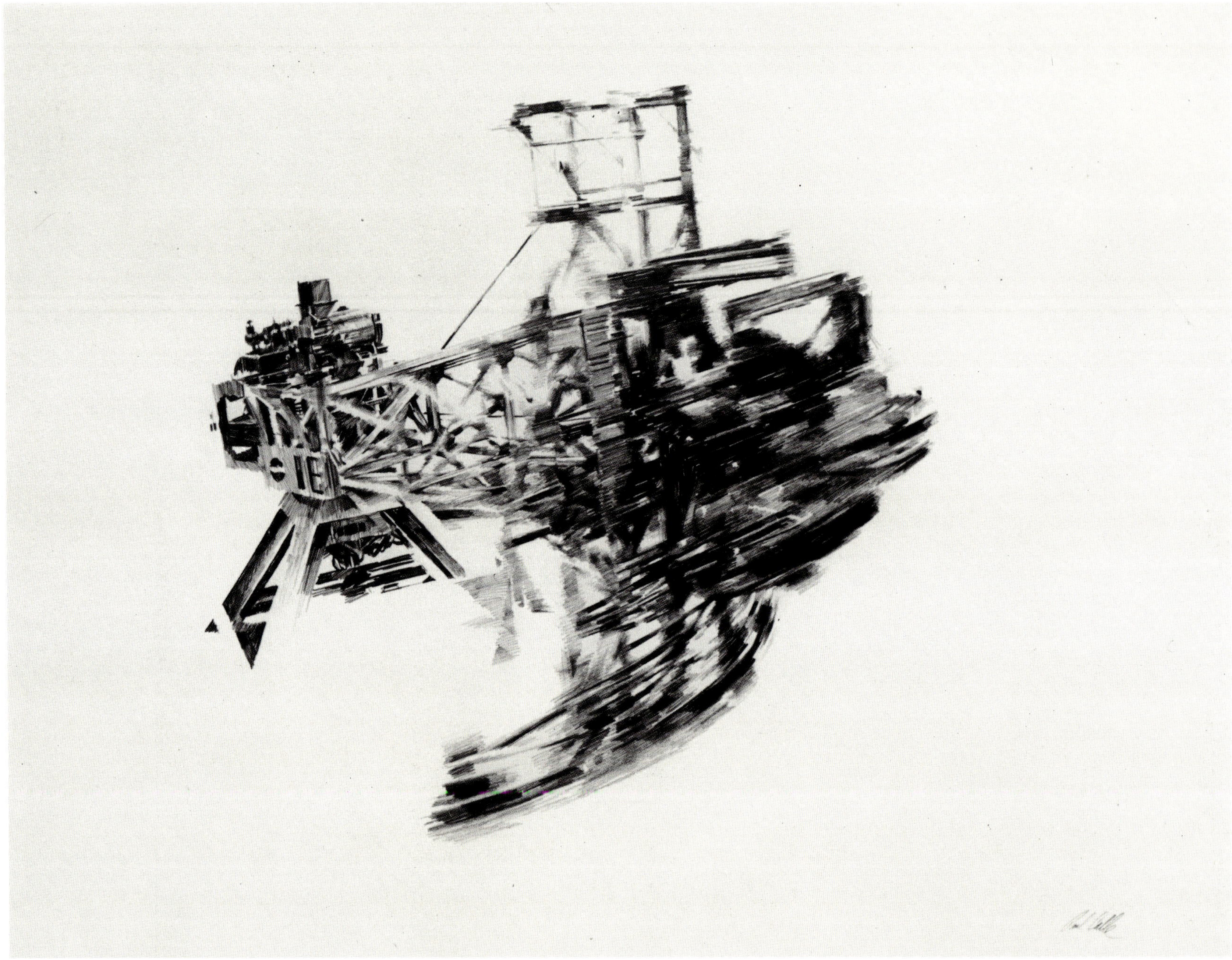

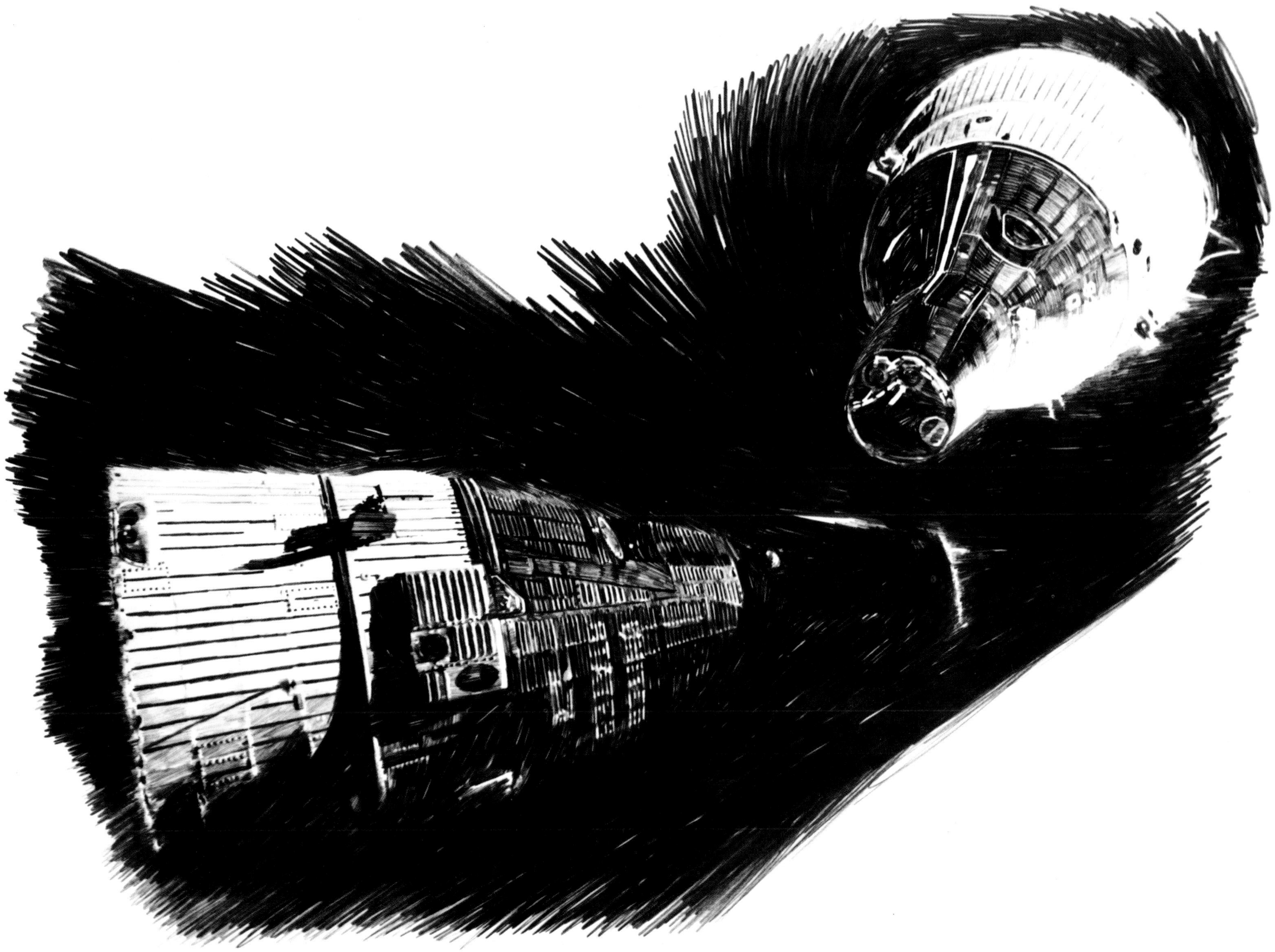

131.
Paul Calle
RENDEZVOUS
Pencil on paper, 22×30″

One of the most critical operations in space, and one in which the United States gained a significant lead on its Russian counterparts, was the technique of rendezvous. This involved incredibly accurate computerized controls of the spacecraft rocket engines.

133.
Norman Rockwell
GRISSOM AND YOUNG
Oil on canvas, 35 × 55″
Reproduced courtesy of *Look* magazine,
© 1964 Cowles Communications, Inc.

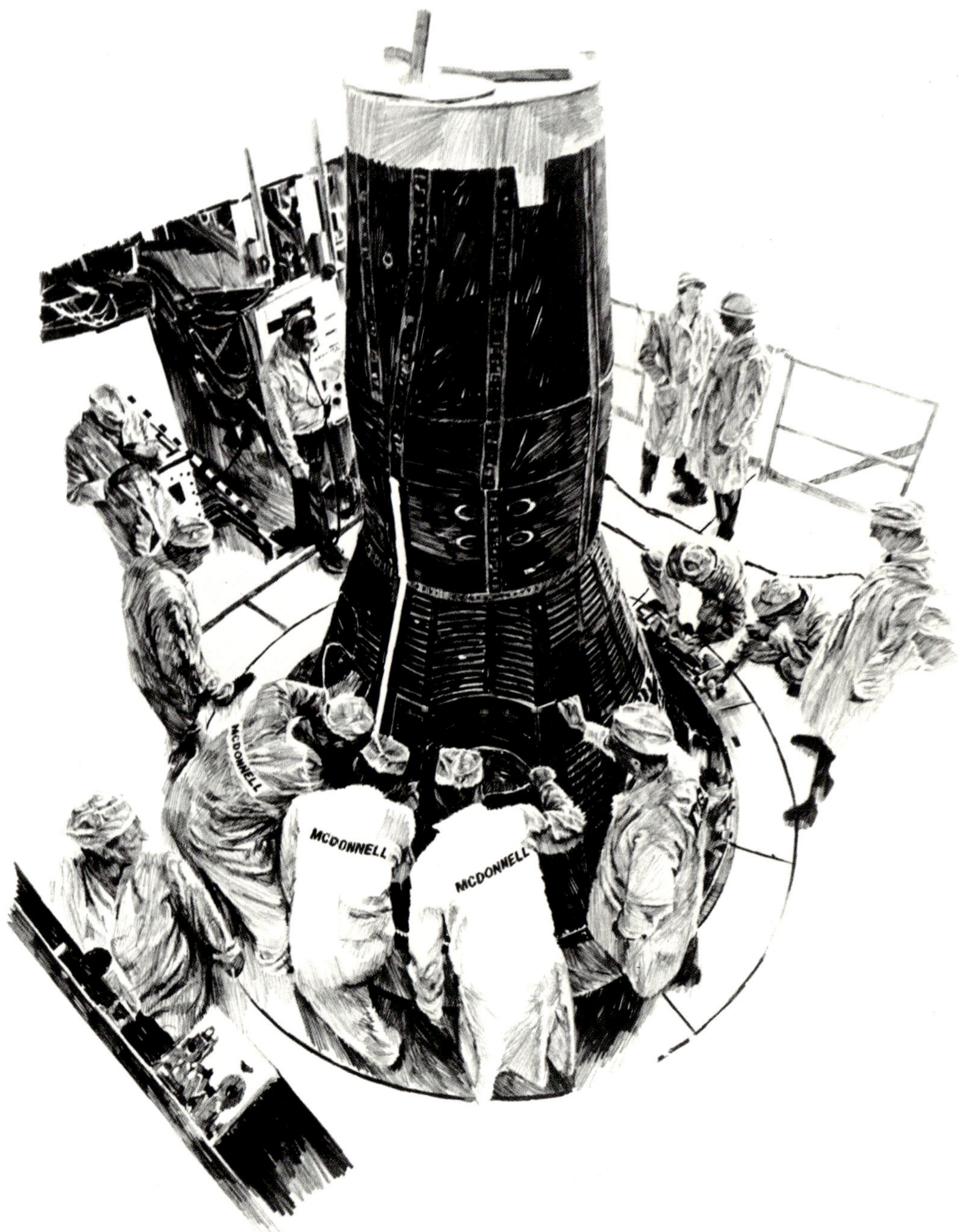

132. *Paul Calle*. GEMINI WHITE ROOM
Pencil on paper, 32 × 23″

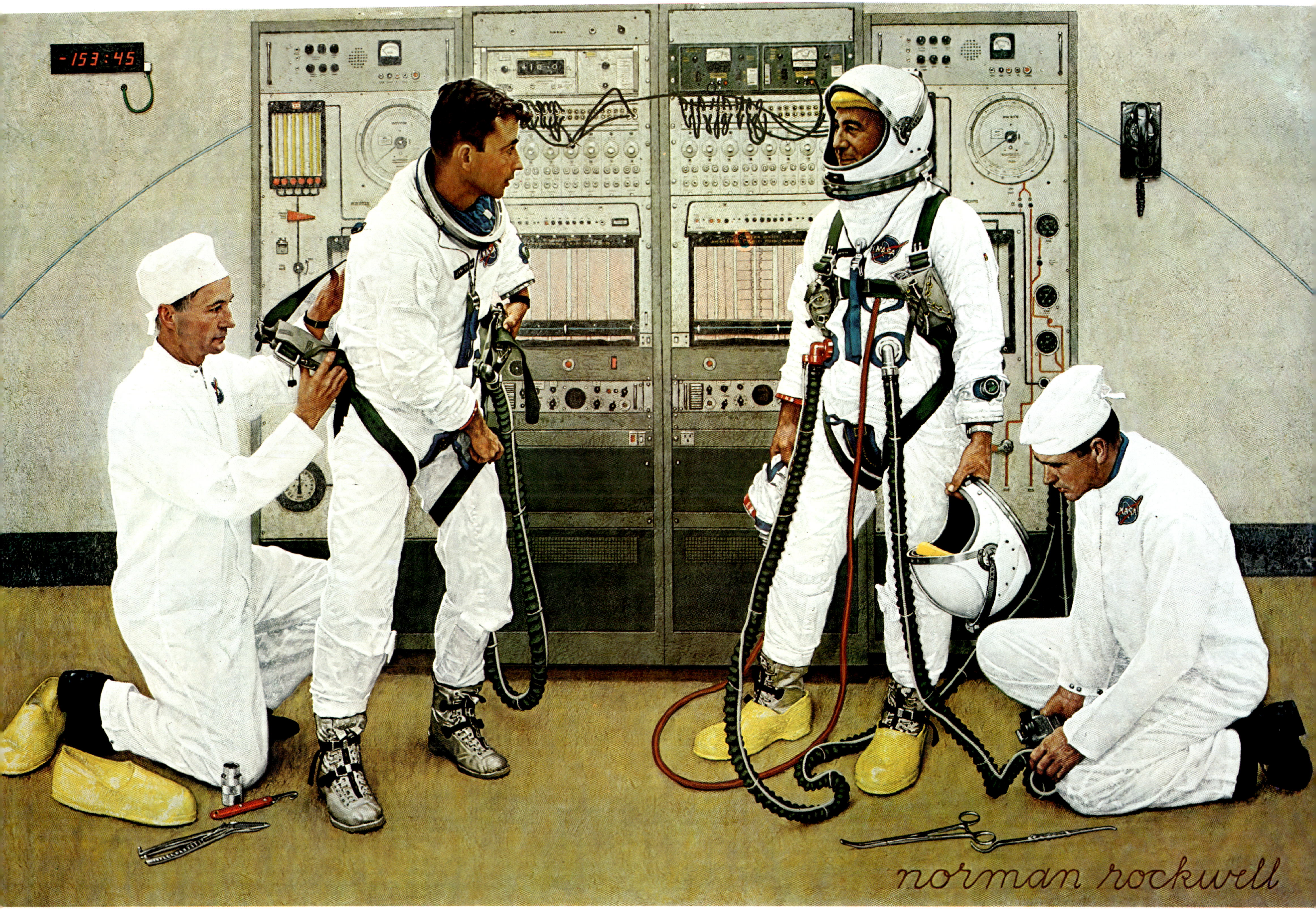
-153:45
norman rockwell

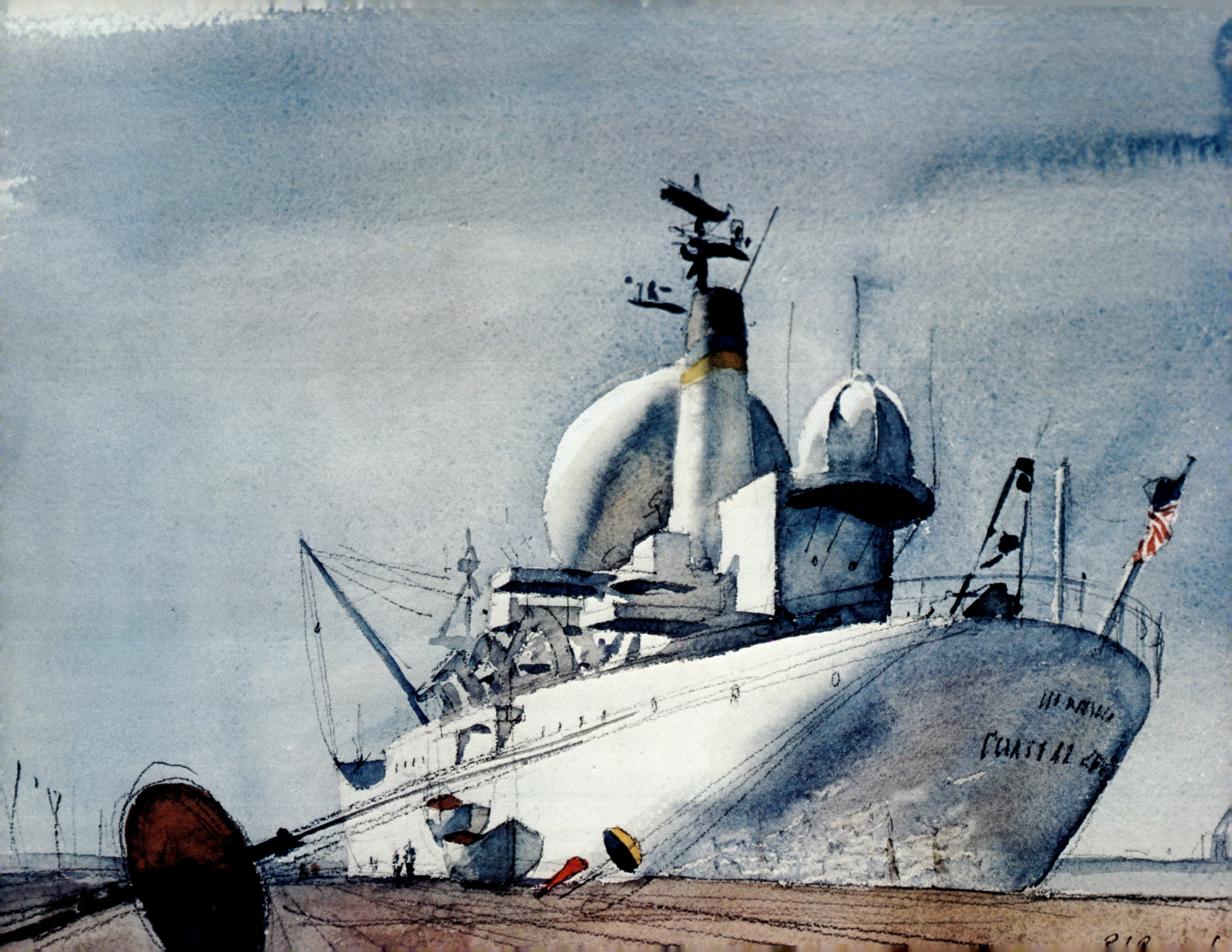

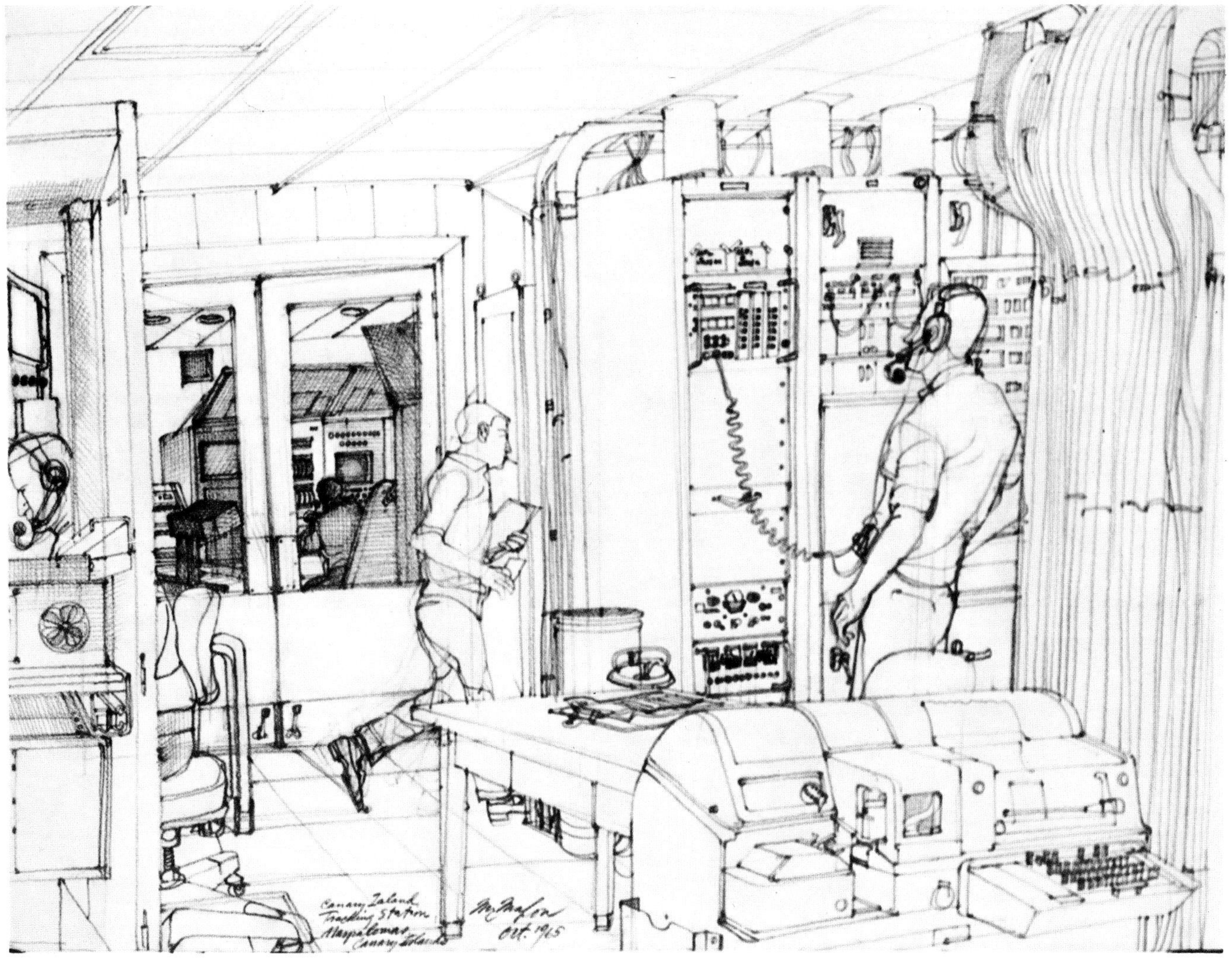

135.
Franklin McMahon
CANARY ISLAND STATION
Pencil on paper, 22 × 30″

134.
Nicholas Solovioff
TRACKING SHIP
Watercolor on paper, 14 × 19″

Tracking vessels, usually converted merchantmen and named after Air Force generals, like many other creations of the Space Age look like nothing ever seen before. The round shapes on board shield radar antennae; the round shapes on the mooring cables keep rats from climbing aboard.

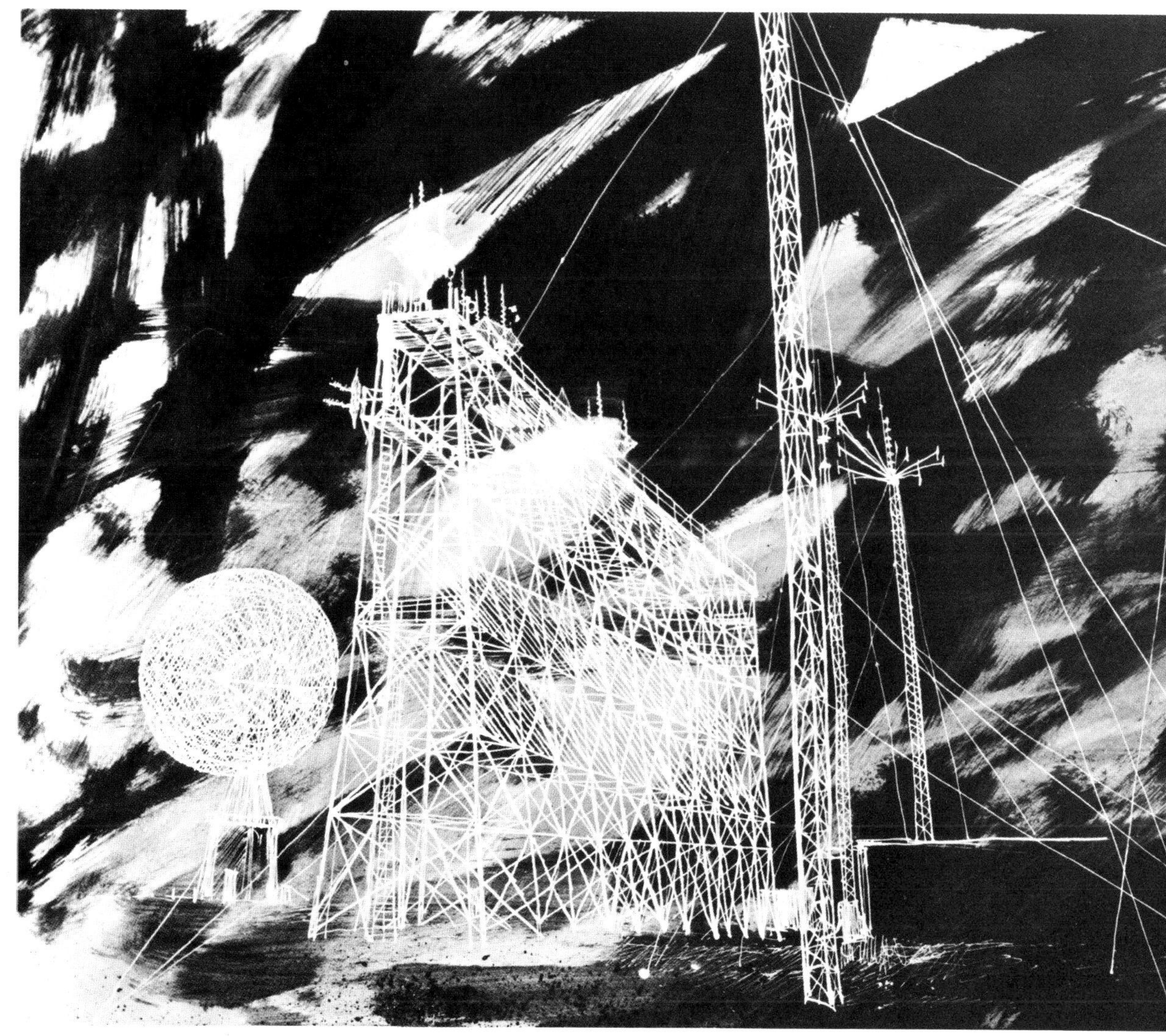

136. *Paul Calle.* GRIDS. Watercolor on paper, 22 × 30″

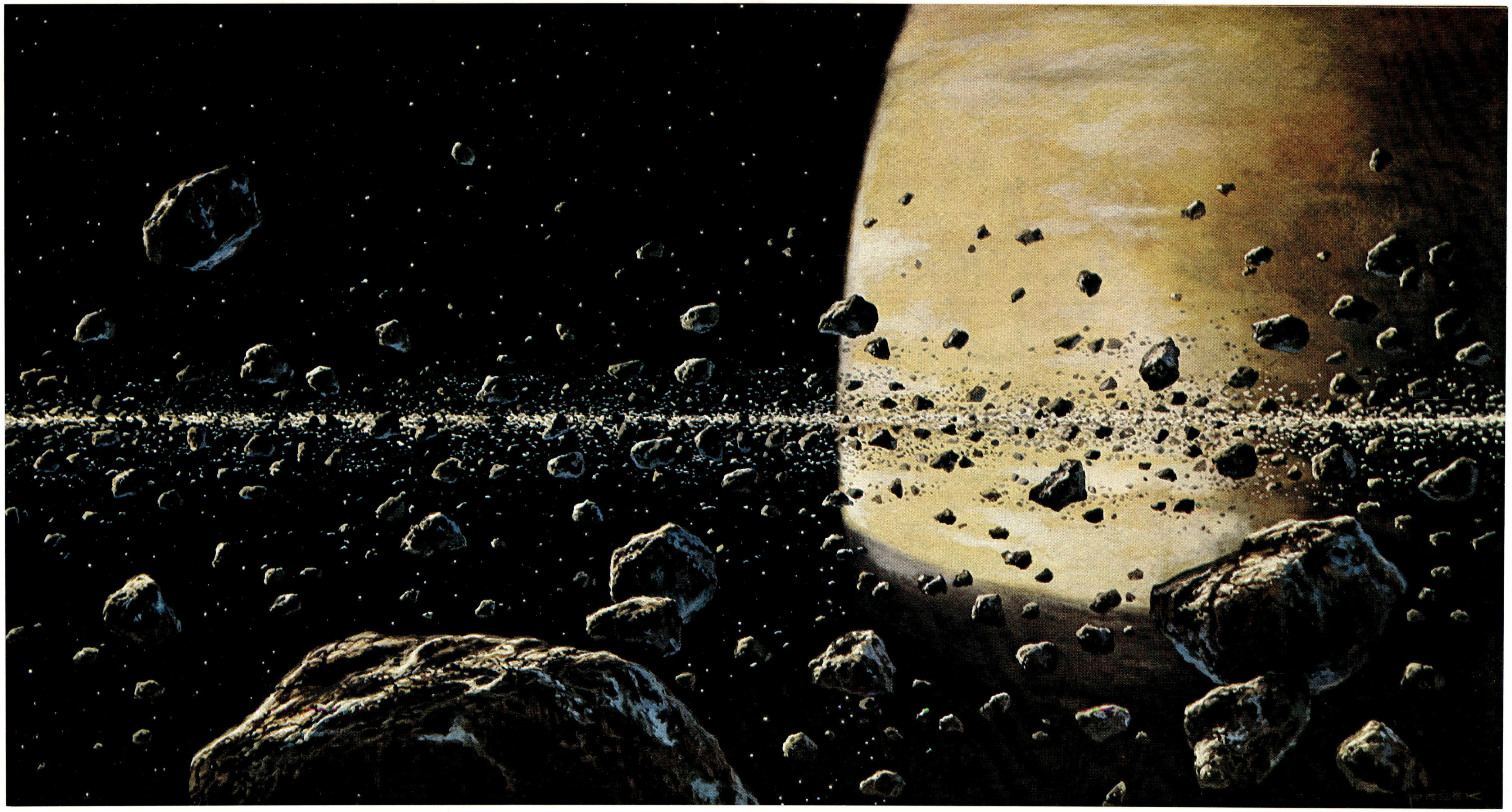

137. *Ludek Pesek.* WITHIN SATURN'S RINGS. Mixed media on panel, 14½ × 22¼". Reproduced courtesy of the Air and Space Museum, Smithsonian Institution

Eyewitness to Space

III

138. Yeffe Kimball *Lunar Gibbous*
139. Mitchell Jamieson *VAB Skeleton*
140. Hugh Laidman *Foundations*
141. Hugh Laidman *The Steel Belief*
142. Hugh Laidman *Sand Mountain*
143. Hugh Laidman *Pilings*
144. Hugh Laidman *Construction at Merritt Island*
145. Hugh Laidman *Unfinished Monster*
146. Paul Arlt *The Crawlerway from VAB*
147. Hugh Laidman *VAB Under Construction*
148. Alden Wicks *The New Olympus*
149. Alden Wicks *VAB Half Built*
150. Hugh Laidman *Construction on Merritt Island*
151. Mario Cooper *Apollo 11 Rollout*
152. Lowell Nesbitt *VAB Bay*
153. Nicholas Solovioff *Interior View VAB*
154. Nicholas Solovioff *Inside VAB*
155. Mario Cooper *Apollo 11 in the VAB*
156. Jack Perlmutter *Moon, Horizon, and Flowers*
157. Tom O'Hara *Spirit of Man in Space*
158. Paul Calle *Saturn 5 First Stage*
159. Mitchell Jamieson *Astronaut Cooper*
160. Ludek Pesek *Mars from Deimos*
161. Lamar Dodd *Lunar Views*
162. Lamar Dodd *Night Before Launch*
163. Lamar Dodd *Saturn—Apollo 10–11*
164. Franklin McMahon *Medical Examination*
165. Paul Calle *Recovery Doctor*
166. Paul Calle *Mike Collins*
167. Hans Cremers *Man's New Dimension*
168. Billy Morrow Jackson *Apollo 4*
169. Paul Calle *Rollout*
170. Billy Morrow Jackson *Press Site*
171. John Pike *Boilerplate*
172. Paul Calle *Gemini Astronaut Borman Recovery*
173. Paul Calle *New Arrival*
174. Tom O'Hara *Checking the Command Module*
175. Yeffe Kimball *Blue Space Concept*

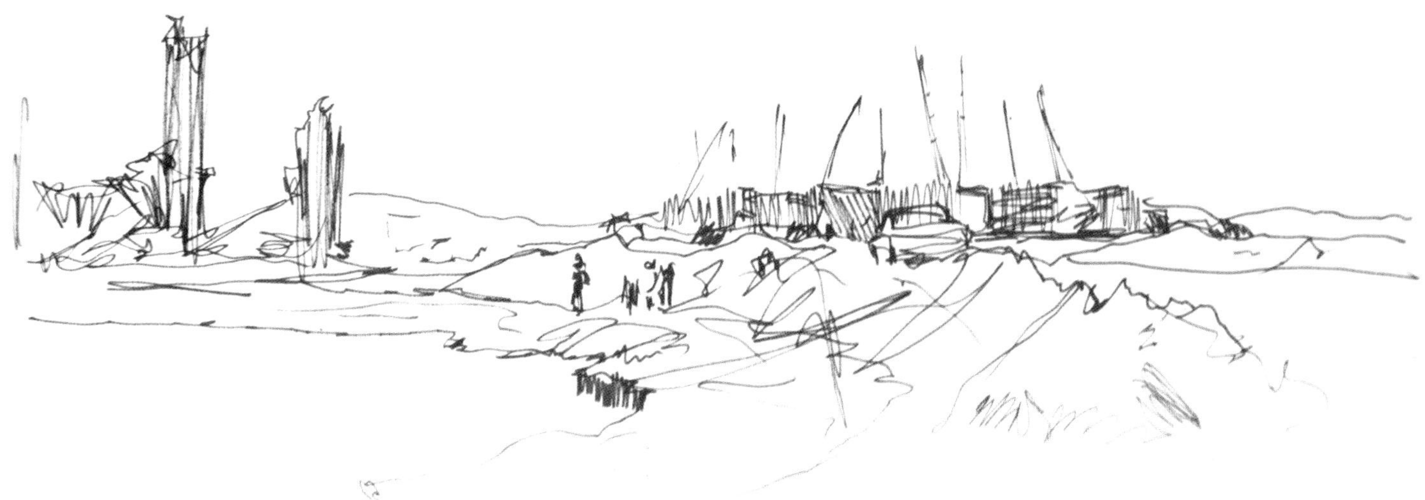

140.
Hugh Laidman. FOUNDATIONS
Felt pen on paper, 12×18″

139.
Mitchell Jamieson. VAB SKELETON
India ink on acetate, 14×17″

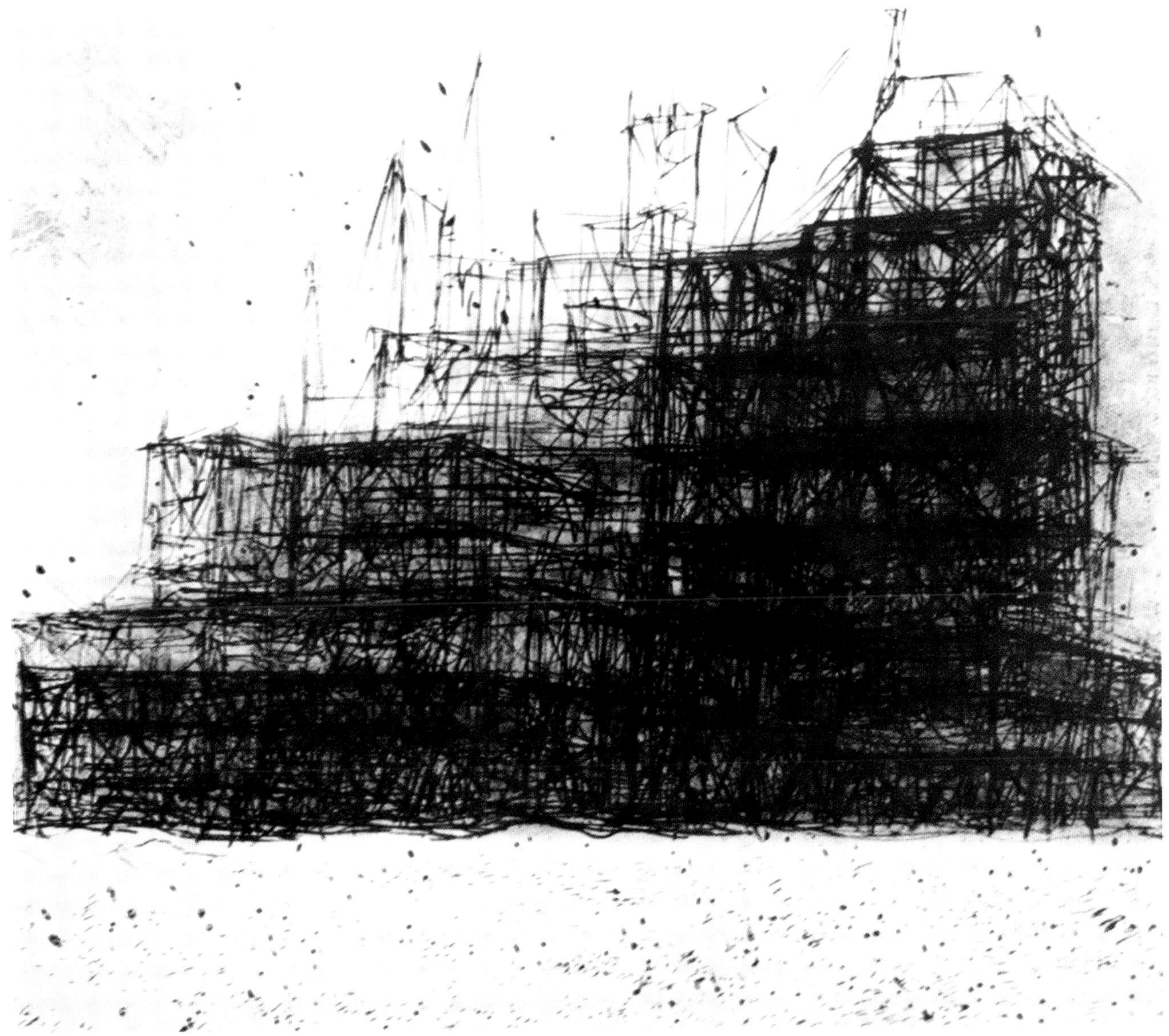

138.
Yeffe Kimball
LUNAR GIBBOUS
Acrylic on canvas, 38″ diameter

Before the Ranger and Surveyor spacecraft photographed the moon, the imagination of the artist had already been there.

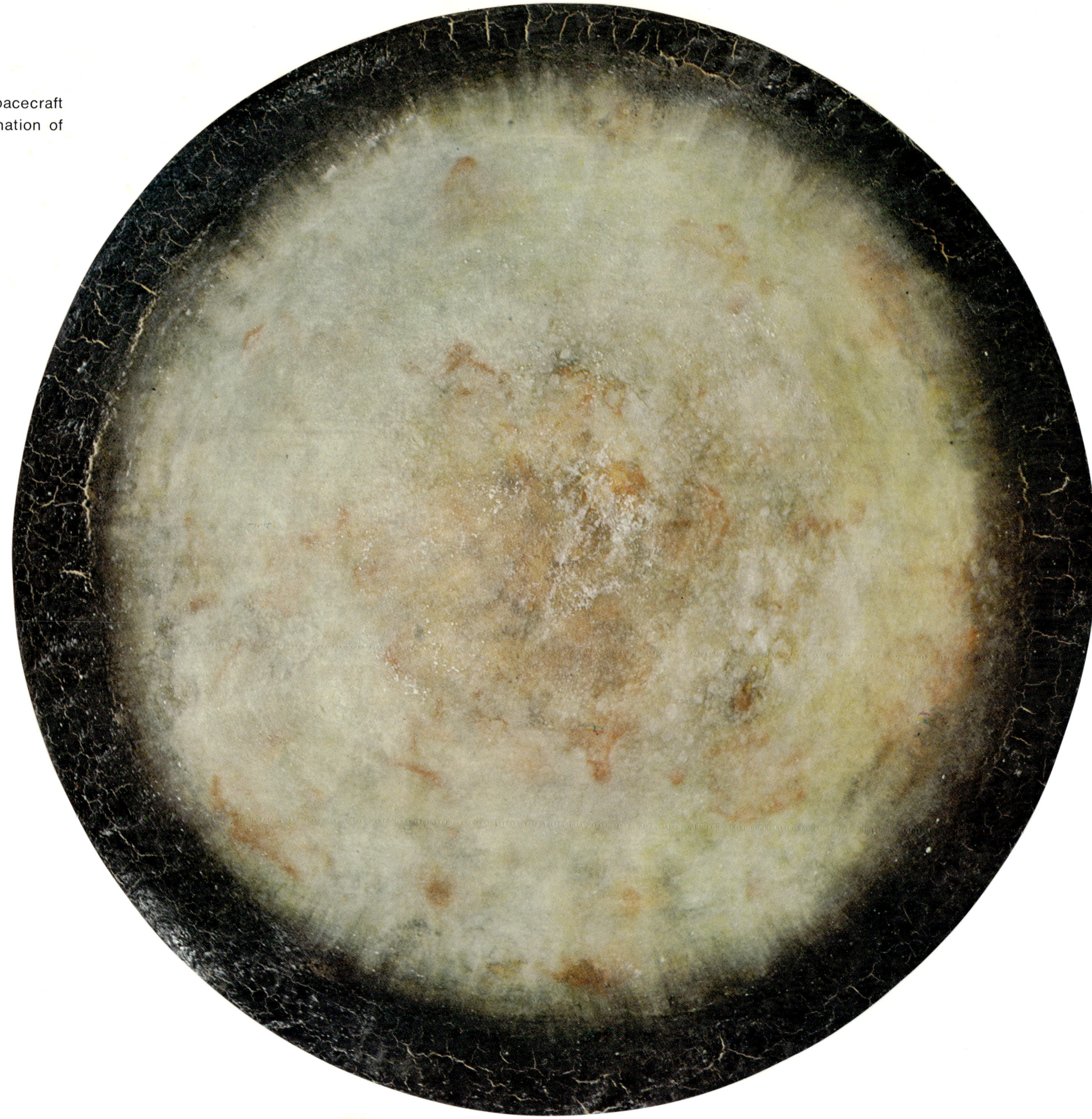

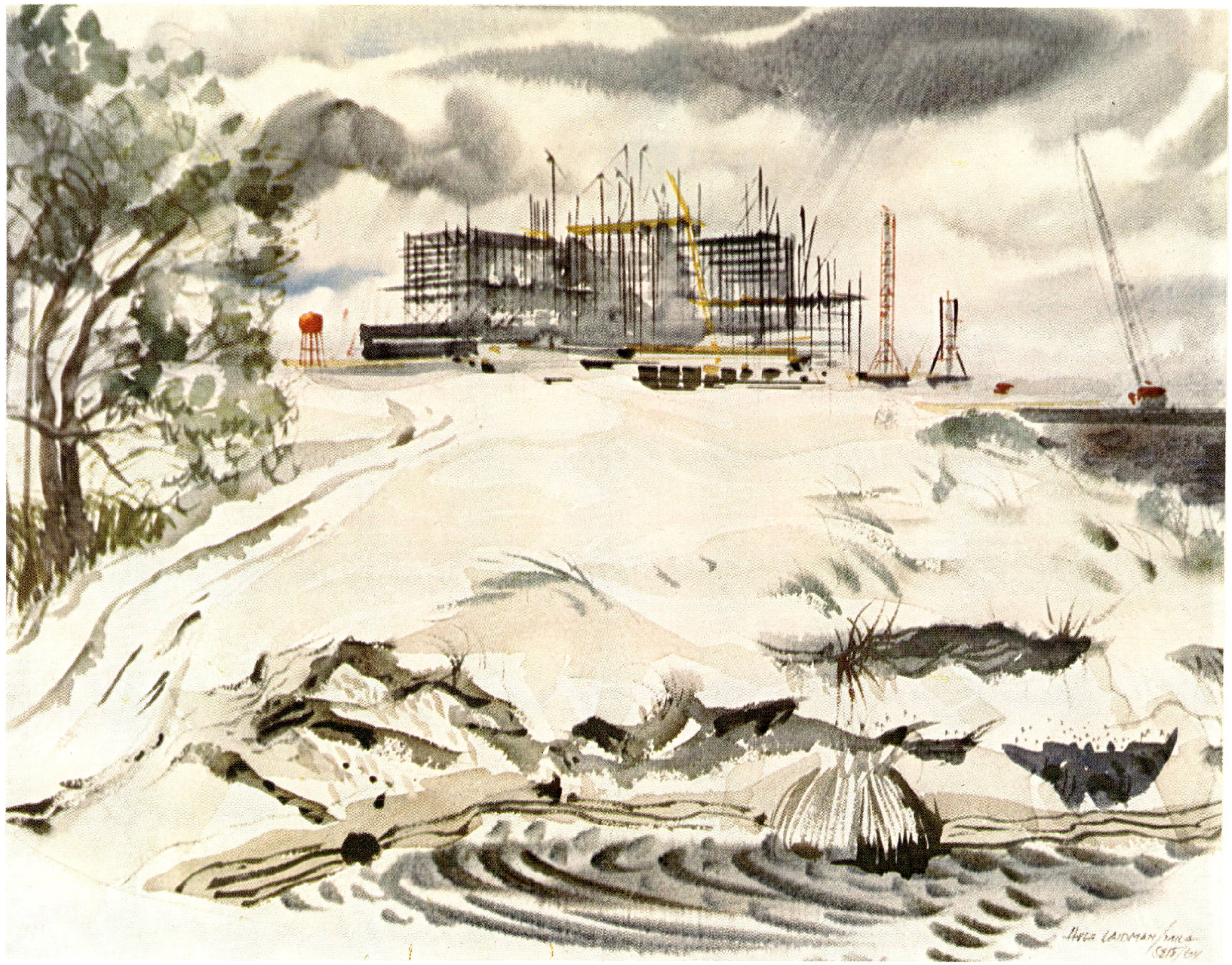

141.
Hugh Laidman
THE STEEL BELIEF
Watercolor on paper, 22×30″

Out of the sand, mud, and scrub of Merritt Island, near Cape Kennedy, the immense framework of the VAB (standing for Vertical, later Vehicle, Assembly Building) arose to dominate the flat wasteland of Cape Kennedy. At this stage, in September, 1964, it looked like the skeletal remains of a prehistoric monster.

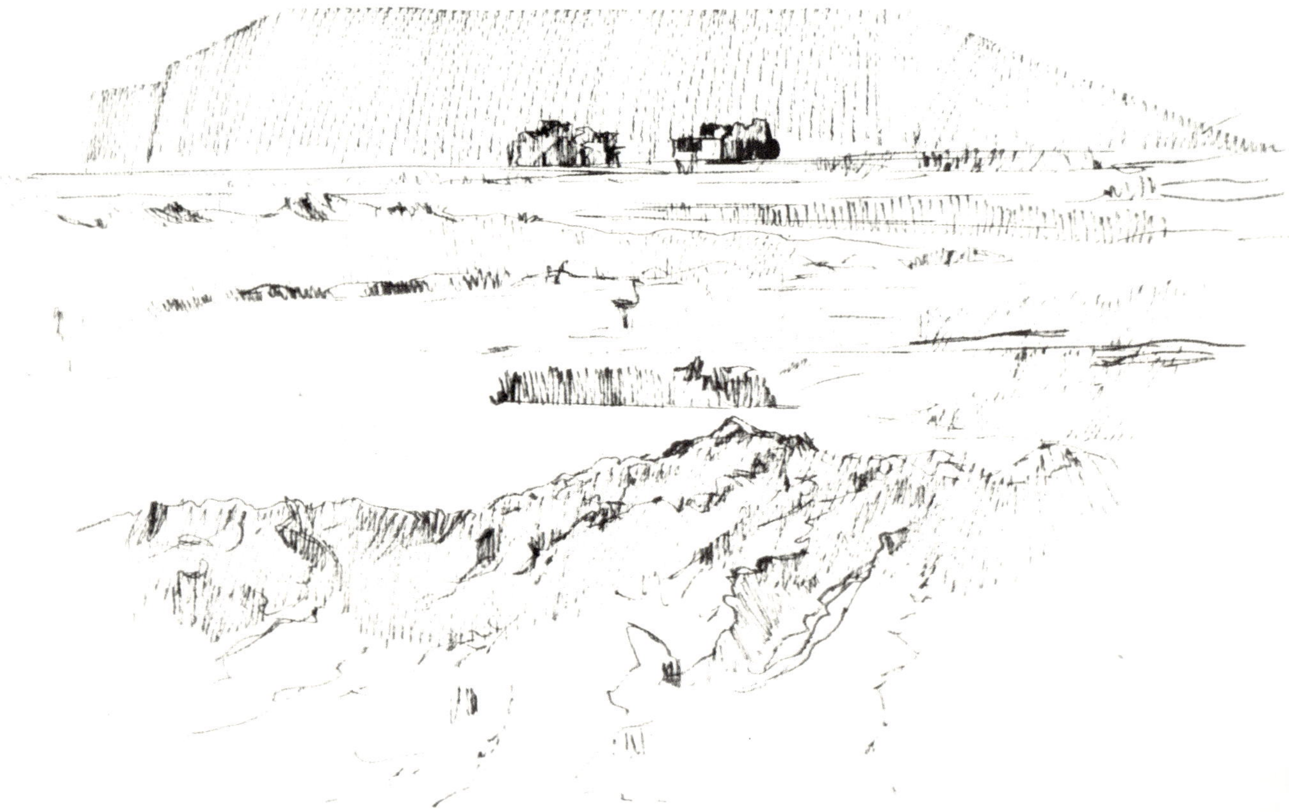

142.
Hugh Laidman. SAND MOUNTAIN
Felt pen on paper, 12×18″

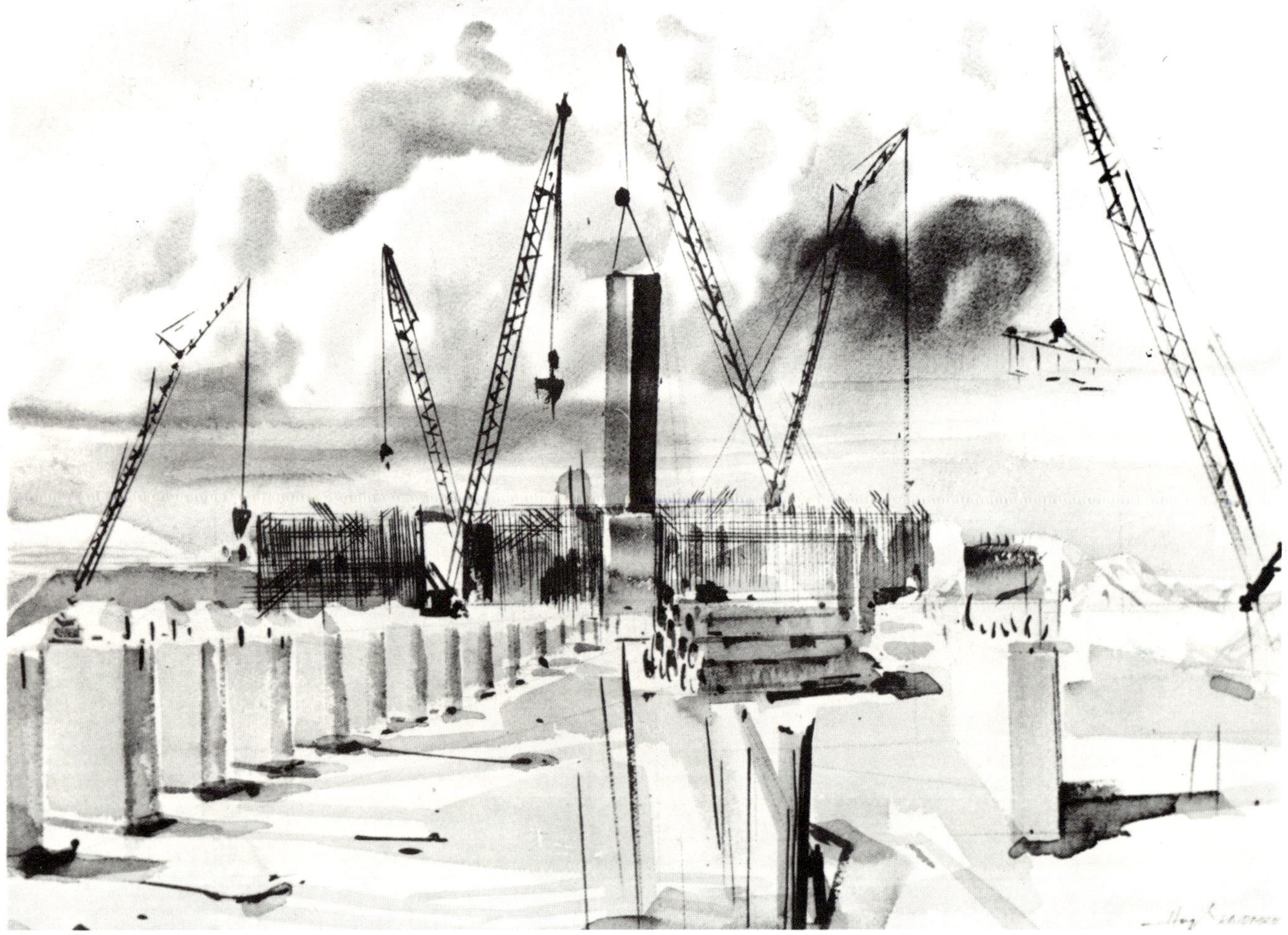

143. *Hugh Laidman.* PILINGS. Watercolor on paper, 22×30″

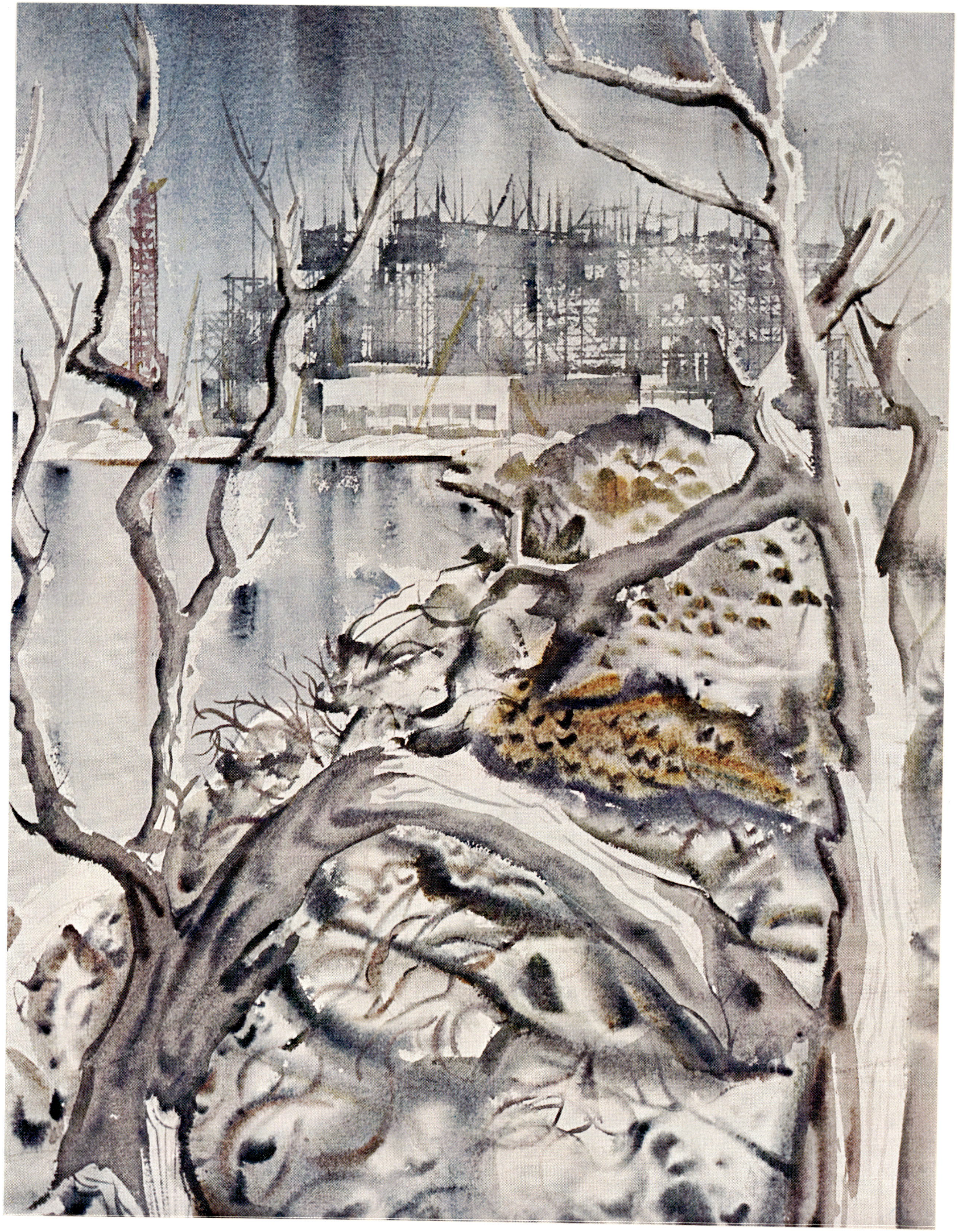

144.
Hugh Laidman
CONSTRUCTION AT MERRITT ISLAND
Watercolor on paper, 30 × 22″

One set of skeletal forms replaces another; the whitened trunks of tropical trees seem to have a common ancestry with the steel girders reaching upward into the subtropical twilight.

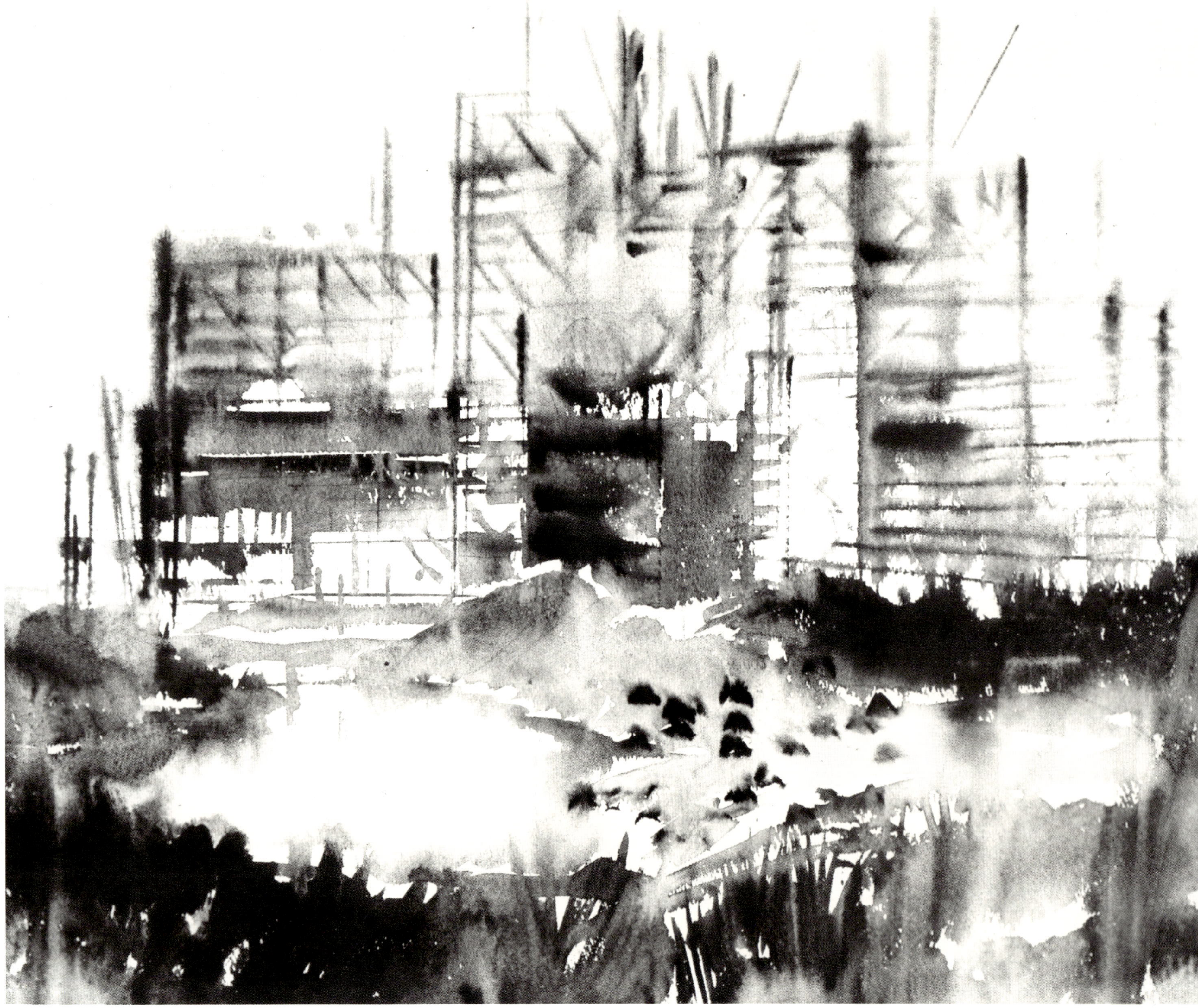

145.
Hugh Laidman
UNFINISHED MONSTER
Watercolor on paper, 22×30″

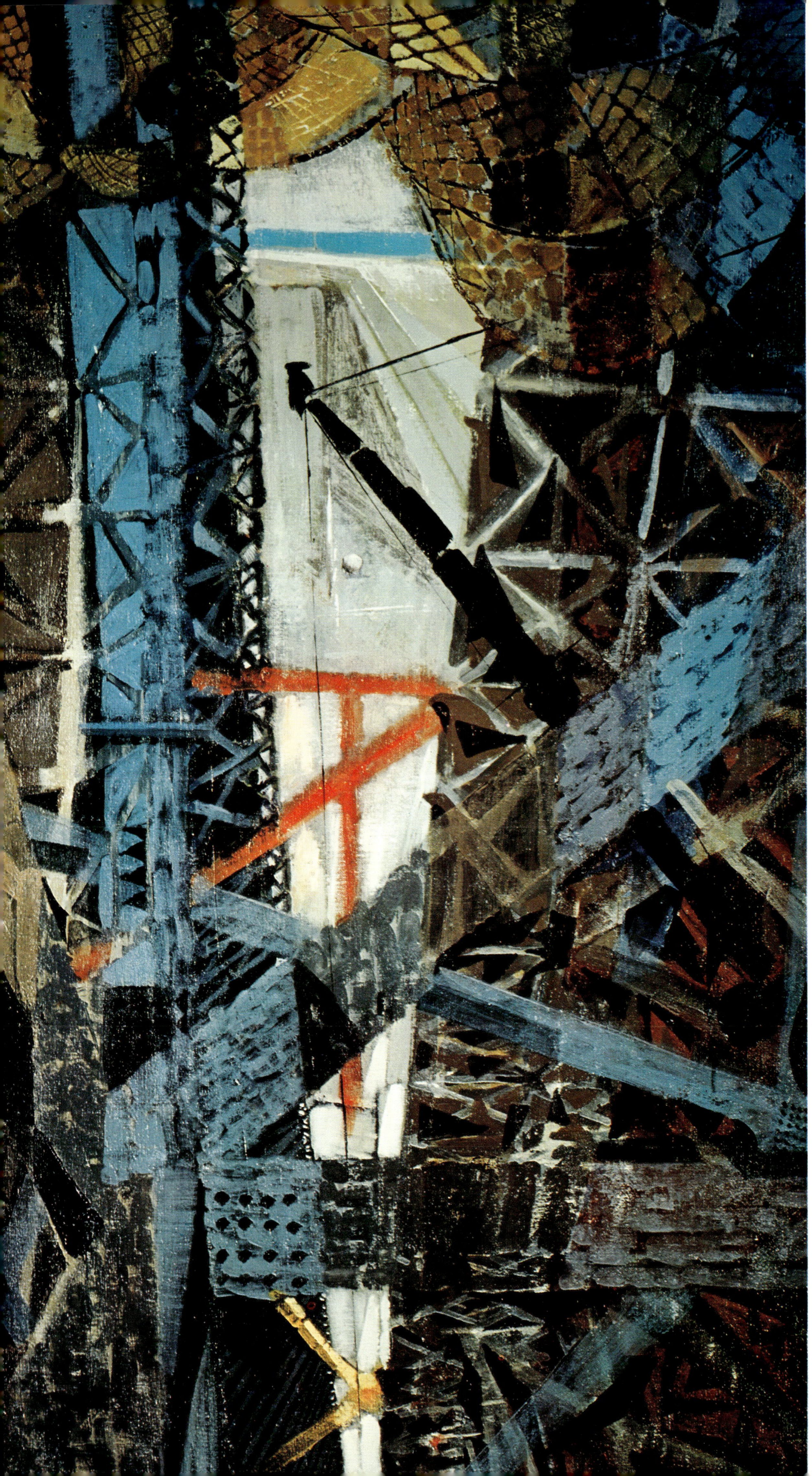

146.
Paul Arlt
THE CRAWLERWAY FROM VAB
Acrylic on canvas, 40×28″

Inside the VAB, the girders, the shafts of sunlight, the dizzy heights, the nets for accidental falls, and the incredible vertical views challenged and excited many artists. Paul Arlt was the first painter to be allowed up inside the half-completed building.

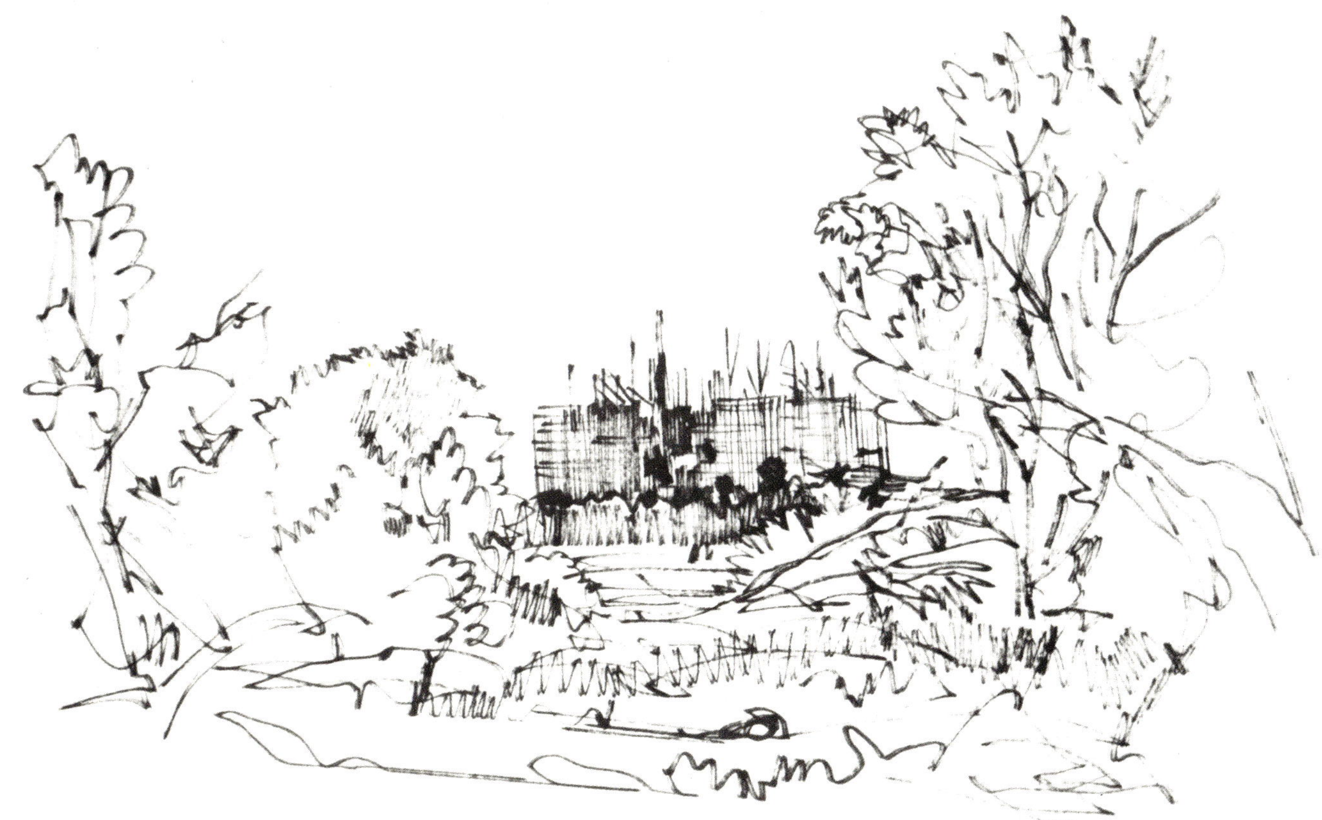

147. *Hugh Laidman*. VAB UNDER CONSTRUCTION. Felt pen on paper, 11×17″

148.
Alden Wicks
THE NEW OLYMPUS
Oil on canvas, 24 × 20″

Classical mythology has been involved in the space program since the first manned flight program was named Mercury. Wicks, noting that the Gemini twins, Castor and Pollux, resulted from the union of Zeus, camouflaged as a swan, and Leda, also gives due honor to the god Apollo. Behind, so huge that thunderstorms rage in the interior, the half-completed VAB makes a suitable temple for the new race of gods.

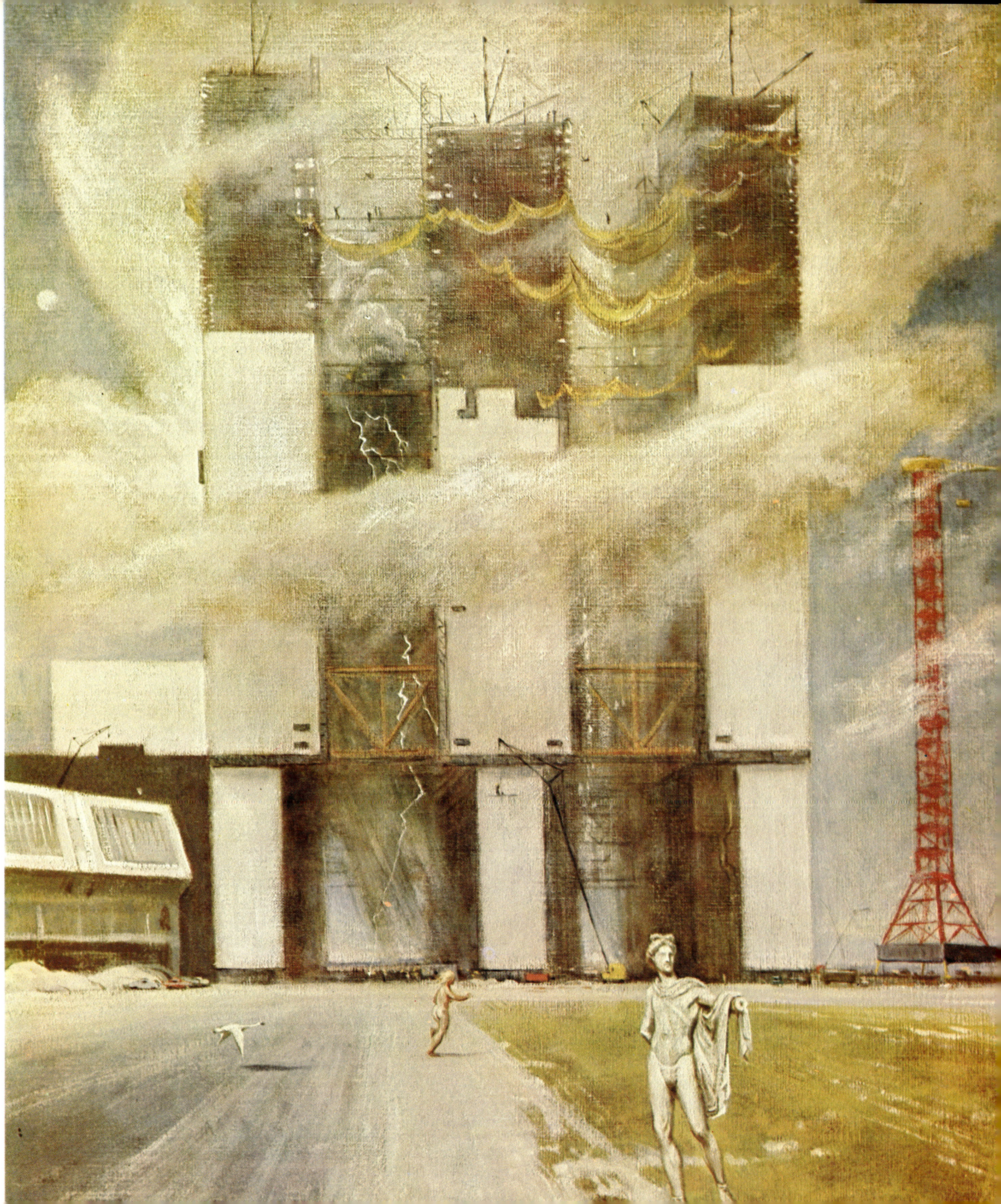

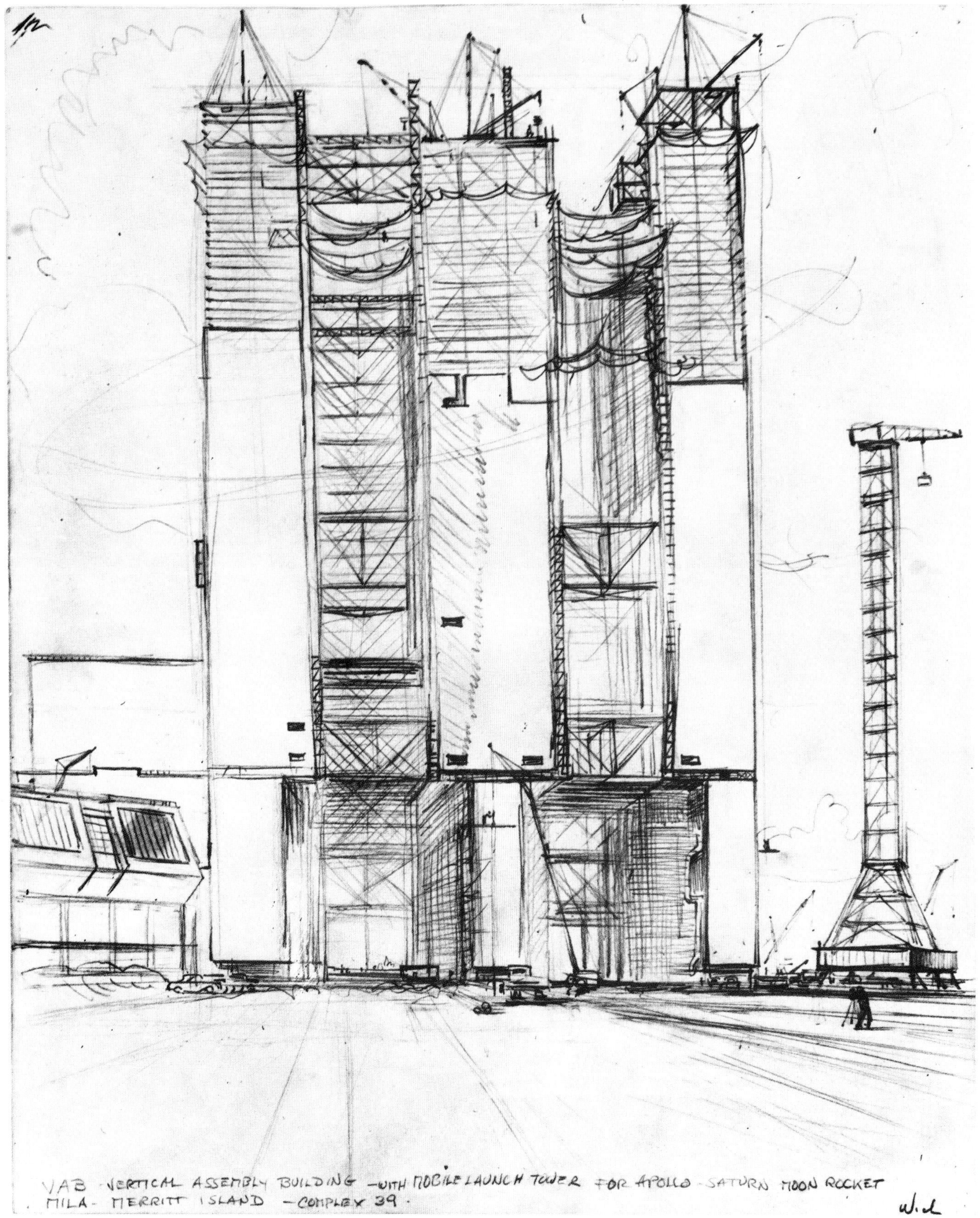

149. *Alden Wicks.* VAB HALF BUILT. Pencil on paper, 14×11″

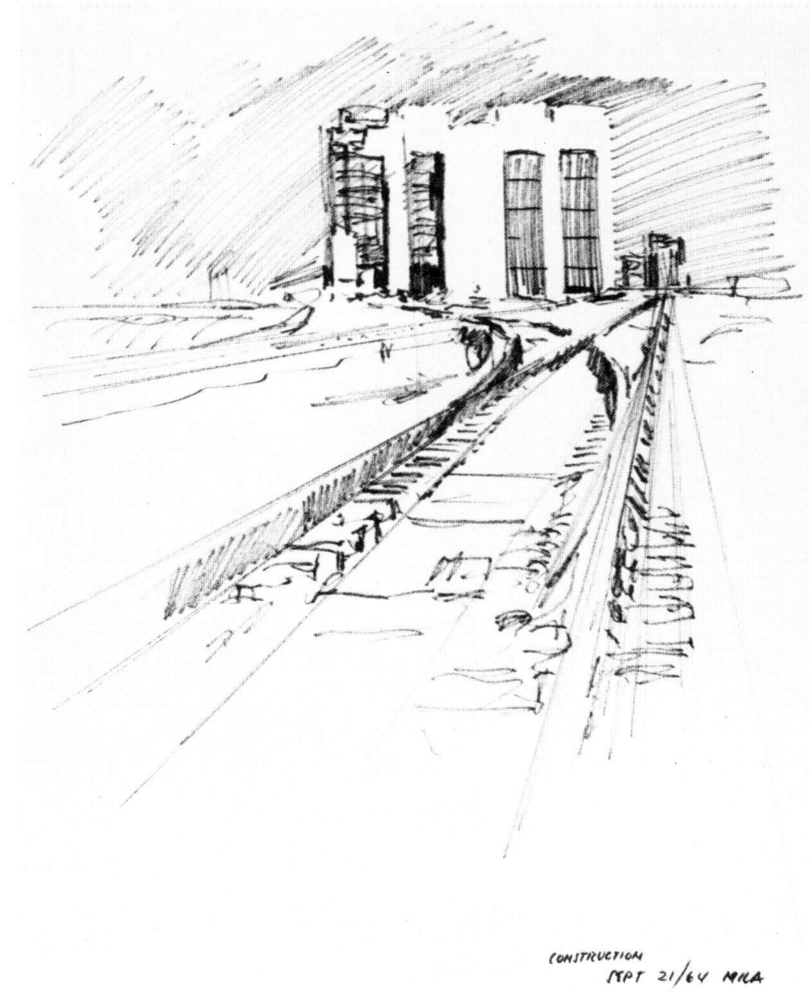

150. *Hugh Laidman.* CONSTRUCTION ON MERRITT ISLAND
Felt pen on paper, 18×12″

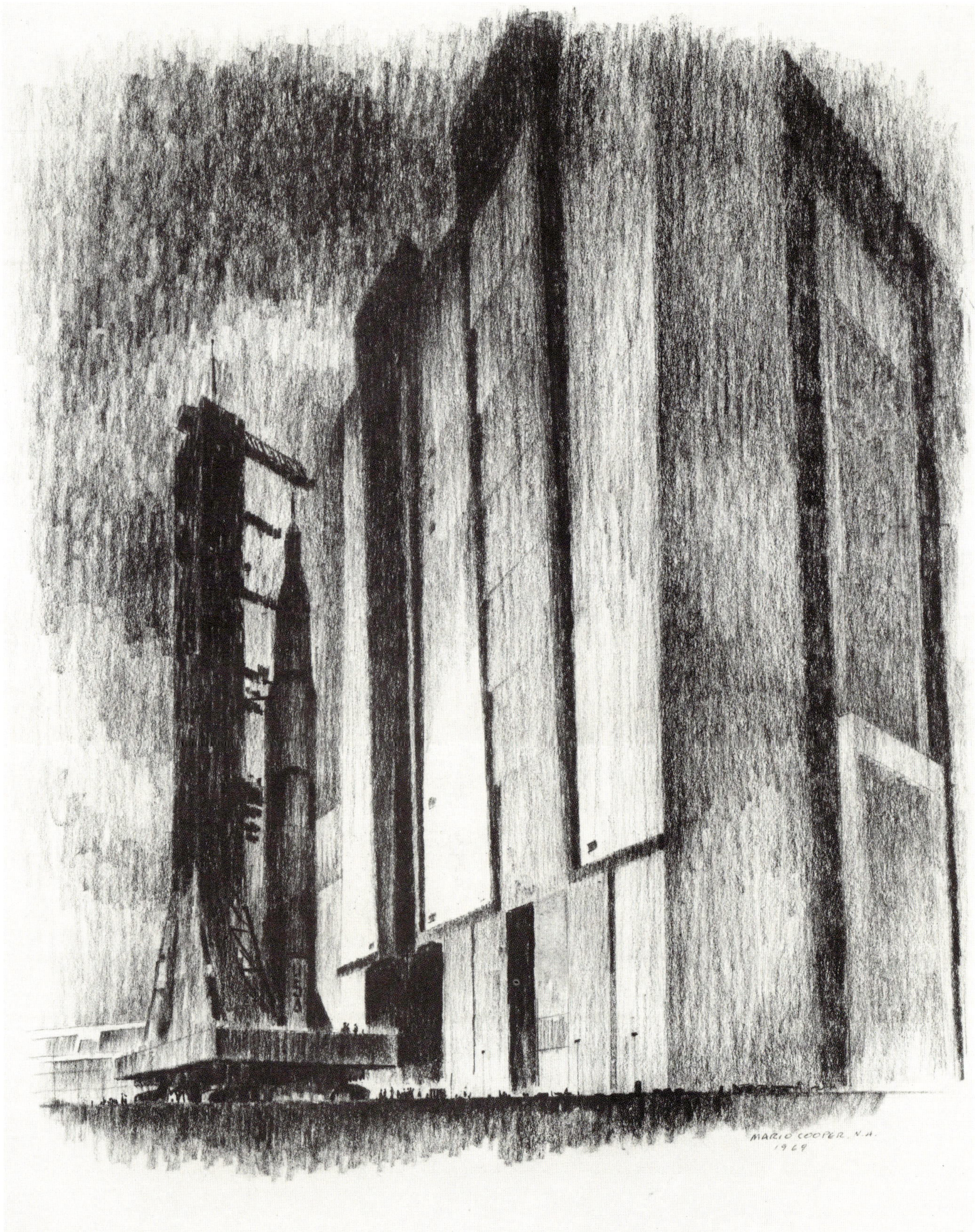

151.

Mario Cooper

APOLLO 11 ROLLOUT

Pencil on paper, 14×11″

The rollout has the majesty and pace of a Byzantine religious ceremony. The vast doors open at dawn and the towering mass of the rocket inch by inch is transported at the speed of a lava flow. Mario Cooper, like most artists, was baffled by the size and has taken artistic license in placing figures on or near the assemblage. Actually, all except a few essential personnel are kept well away from the moving mass.

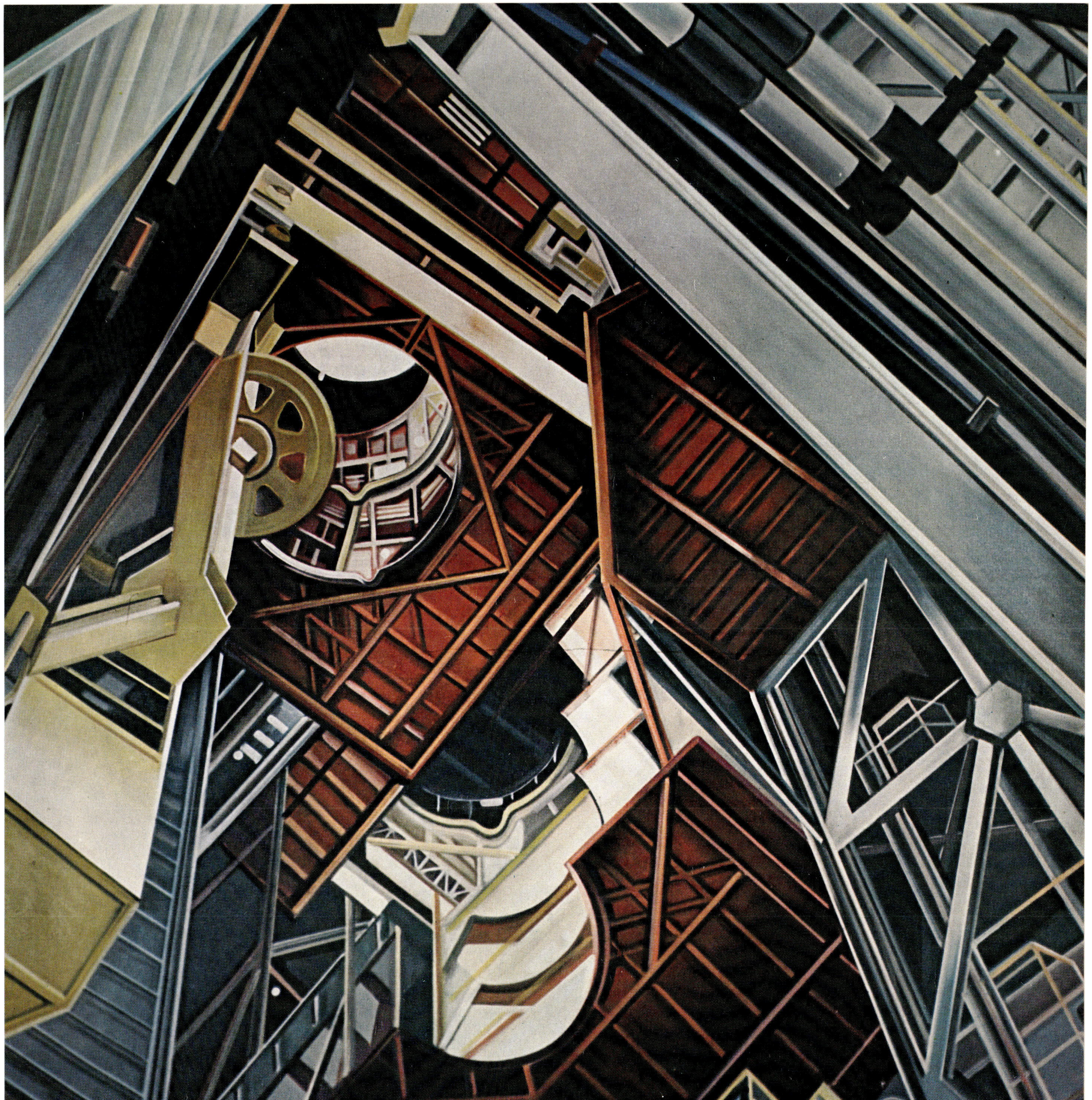

152.
Lowell Nesbitt
VAB BAY
Oil on canvas, 5×5′

Inside the VAB are incredible vistas. The one looking upward into the bays designed to house a Saturn V rocket presents the artist with a maze of interlocking planes, platforms, girders, shafts, catwalks, monstrous wheels, and, at the end, a segment of open sky.

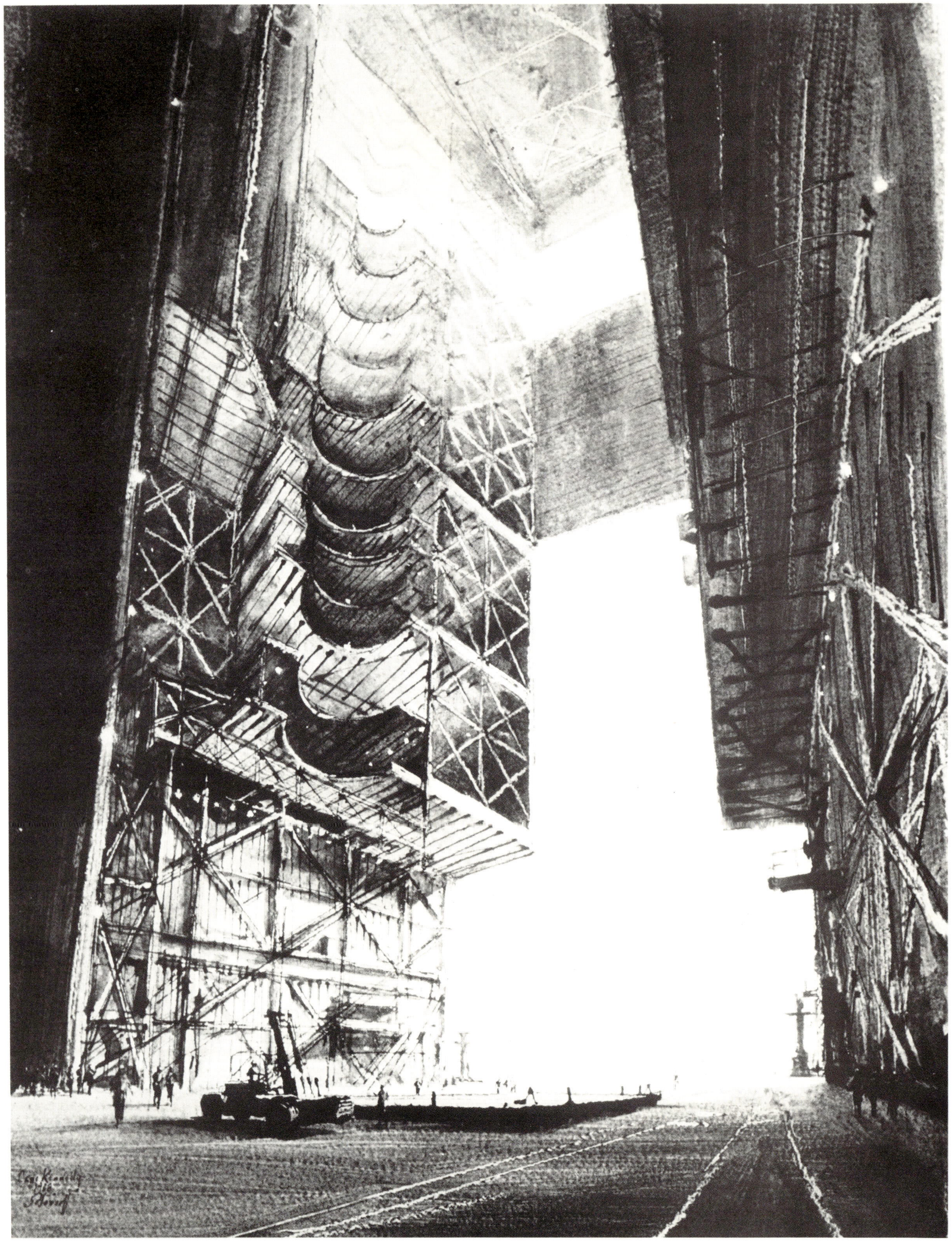

153.
Nicholas Solovioff
INTERIOR VIEW VAB
Watercolor on paper, 30×22″

The Vehicle Assembly Building was the biggest building in the world, but no one would have guessed it from the outside. The view from the inside, however, looking through one of the doors, which is big enough to encompass the United Nations building in New York, gives some idea of the gargantuan proportions.

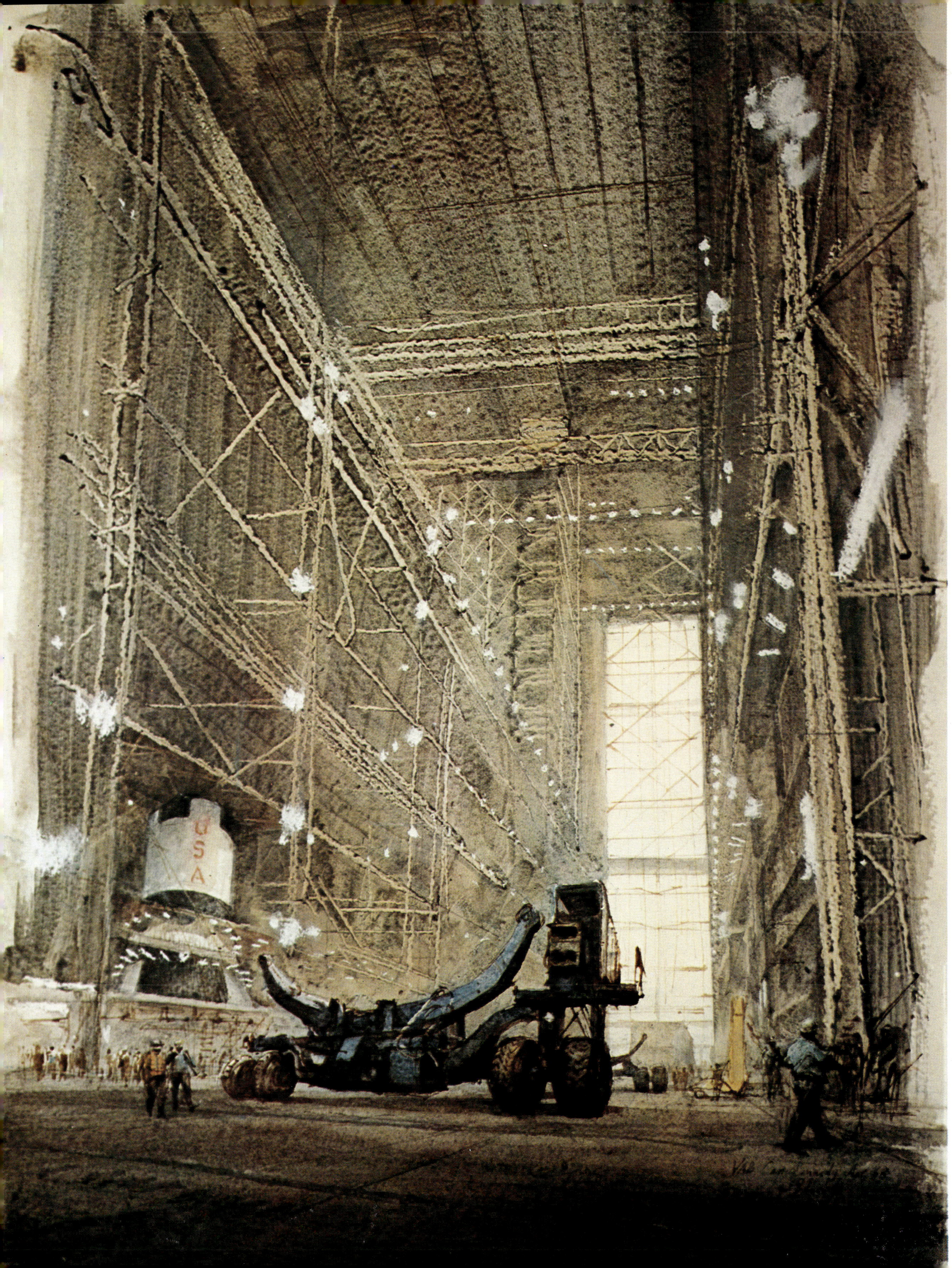

154.
Nicholas Solovioff
INSIDE VAB
Watercolor on paper, 30 × 22″

The echoing vastness of the VAB is reminiscent of a fantasy by Piranesi.

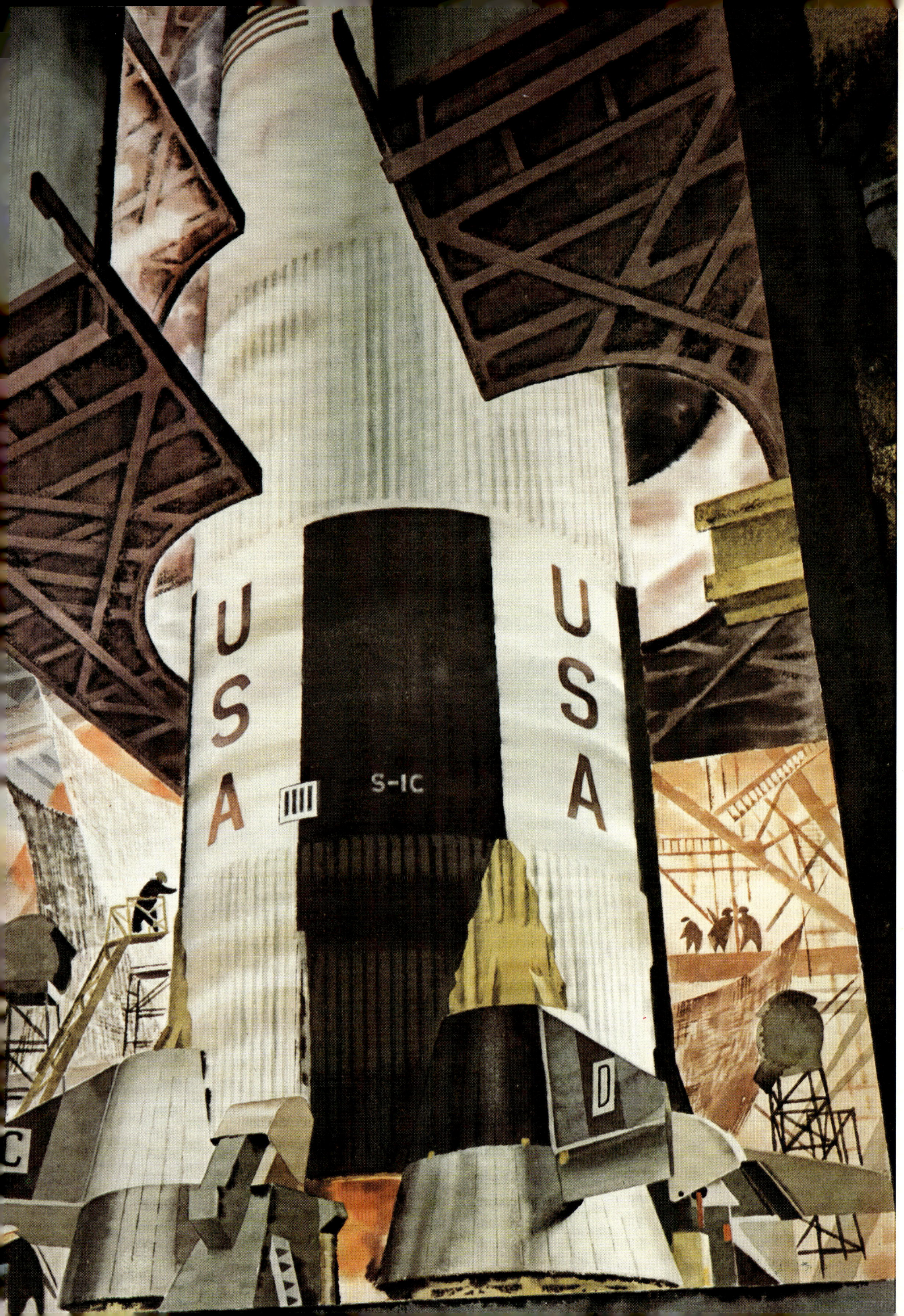

155.
Mario Cooper
APOLLO 11 IN THE VAB
Watercolor on paper, 29×21″

To the engineer, the final checking of the engines of the Saturn–Apollo 11 may have been an exacting chore, but to Mario Cooper, president of the American Watercolor Society, it was a Baroque stage setting of transcendent beauty. Actors in the drama, like high priests, minister to the majestic idol on center stage.

156.
Jack Perlmutter
MOON, HORIZON, AND FLOWERS
Oil on canvas, 38×52″

The rollout is the agonizingly slow journey of the Saturn-Apollo rocket from the service area inside the VAB to the launch pad more than three miles away. Perlmutter records the measured progress in an artist's terms. Overhead, the lacelike metal of the arming tower is silhouetted against the rising moon.

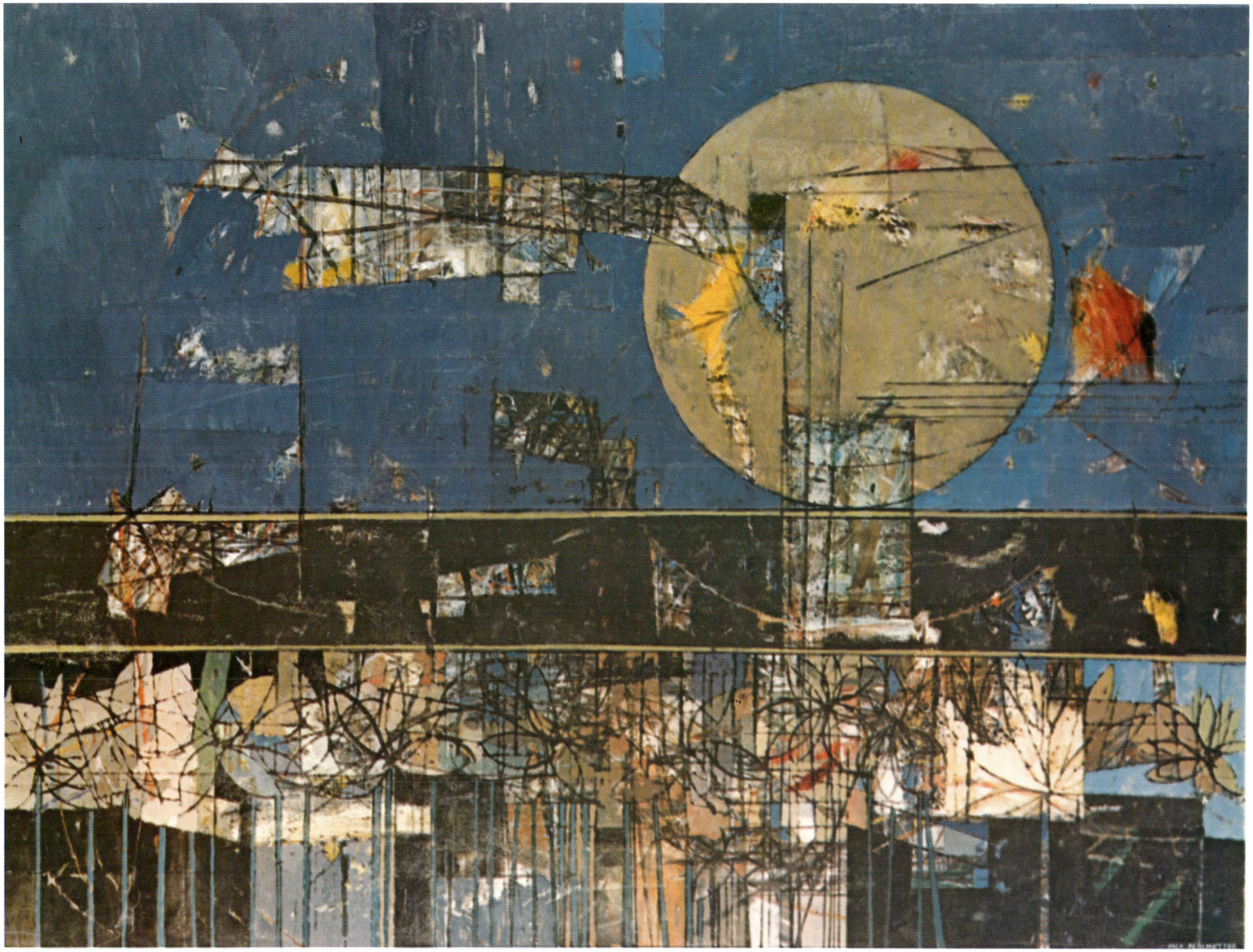

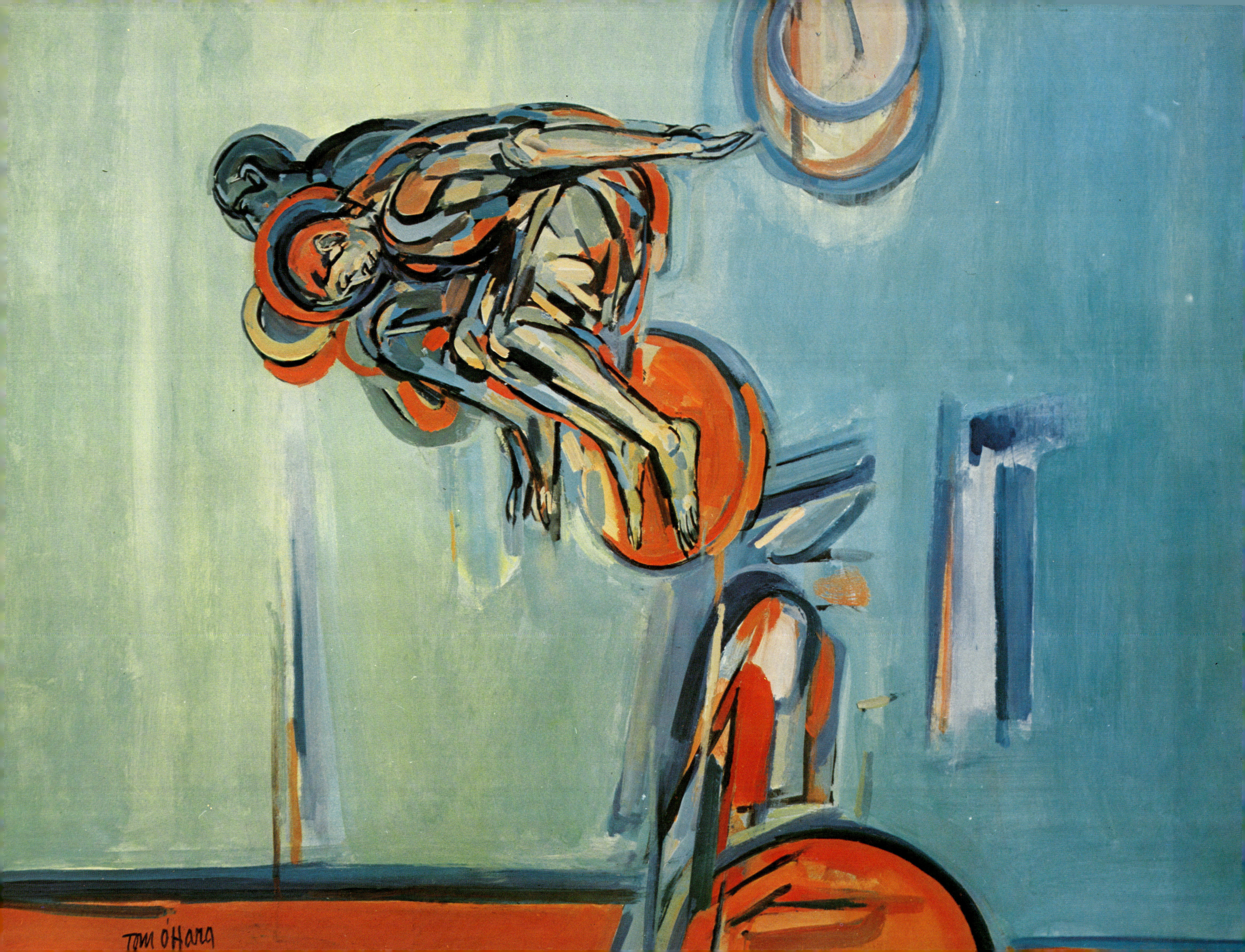
Tom O'Hara

157.
Tom O'Hara
SPIRIT OF MAN IN SPACE
Acrylic on paper, 22×30″

In space there is no up, no down, no front, no back. Man wears his helmet like a halo and trusts that the dangers of his mission will not be more than he can overcome.

158. *Paul Calle*. SATURN 5 FIRST STAGE. Pen and ink on paper, 18×22″

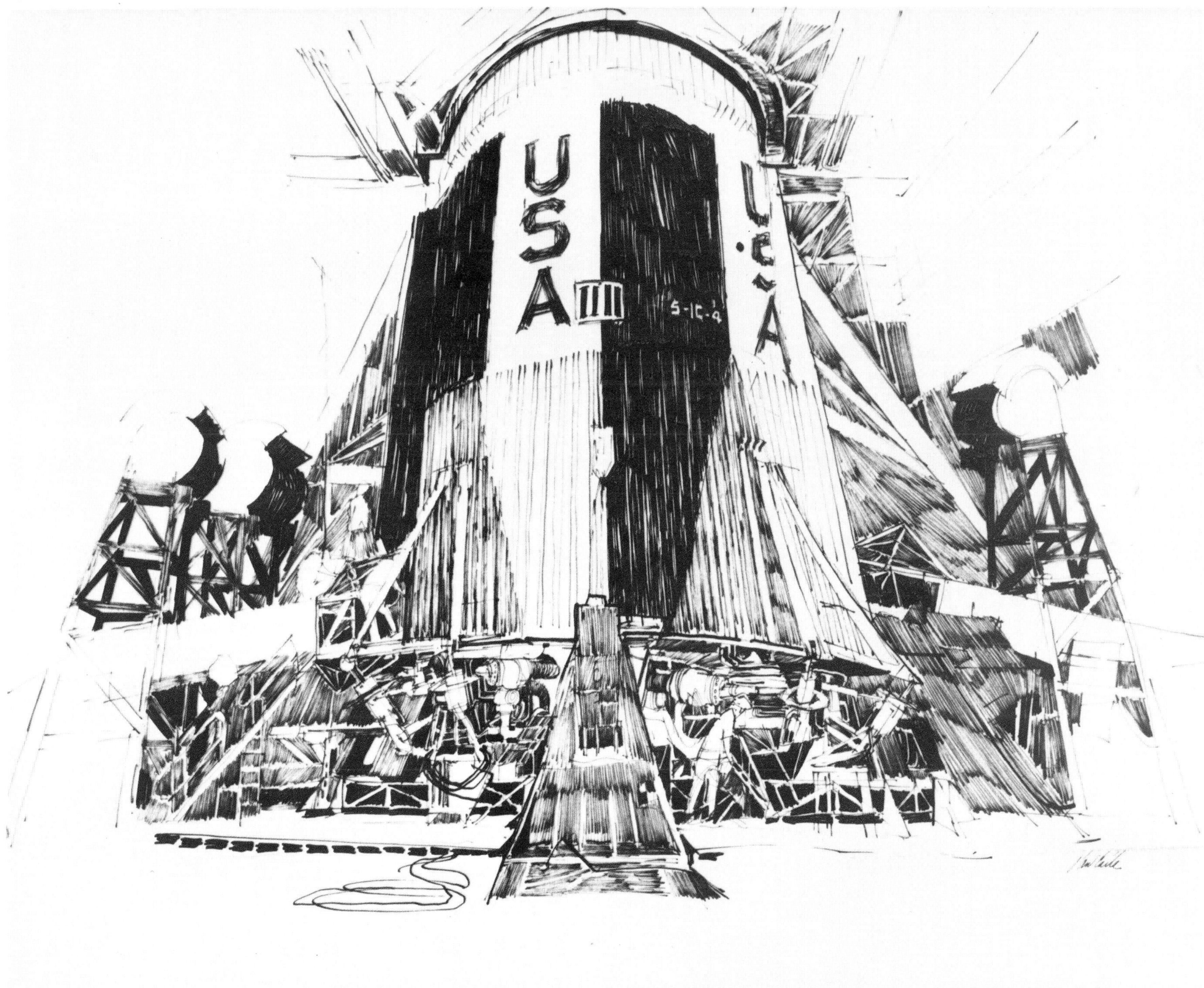

159.
Mitchell Jamieson
ASTRONAUT COOPER
Ink on printed map, 33×9½″

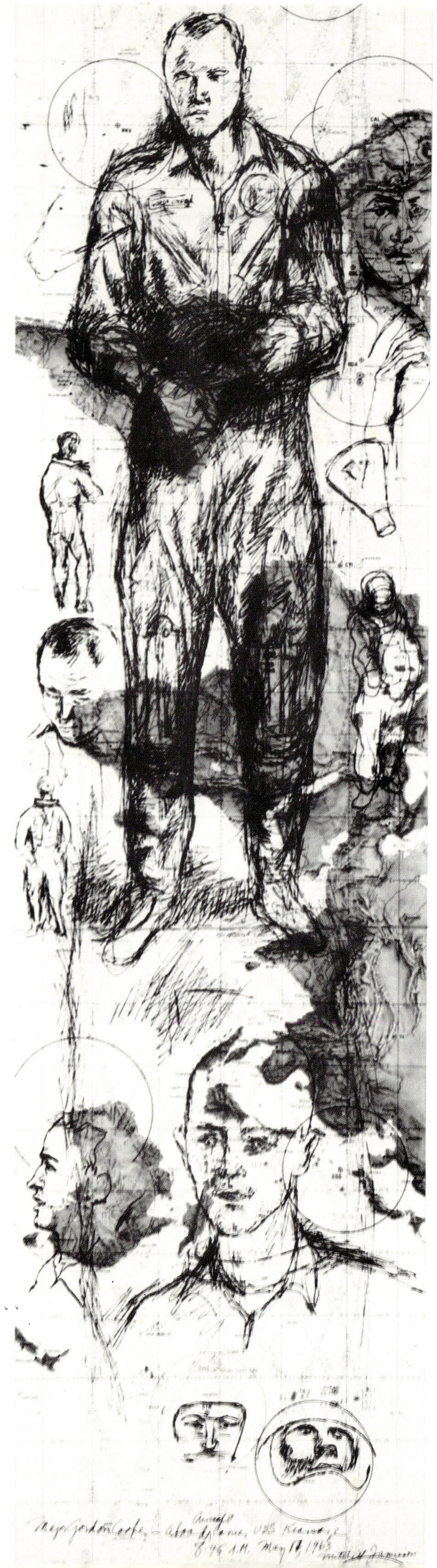

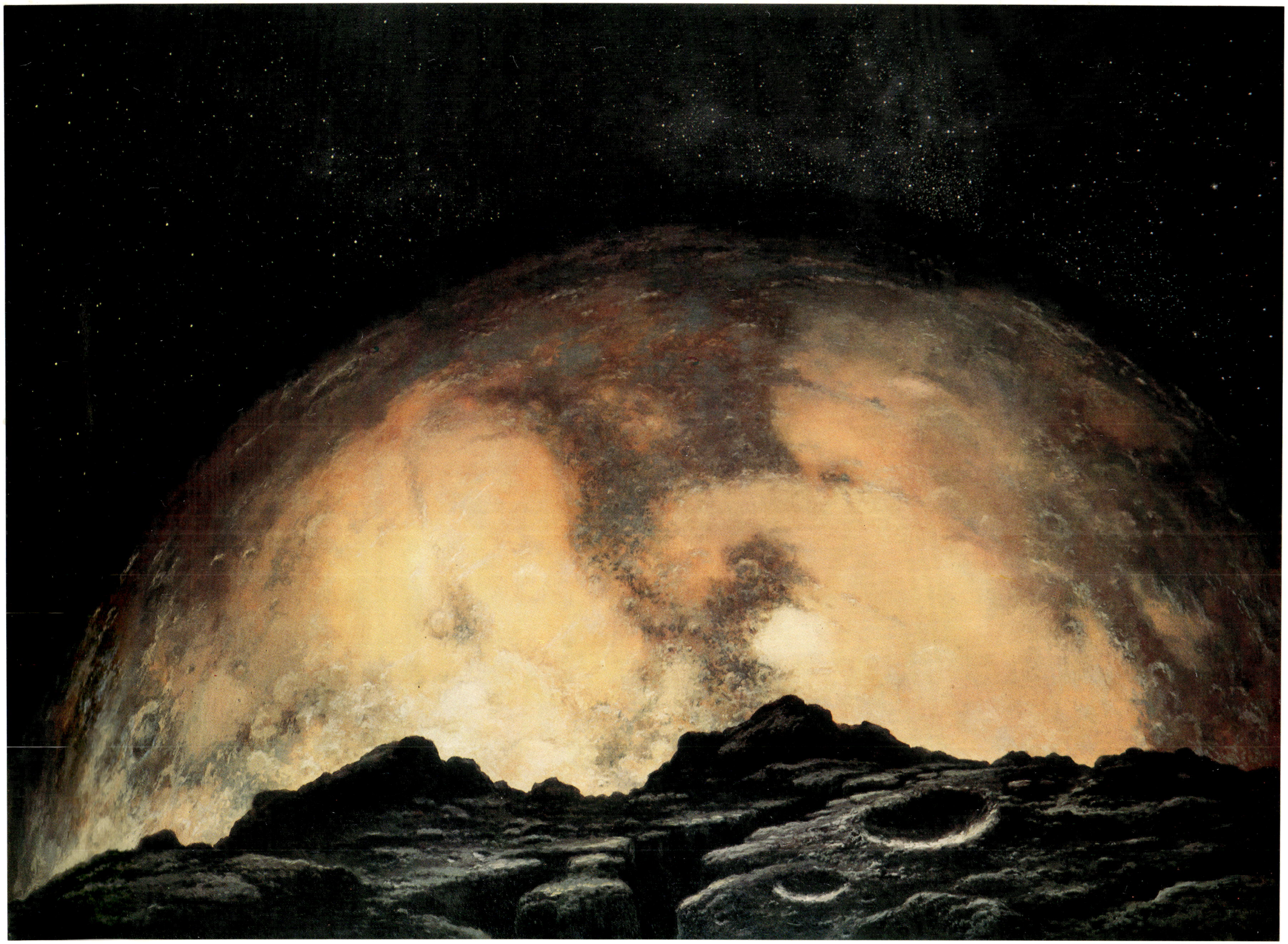

160.
Ludek Pesek
MARS FROM DEIMOS
Mixed media on panel, 14½ × 22¼″
Reproduced courtesy of the Air and Space Museum, Smithsonian Institution

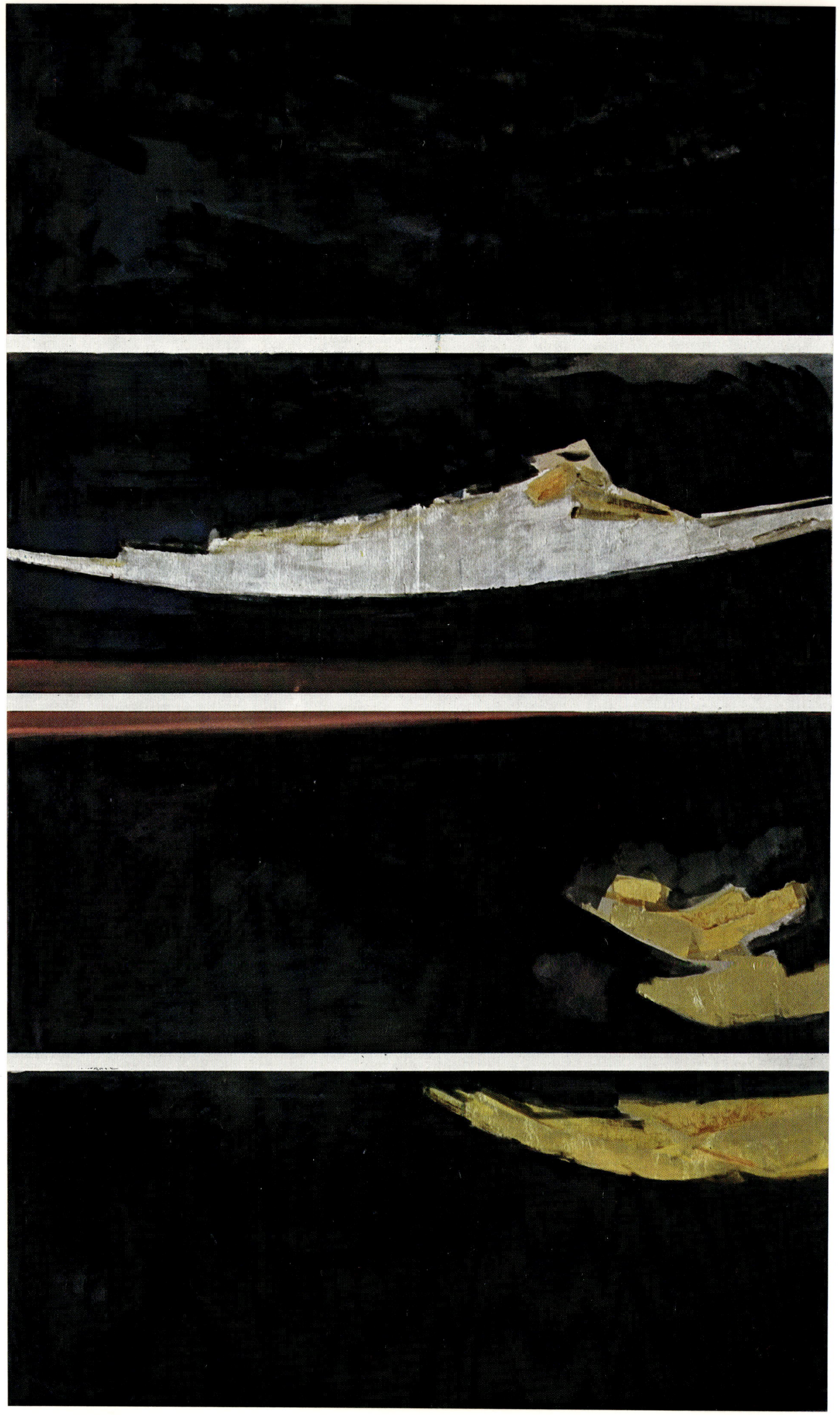

161.
Lamar Dodd. LUNAR VIEWS
Oil on canvas, 82 × 60″

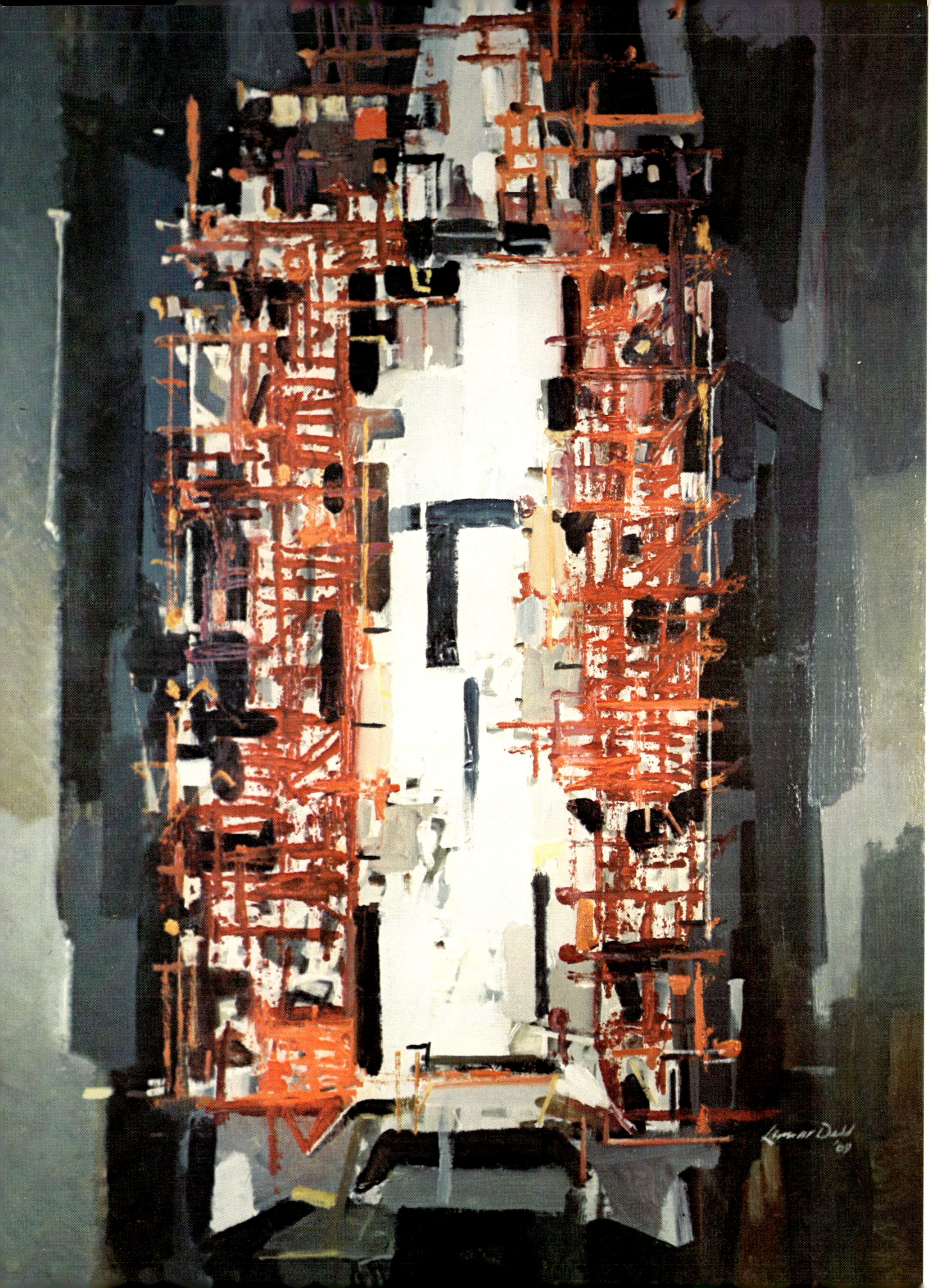

162.
Lamar Dodd
NIGHT BEFORE LAUNCH
Oil on canvas, 50 × 36″

The Saturn encased in its shimmering ganglion of red supports, the arming tower on one side, the service gantry on the other, is a sight that no painter could ever forget. Many artists have dealt with this theme, each in his own way. Lamar Dodd, perhaps more than any other, has seen and captured the mystic beauty of the theme and transformed it into a fantasy from the *Arabian Nights*.

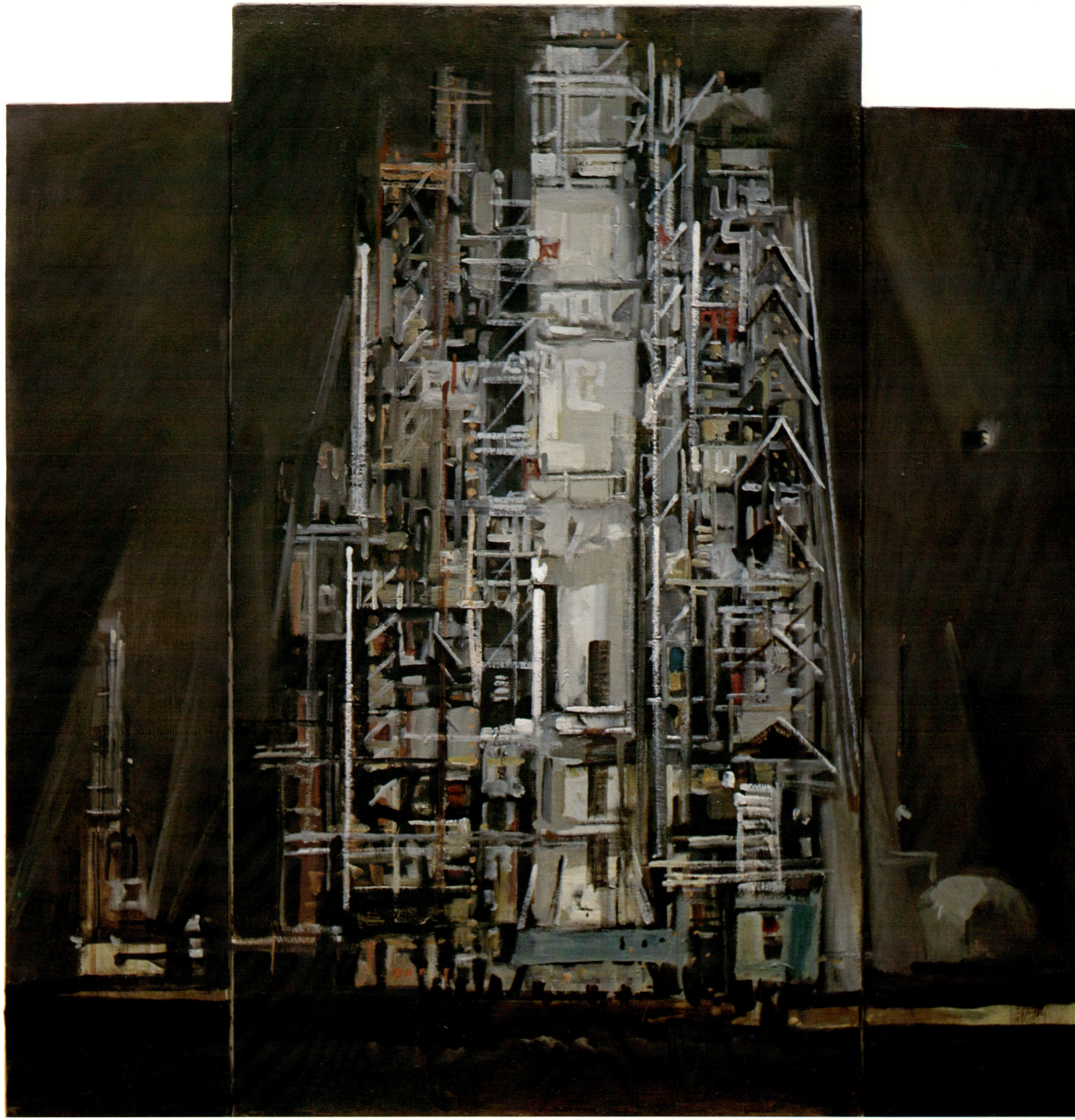

163.
Lamar Dodd
SATURN–APOLLO 10–11.
Oil on canvas, 60×52″

164. *Franklin McMahon.* MEDICAL EXAMINATION. Pencil on paper, 22×14″

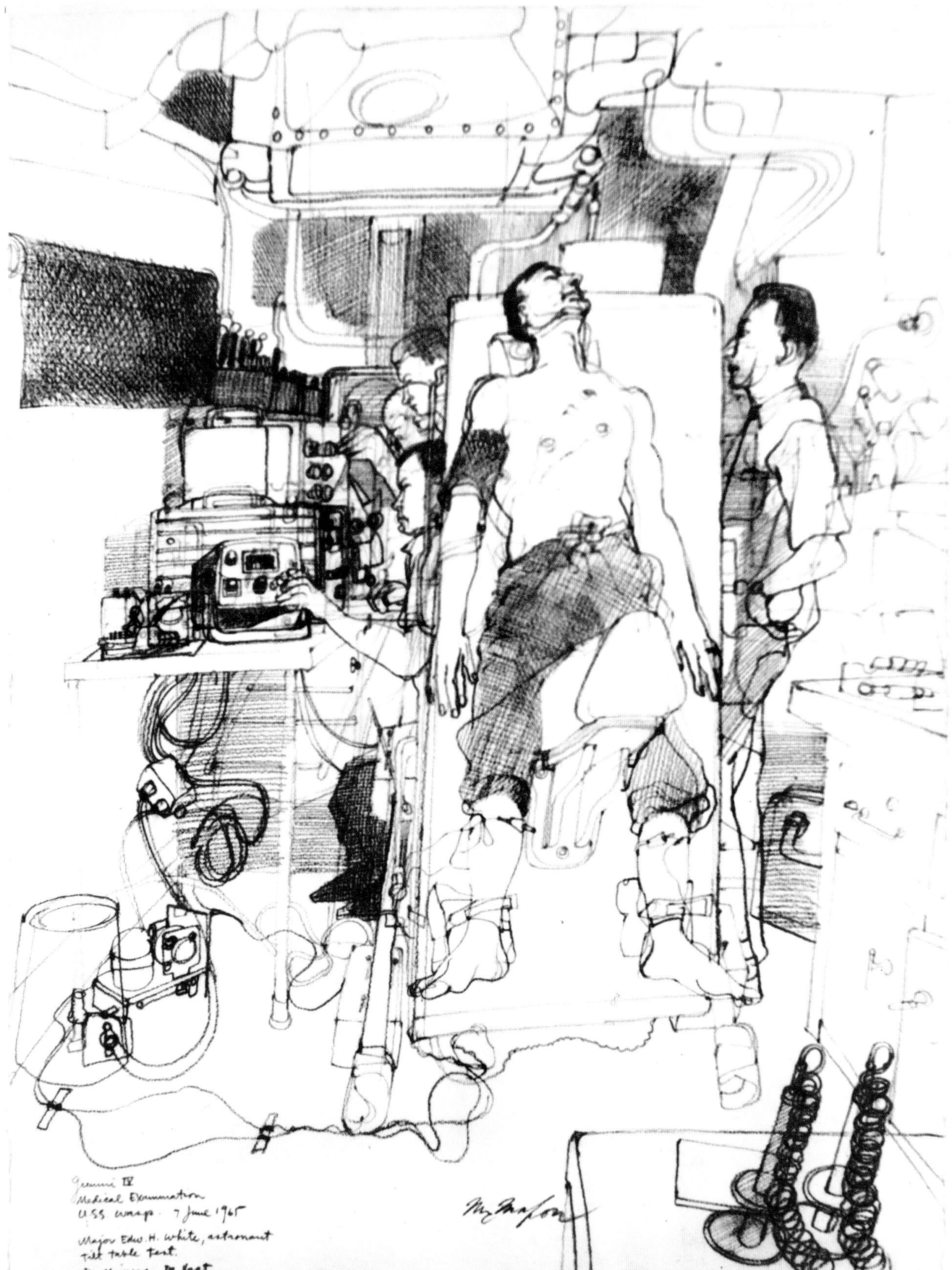

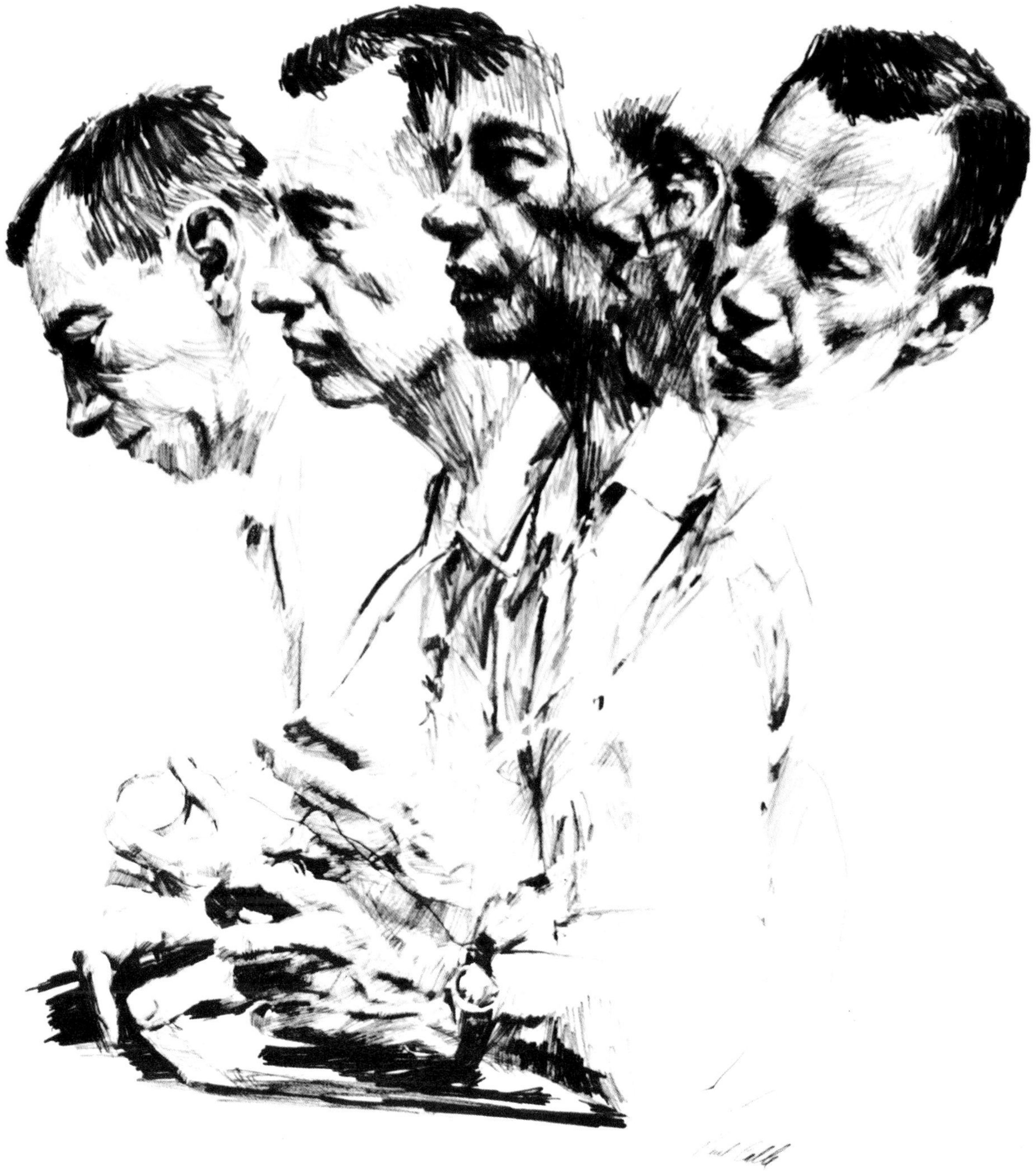

165. *Paul Calle.* RECOVERY DOCTOR
Pencil on paper, 22×30″

166.
Paul Calle
MIKE COLLINS
Pen and ink on paper, 23×18″

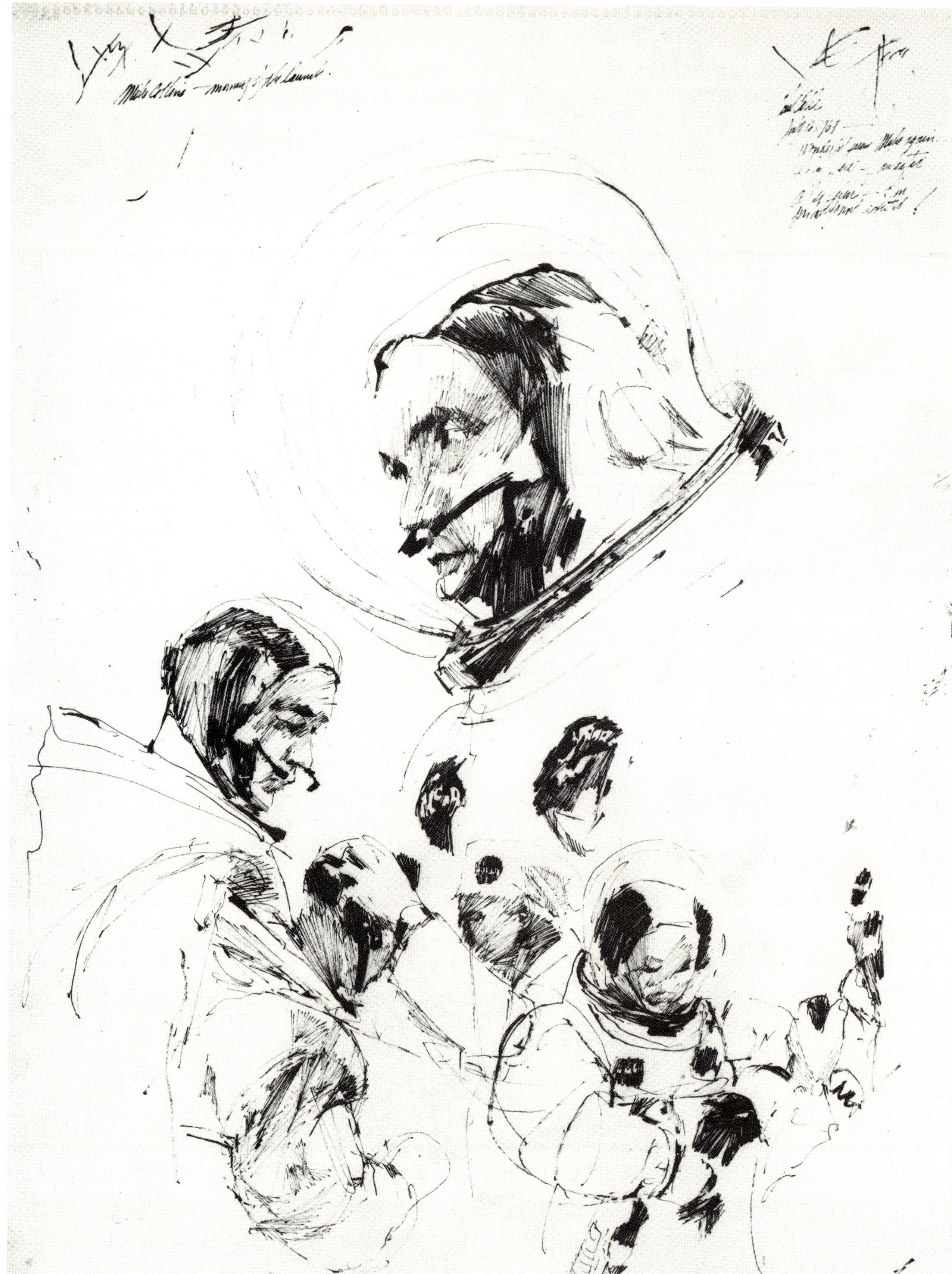

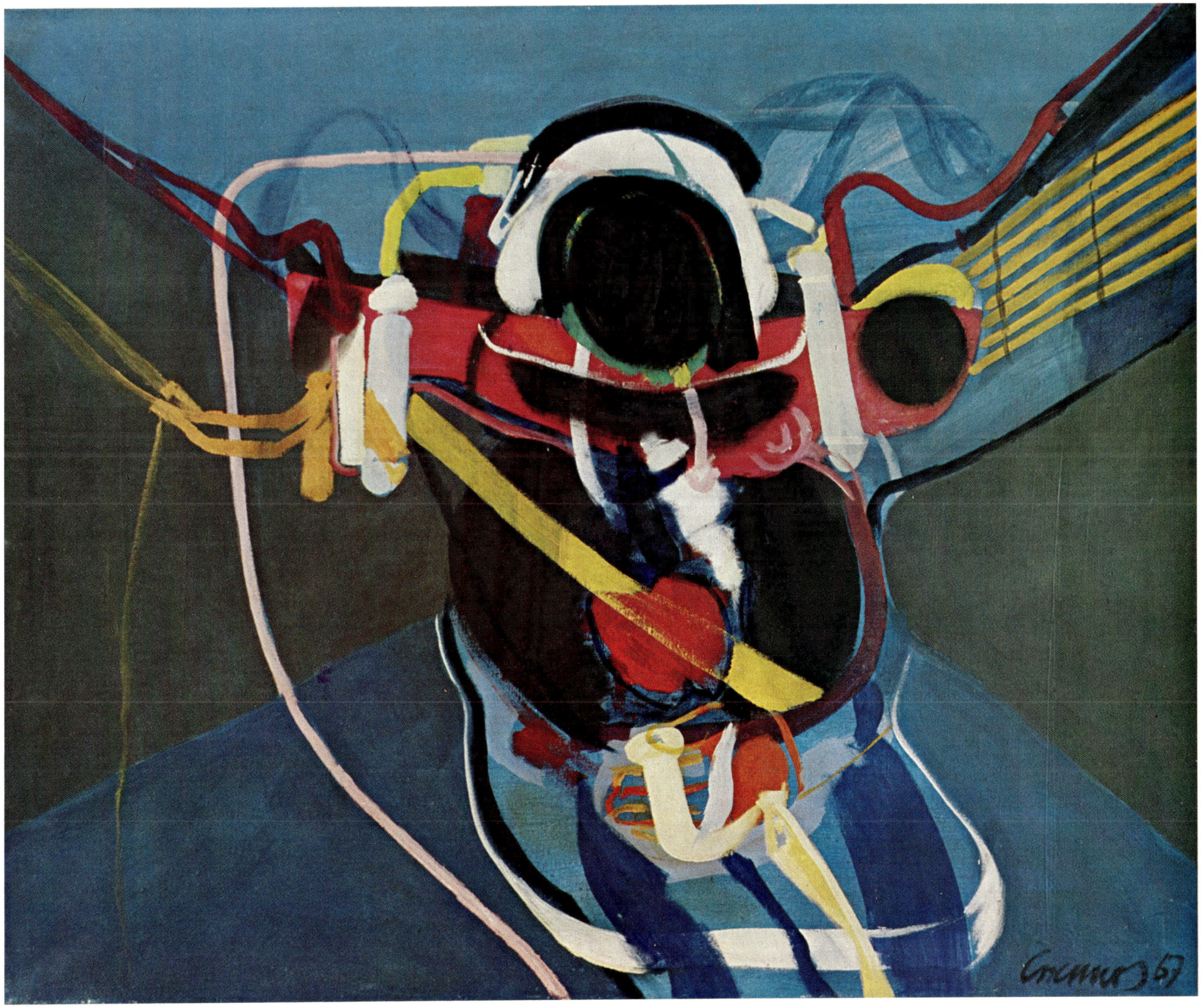

167.
Hans Cremers
MAN'S NEW DIMENSION
Acrylic on canvas, 33 × 42″

Although man in space is freer than any creature has ever been, his absolute liberty is not his own. Encased in wonderfully complex armor to shield him from his lethal surroundings, he is also linked by a thousand command radio systems to distant masters and monitors. In a sense, he is a flying laboratory whose face is a blank.

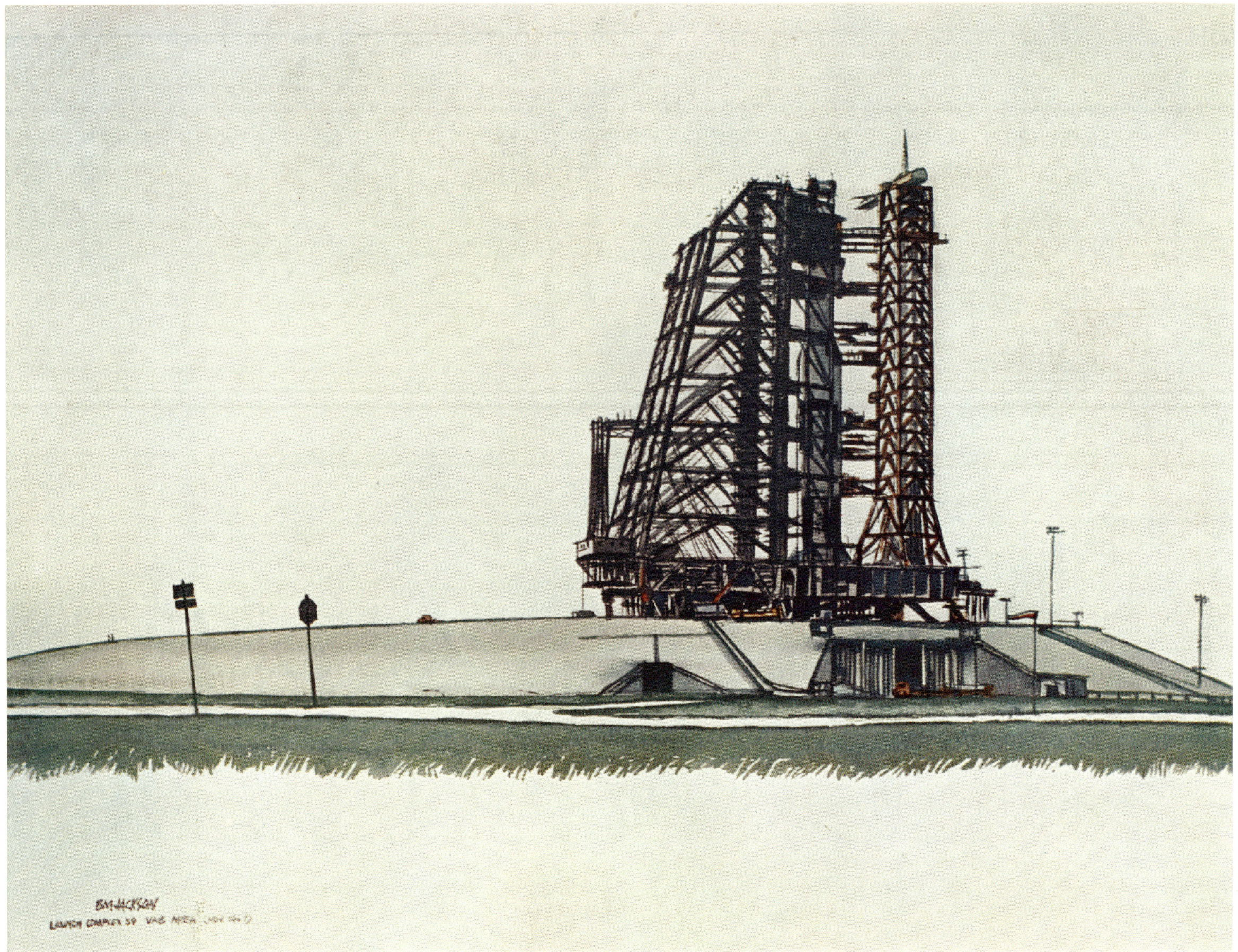

168. *Billy Morrow Jackson.* APOLLO 4
Watercolor on paper, 22×30″

In the hard light of day, the Saturn V on Pad 39 becomes structurally understandable and loses much of its mystic glamour.

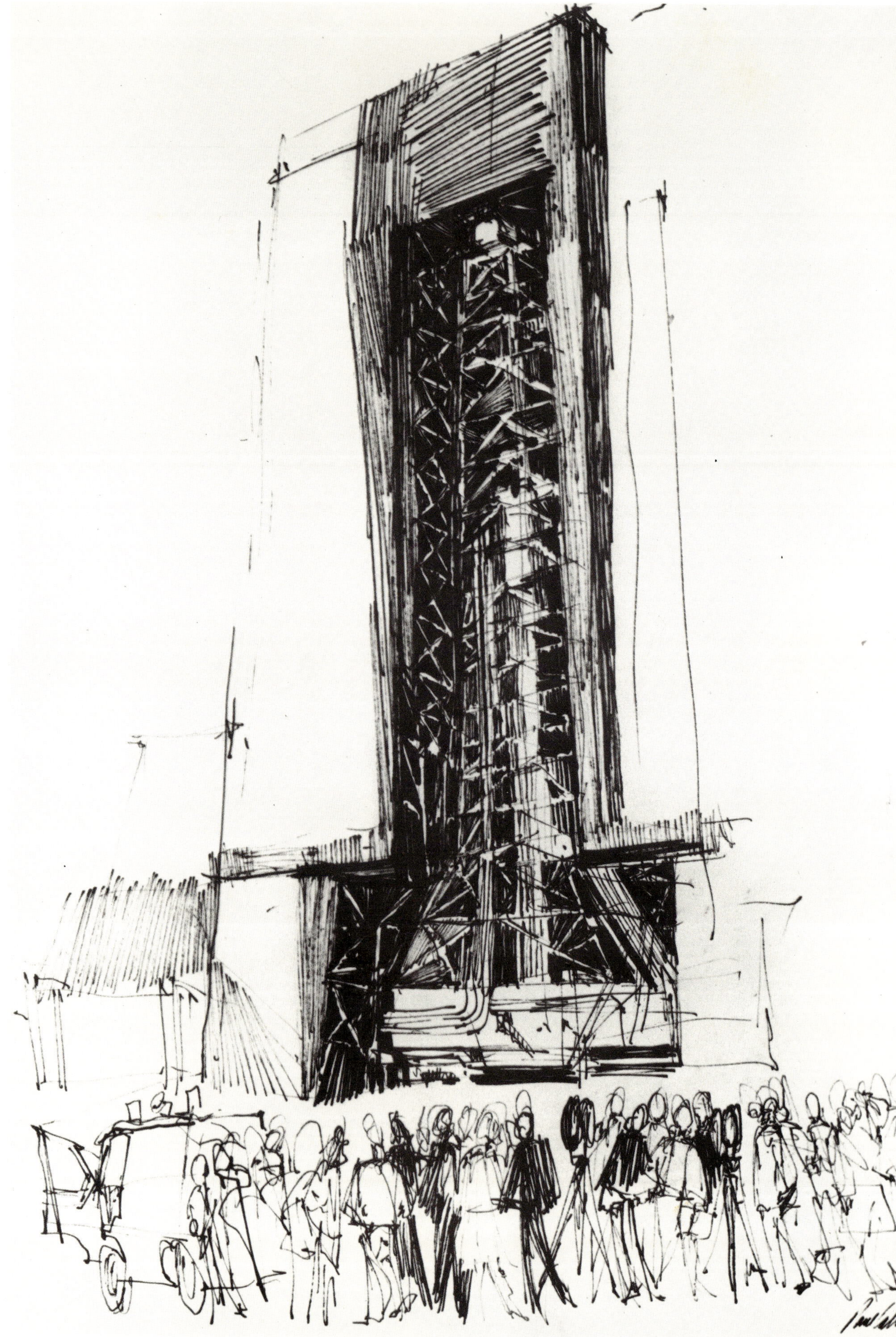

169. *Paul Calle.* ROLLOUT. Ink on paper, 24×18″

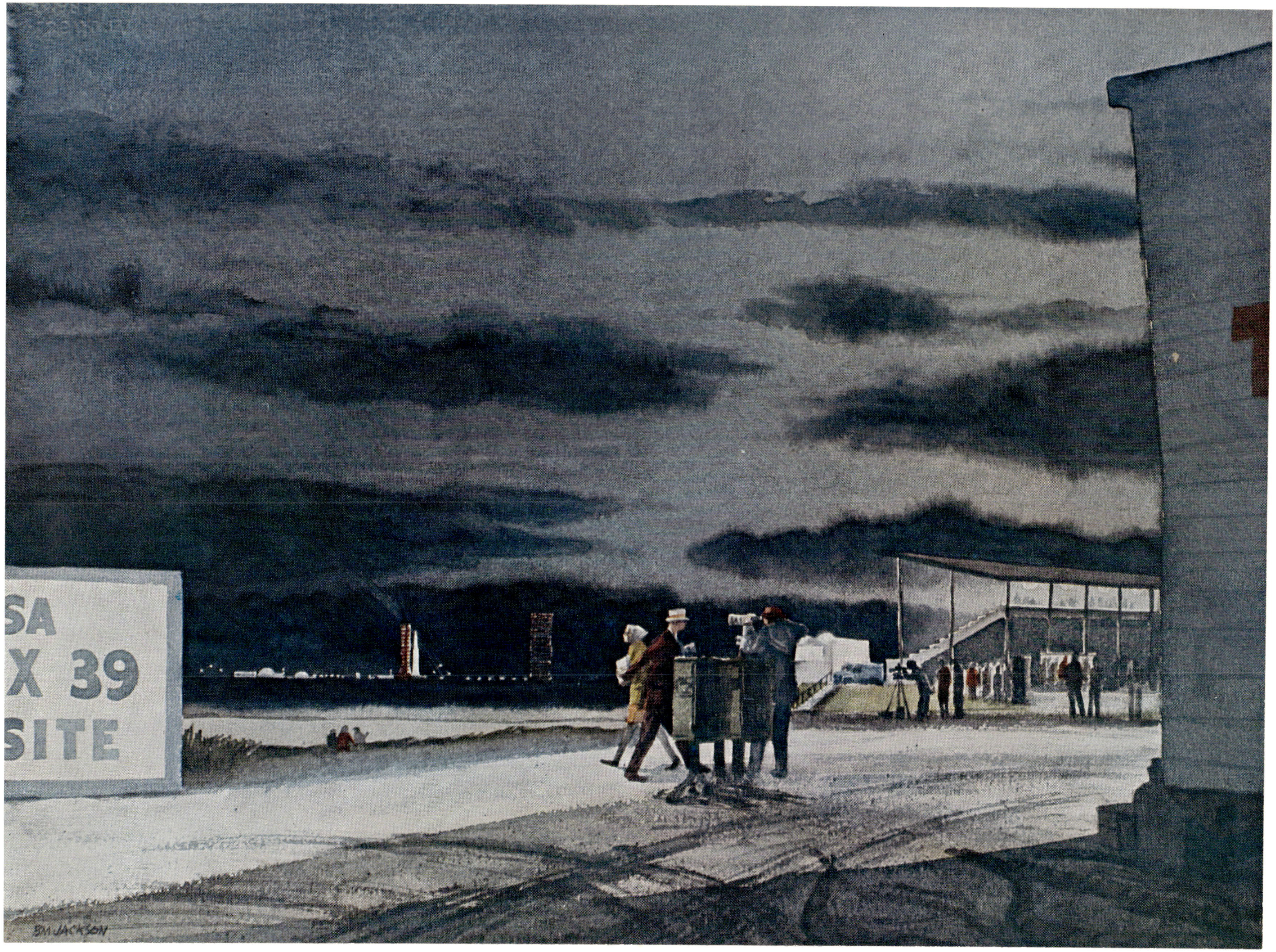

170.
Billy Morrow Jackson
PRESS SITE
Watercolor on paper, 22×30″

TV viewers usually are not aware that the news commentators, whose breathless bulletins tell them of history on the march, are more than three miles away from a Saturn V when it is fueled and ready for flight. This is a safety measure. If a rocket exploded on the pad the radius of utter destruction would be at least three miles. The press is accommodated in a viewing stand reminiscent of a country race track. There is a difference, however; each seat has a private telephone jack.

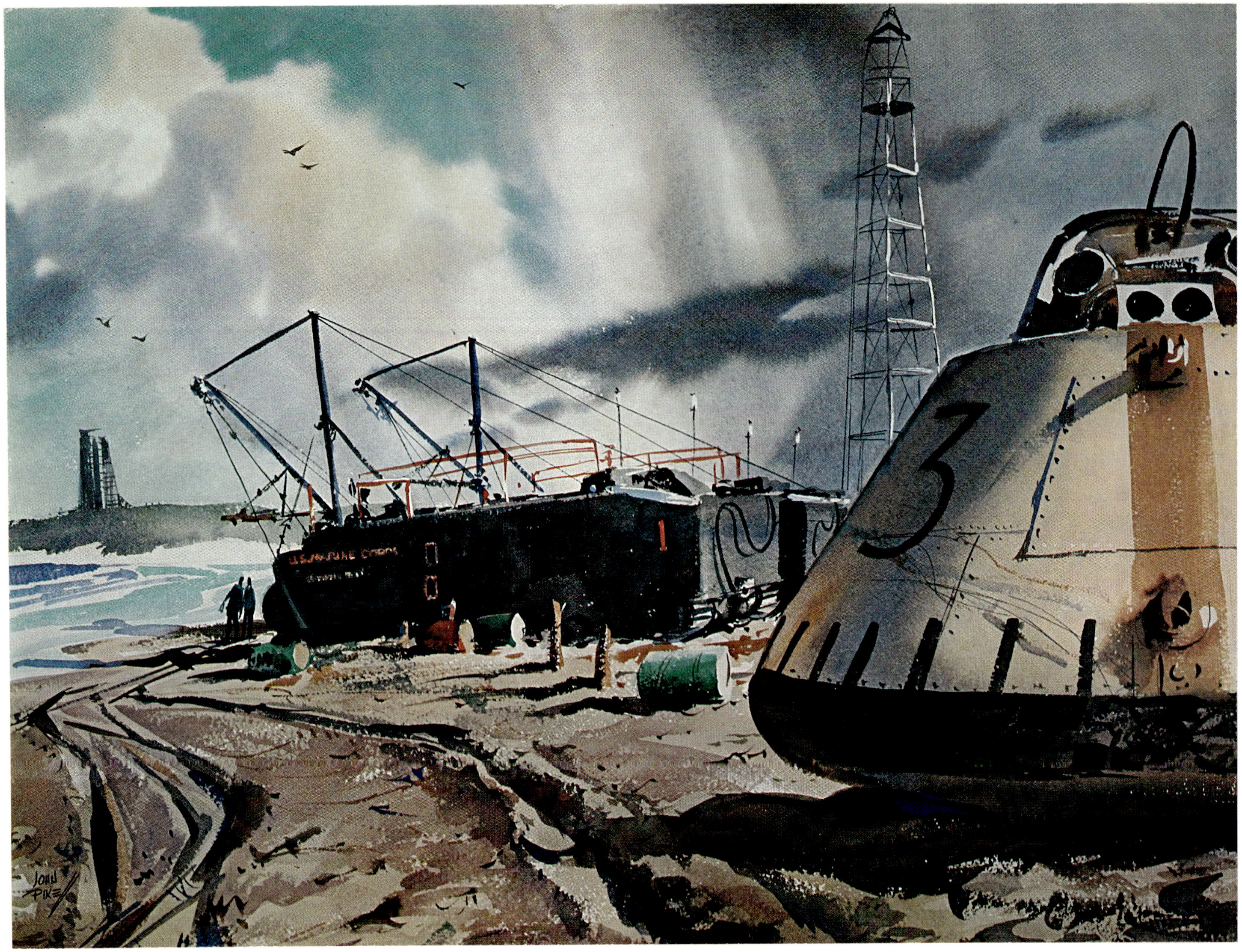

171. *John Pike*. BOILERPLATE. Watercolor on paper, 28×36″

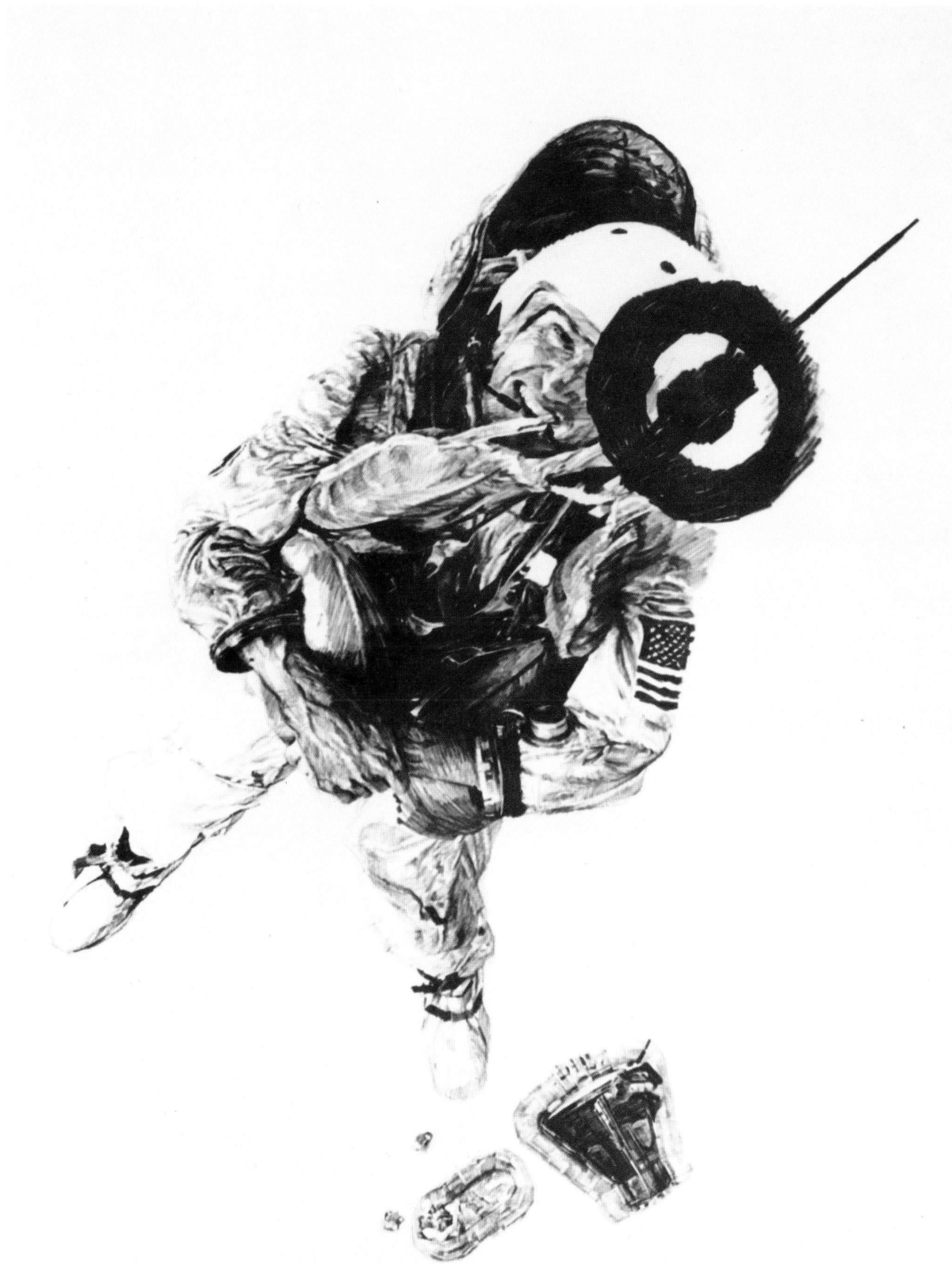

172.
Paul Calle
GEMINI ASTRONAUT BORMAN RECOVERY
Pencil on paper, 30×22″

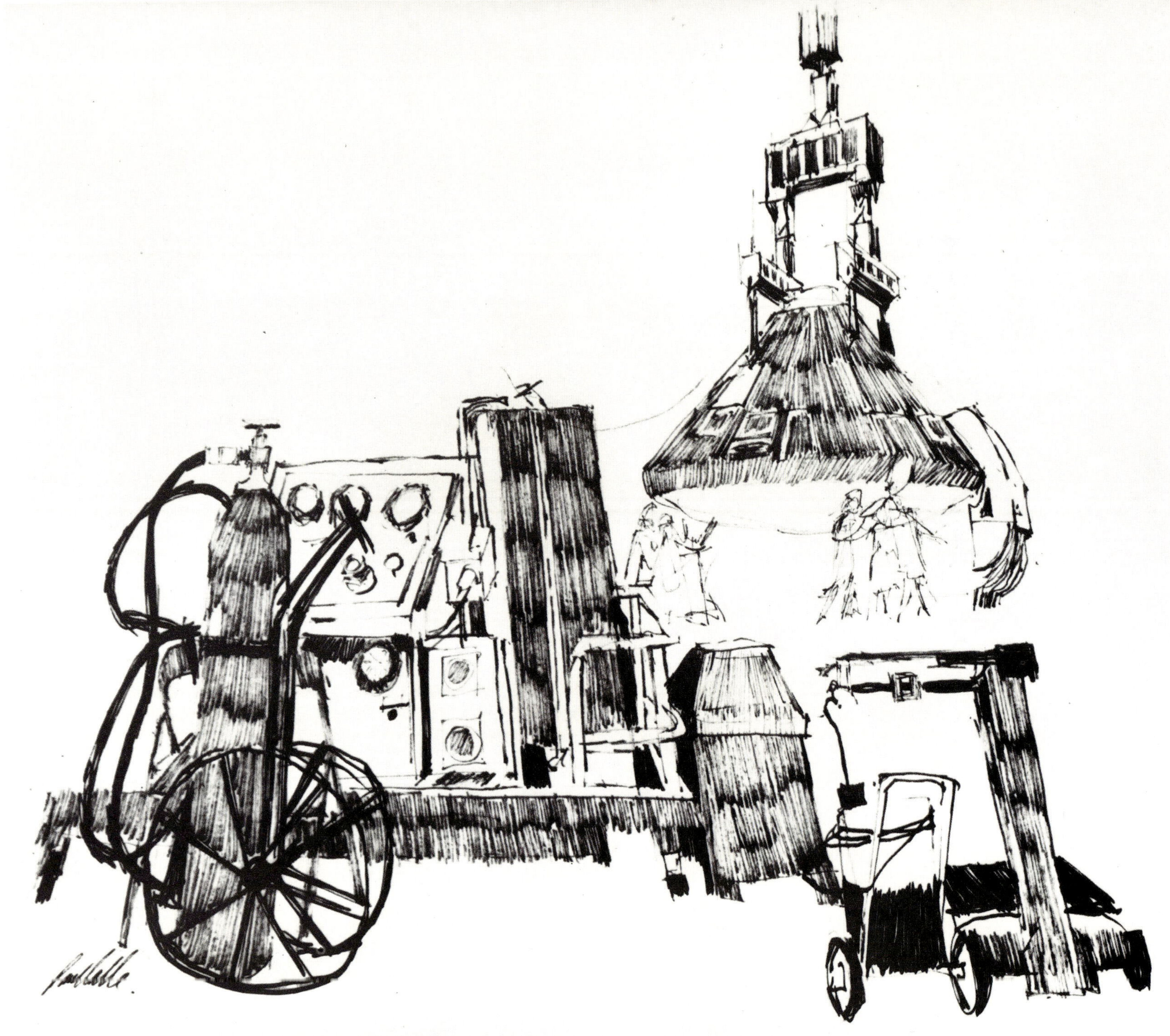

173. *Paul Calle*. NEW ARRIVAL. Ink on paper, 12×18″

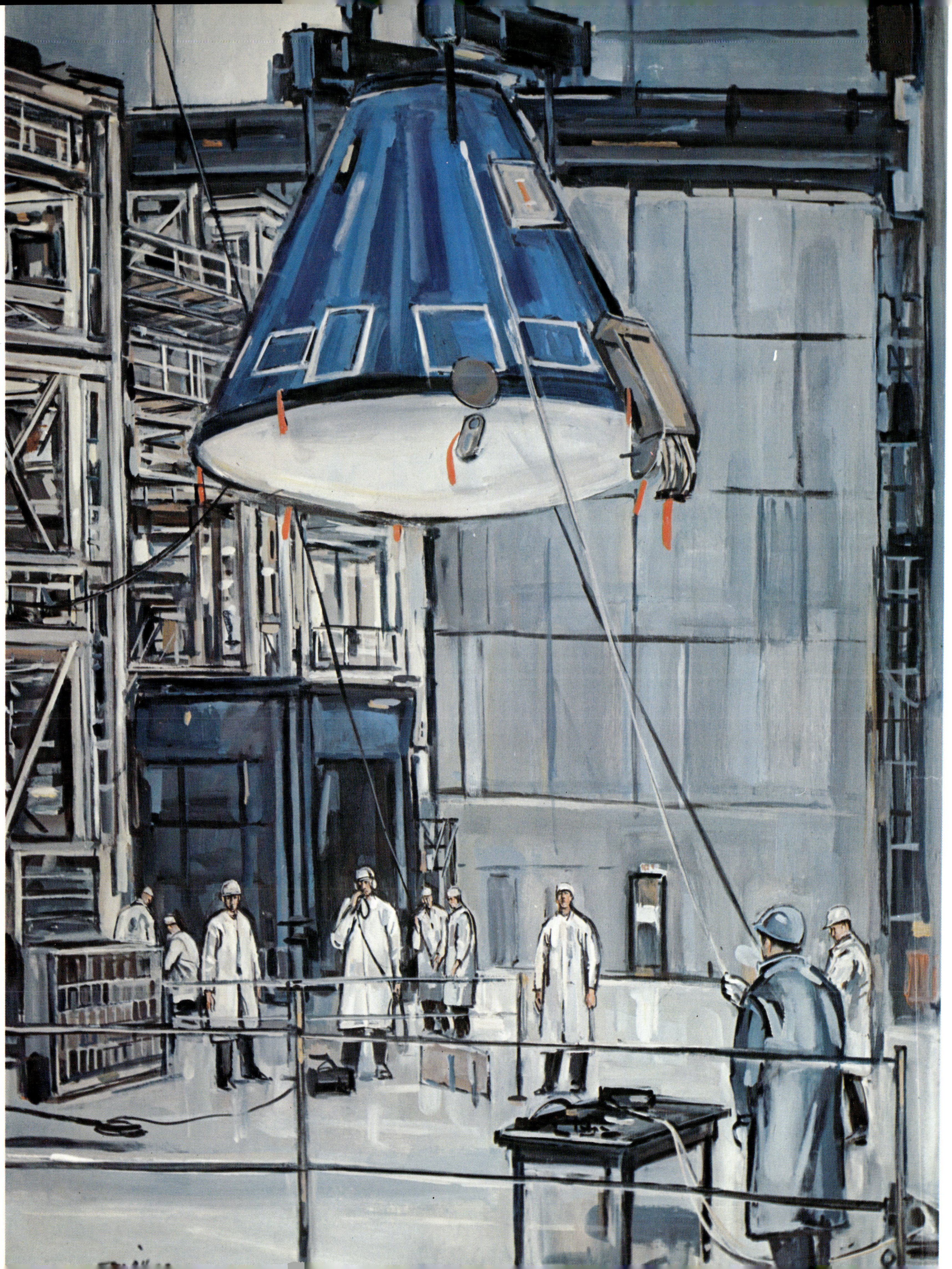

174.
Tom O'Hara
CHECKING THE COMMAND MODULE
Acrylic on paper, 30×22″

When a new spacecraft is received at Cape Kennedy, it is unwrapped and inspected like a crown jewel and handled like a newly laid egg. In this case the craft was Apollo 11, destined for an immortal place in history.

Overleaf:
175. *Yeffe Kimball.* BLUE SPACE CONCEPT
Acrylic on panel, 7×10′

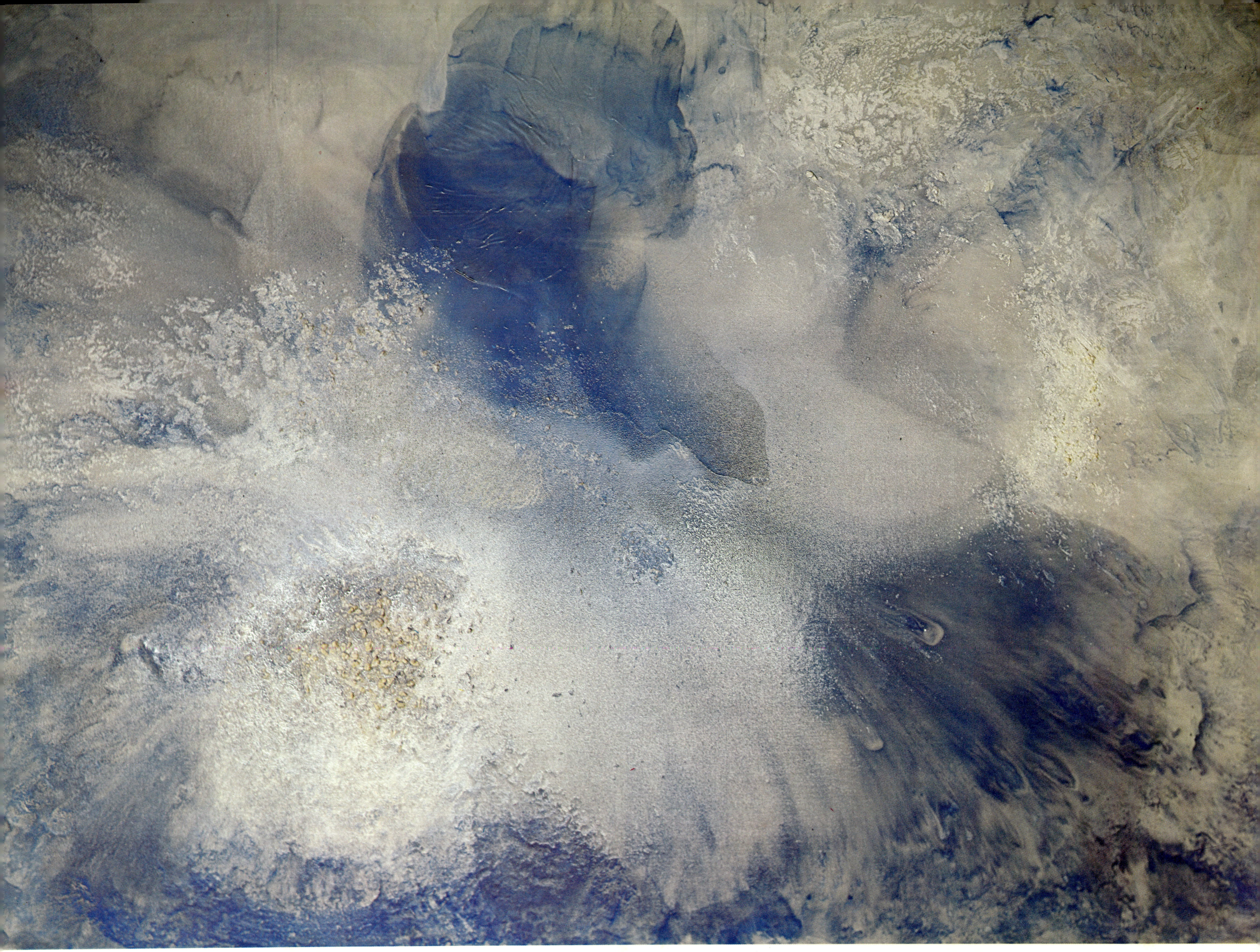

Eyewitness to Space

IV

176. Robert McCall *Apollo 8 Coming Home*
177. Fred Messersmith, A. W. S. *Prelude to Liftoff*
178. Chrystal Jackson *Cocoa Beach Footnootes*
179. Franklin McMahon *Waiting for Words from the Moon*
180. Franklin McMahon *Stacking the Saturn*
181. Tom O'Hara *Top of the Stack*
182. Henry C. Pitz *Last Check*
183. John Meigs *To the Moon*
184. Lamar Dodd *Celestial Journey*
185. Lamar Dodd *Eagle*
186. Paul Calle *LM Ascent Stage*
187. Billy Morrow Jackson *Lunar Module White Room*
188. Nicholas Solovioff *Training Simulator*
189. Nicholas Solovioff *Command Module Simulator*
190. Nicholas Solovioff *Goldstone*
191. Robert McCall *Practice Recovery of Boilerplate Spacecraft*
192. Billy Morrow Jackson *Gemini Pad*
193. Nicholas Solovioff *Apollo Launch*
194. Robert Rauschenberg *Sky-Rite*
195. James Wyeth *Firing Room*
196. Vincent Cavallaro *Ignition Sequence*
197. Paul Calle *T Plus 50 Seconds*
198. Paul Calle *Red Carpet*
199. Tom O'Hara *Apollo 10 Is Hauled Aboard*
200. Leonard Dermott *End of the Voyage*
201. Leonard Dermott *Coming Aboard*
202. Tom O'Hara *Medical Examination*
203. Tom O'Hara *Astronaut Returns to Houston*
204. Tom O'Hara *Apollo 10 Astronauts Sleep on Plane Back to Houston*
205. Mitchell Jamieson *First Look*
206. Jack Perlmutter *Sun and Moon*
207. Mario Cooper *Apollo 11*
208. Dale Meyers *Apollo 11—Firefly to the Moon*
209. Tom O'Hara *Checking the Engines*
210. Paul Calle *Breakfast for the Apollo 11 Crew*
211. Paul Calle *Armstrong Being Helped with His Spacesuit*
212. Paul Calle *Suiting up for Apollo 11*
213. Paul Calle *Launch Morning*
214. William Thon *Dawn for Apollo 11*
215. Paul Calle *Gemini Astronauts*
216. James Wyeth *T Minus 3 Hours 30 Minutes and Counting*
217. Lamar Dodd *Destination*
218. Robert McCall *Sunday Service*
219. Mitchell Jamieson *Earth Terminal*
220. Julio Fernandez *Apollo 11*
221. William Thon *"We Have Liftoff"*
222. Vincent Cavallaro *Power to Go*
223. Franklin McMahon *Astronaut Manager*
224. Franklin McMahon *LM on the Moon*
225. Franklin McMahon *Houston Control*
226. Franklin McMahon *"The Eagle Has Landed"*
227. Norman Rockwell *Astronauts on the Moon*
228. M. McCaffrey *Chris Kraft and Dr. Gilruth*
229. Franklin McMahon *Space Conversation*
230. Franklin McMahon *Advice for a Man on the Moon*
231. Mitchell Jamieson *On the Moon*
232. Lamar Dodd *The Cosmos Awaits*
233. Franklin McMahon *Home Town Cheese*
234. Lamar Dodd *Lunar Module*
235. Lamar Dodd *Celestial Body*
236. Yeffe Kimball *Copernicus*
237. Hans Cremers *Metamorphoses*
238. Mitchell Jamieson *Predawn Pickup*
239. Nicholas Solovioff *Apollo 8 Returned*
240. Robert McCall *Spacecraft on Board*
241. Robert McCall *Moonlit Pacific*
242. Mitchell Jamieson *Frogmen*
243. Leonard Dermott *Apollo 9 Coming Aboard*
244. Mitchell Jamieson *Quarantine Facility*
245. Mitchell Jamieson *Equator Ceremonies*
246. Mitchell Jamieson *Germ-Proof Passage*
247. Robert Rauschenberg *Sack*
248. Robert Rauschenberg *Medallion*
249. Robert Rauschenberg *Sky Garden*
250. Robert Rauschenberg *Moon Rose*
251. Robert Rauschenberg *Bait*
252. Robert Rauschenberg *Hybrid*
253. Lamar Dodd *Moments of Decision*
254. Lamar Dodd *Future Target*
255. Robert McCall *Space Station #1*
256. Francis J. Krasyk *Man in Copernicus*
257. John Willenbecher *Allegorical System: Night*
258. Chesley Bonestell *A New World*

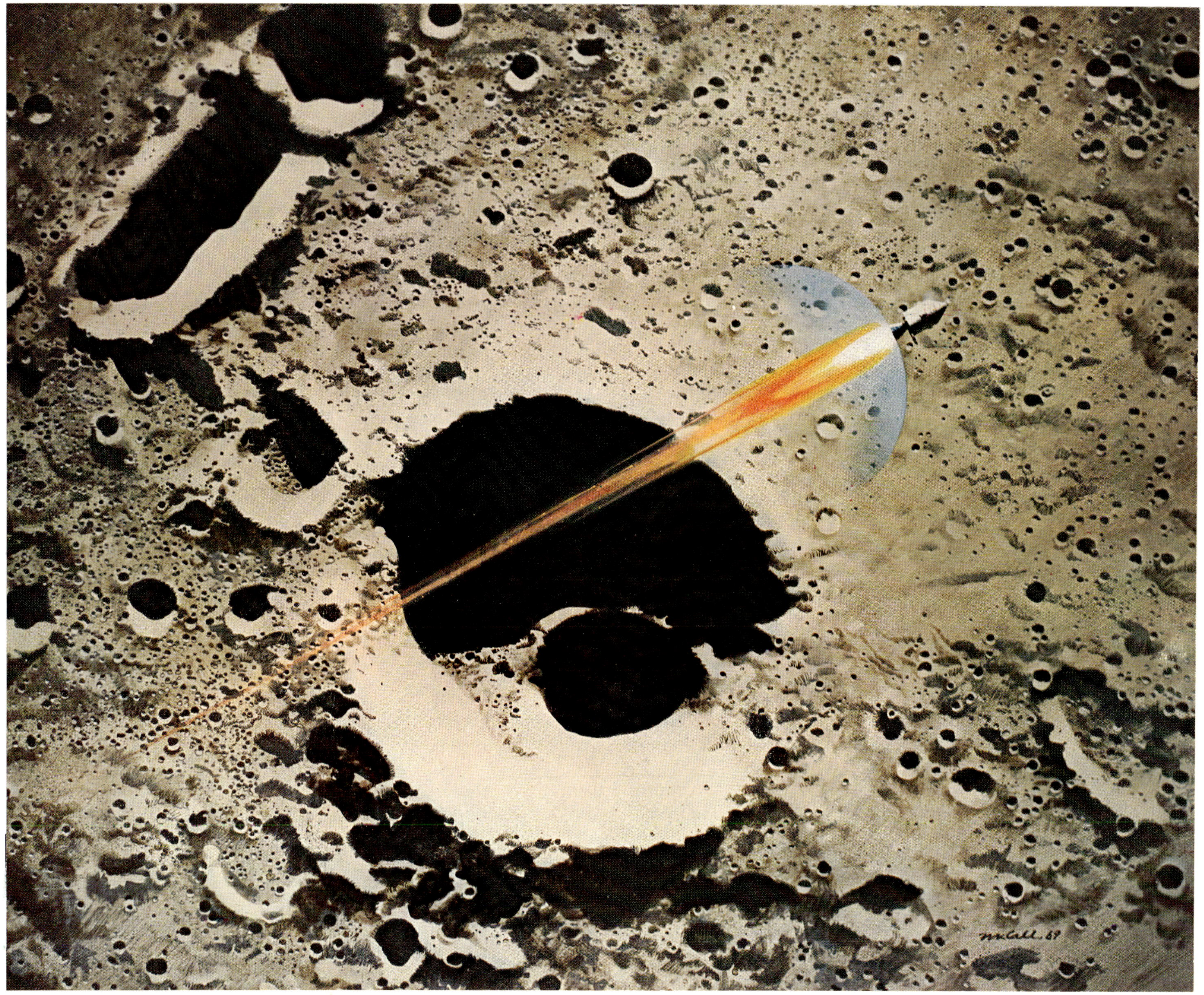

176.
Robert McCall
APOLLO 8 COMING HOME
Oil on panel, 39×48″

On Christmas eve, 1968, man for the first time saw with his own eyes the far side of the moon. It was like "dirty beach sand with lots of footprints in it," more mountainous and rugged than the side always facing toward the earth. A crucial maneuver during this flight was the firing of the service module engine, which increased the speed and thus released the astronauts from lunar orbit for the journey back to earth. One of the many unexpected phenomena of space has been the discovery that a rocket engine fired in a vacuum produces a beautiful peacock-tail flame pattern, as McCall has recorded in this painting.

177.
Fred Messersmith, A.W.S.
PRELUDE TO LIFTOFF
Watercolor on paper, 22×28″

Those who did not have passes to enter the base for the great event watched from the banks of the Banana River about six miles away. Trailers, campers, and tents lined the roads. Even at this range, the spectacle had an element of magic.

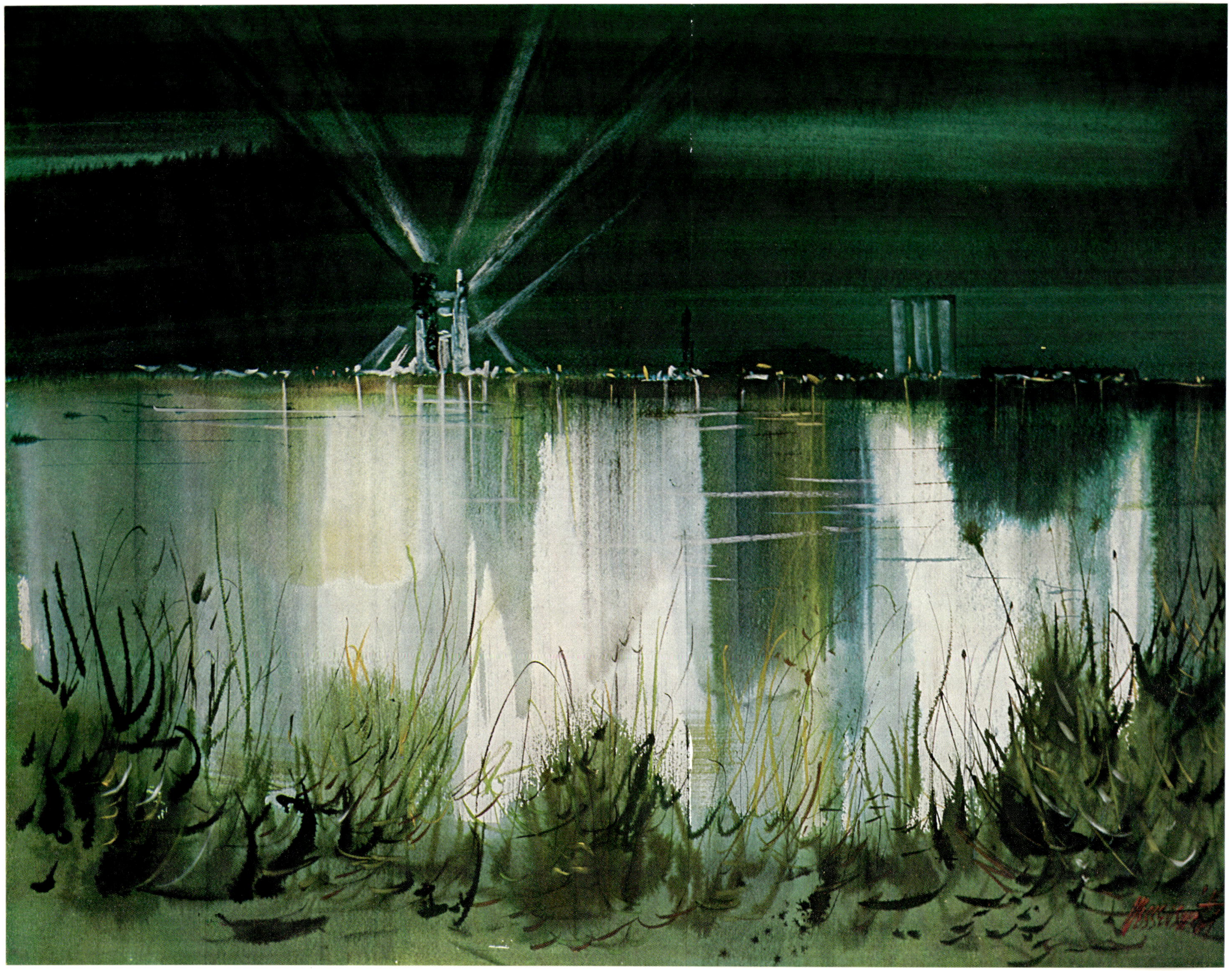

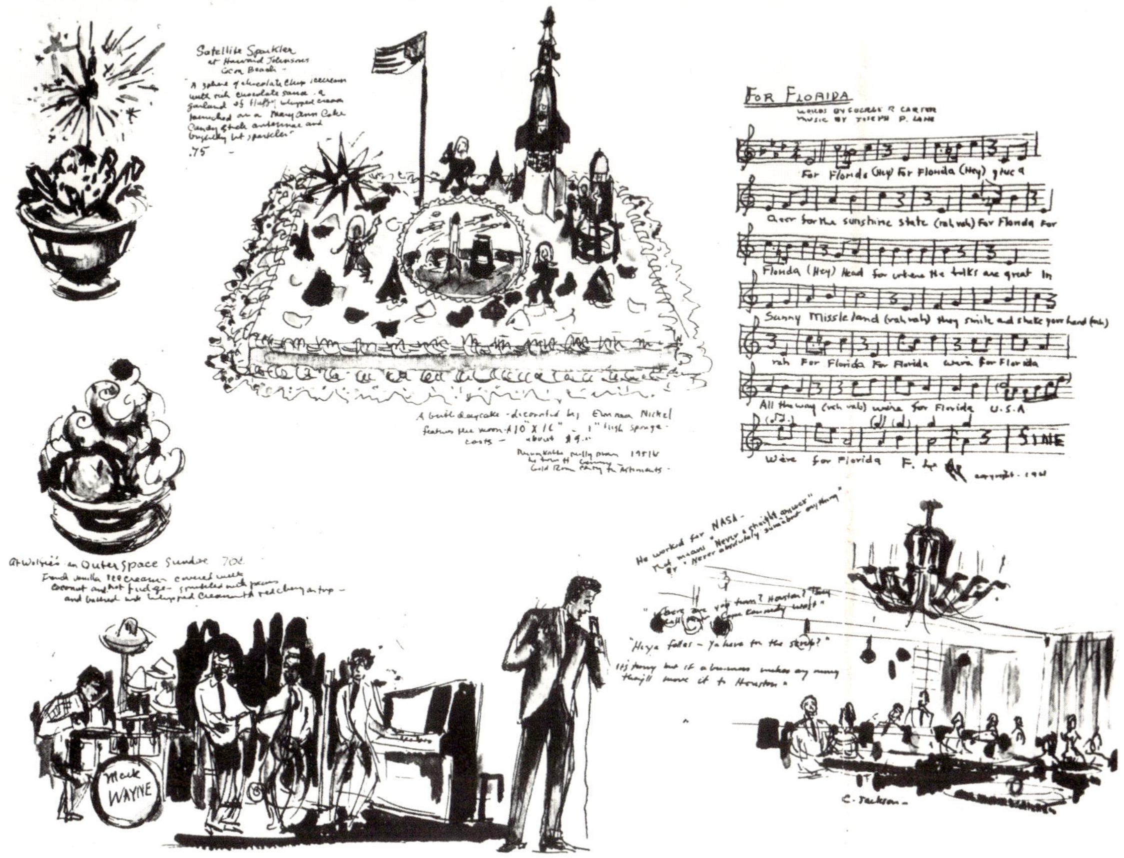

178. *Chrystal Jackson.* COCOA BEACH FOOTNOTES
Watercolor on paper, 6×10″

Cocoa Beach, just outside the gates of Cape Kennedy, produced for a brief period an American subculture, derived entirely from Space Age associations. Chrystal Jackson, whose witty, charming vignettes are known to millions of air travelers, observed and recorded these spin-off trivia.

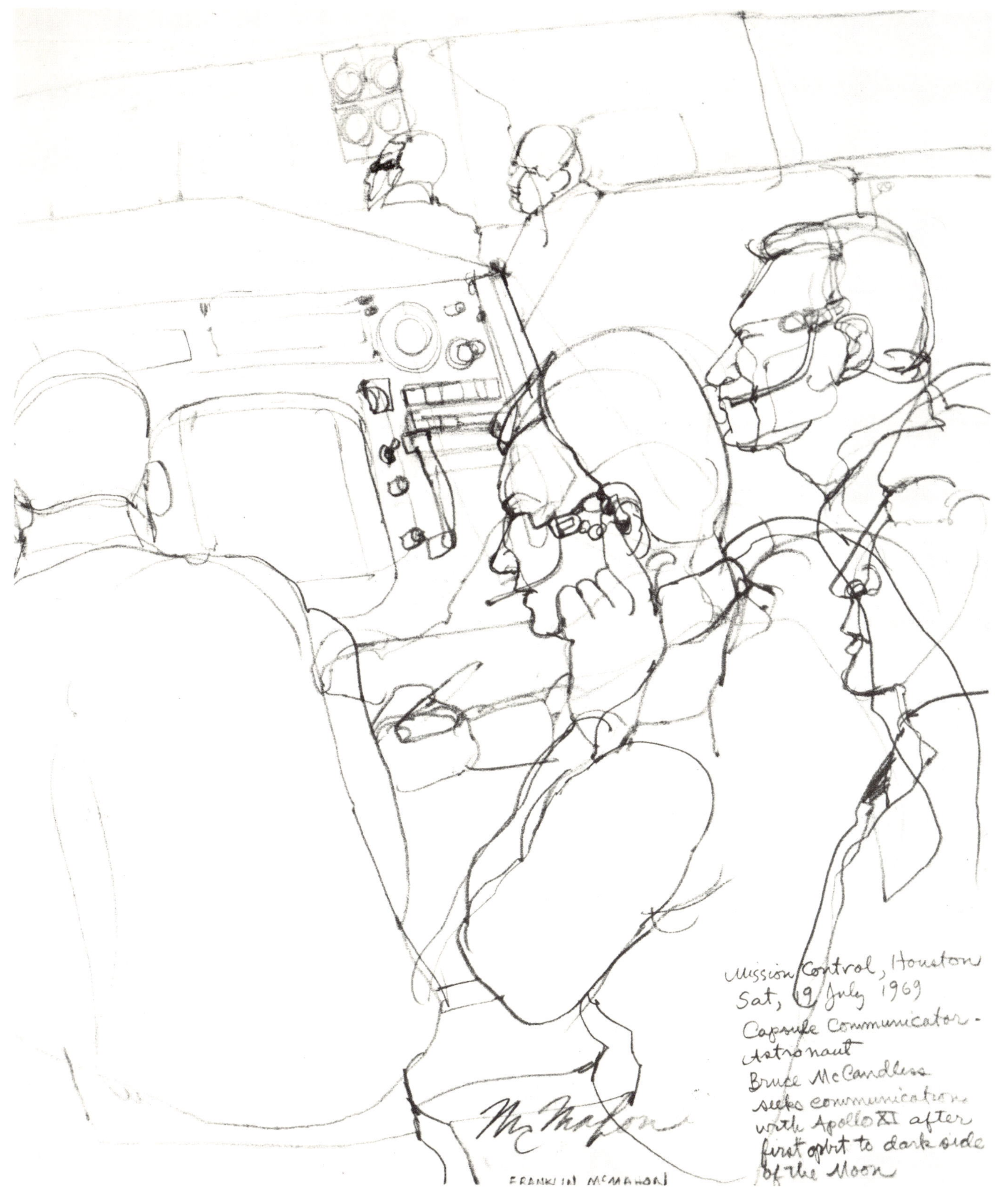

179. *Franklin McMahon.* WAITING FOR WORDS FROM THE MOON. Pencil on paper, 17×14″

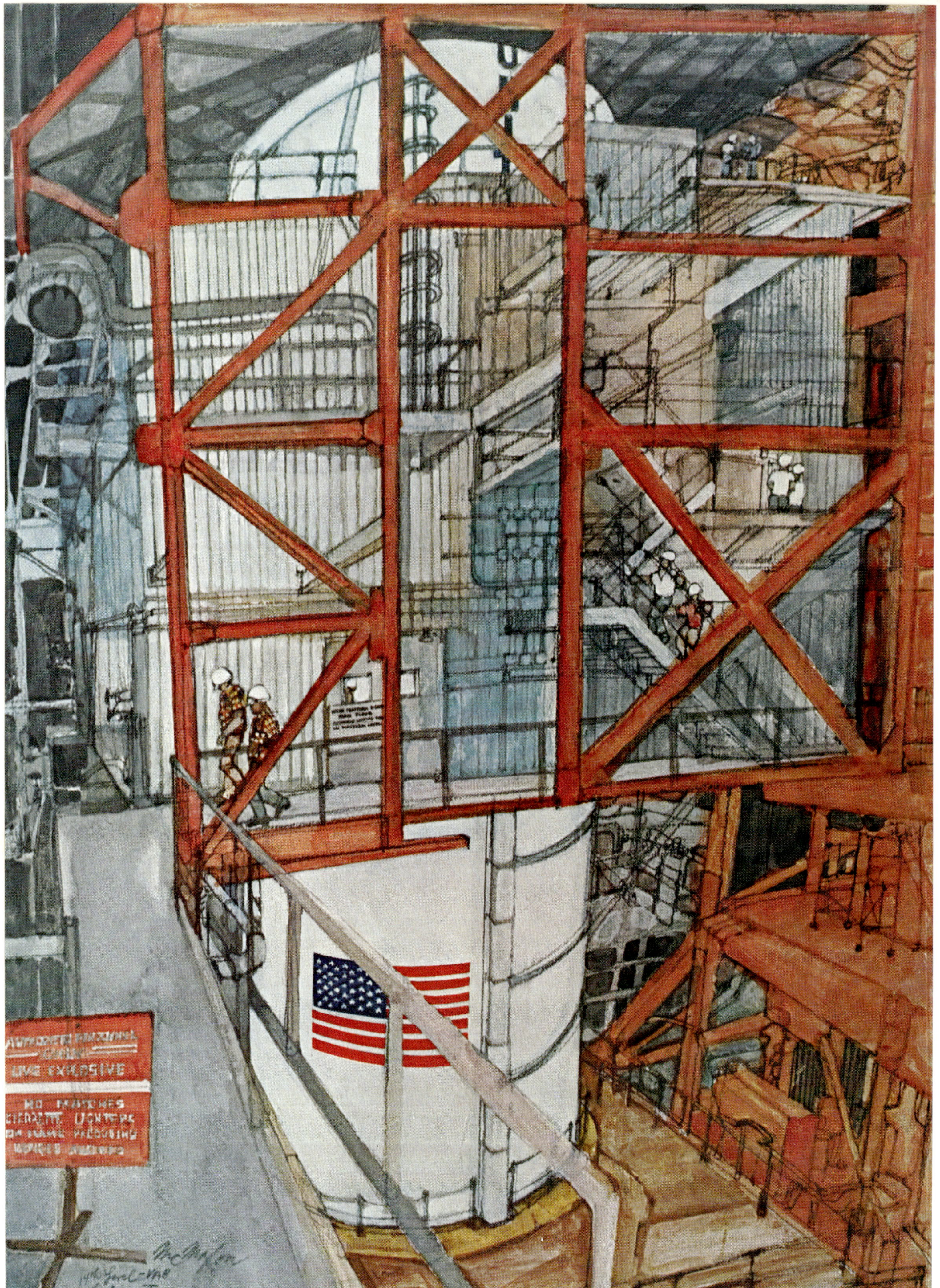

180.
Franklin McMahon
STACKING THE SATURN
Watercolor on paper, 30 × 22″

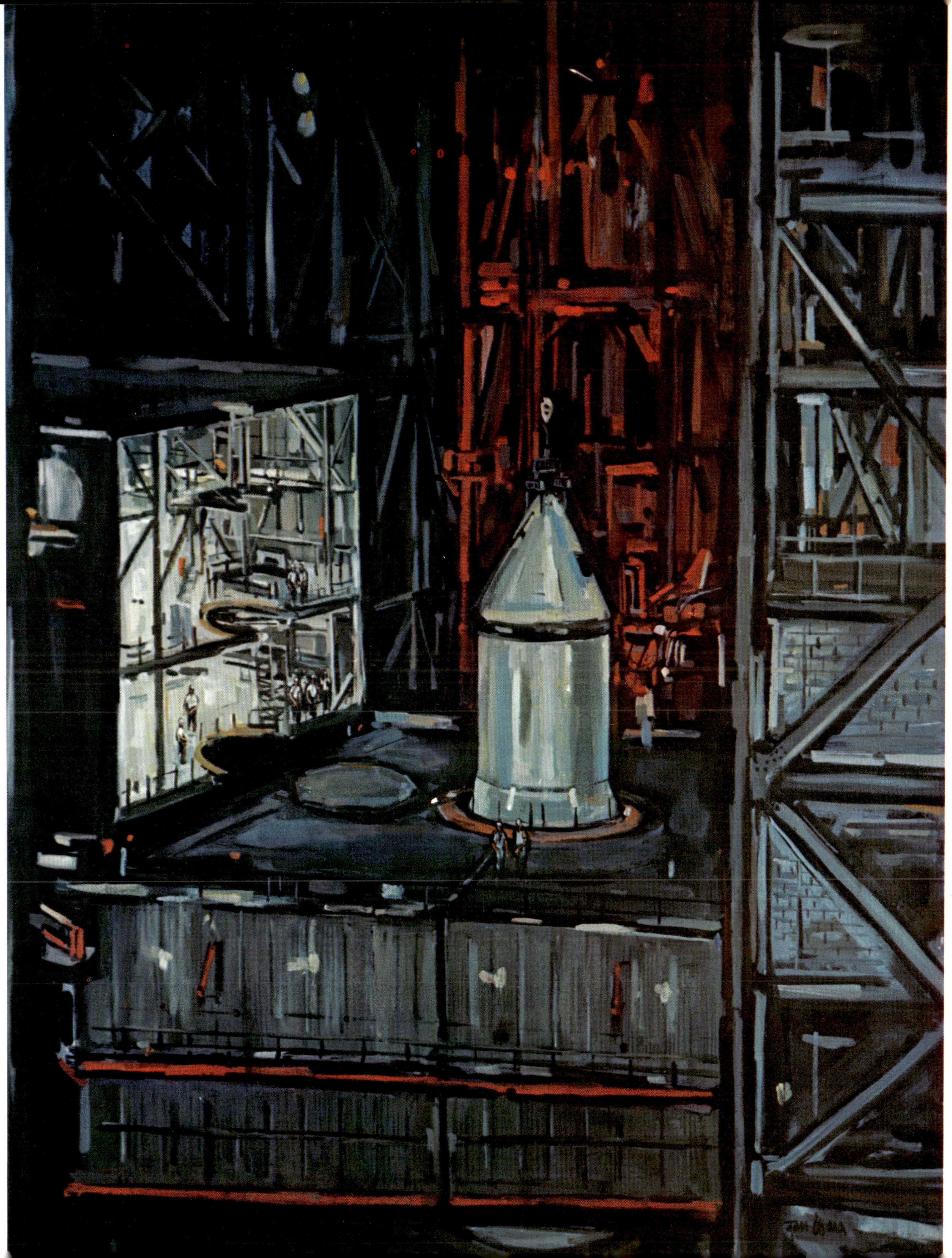

181.
Tom O'Hara. TOP OF THE STACK
Watercolor on paper, 30×22″

When the work platforms are in place around a rocket, only one segment is accessible from one level. Here the Apollo command module at the top of Saturn V is undergoing a final check.

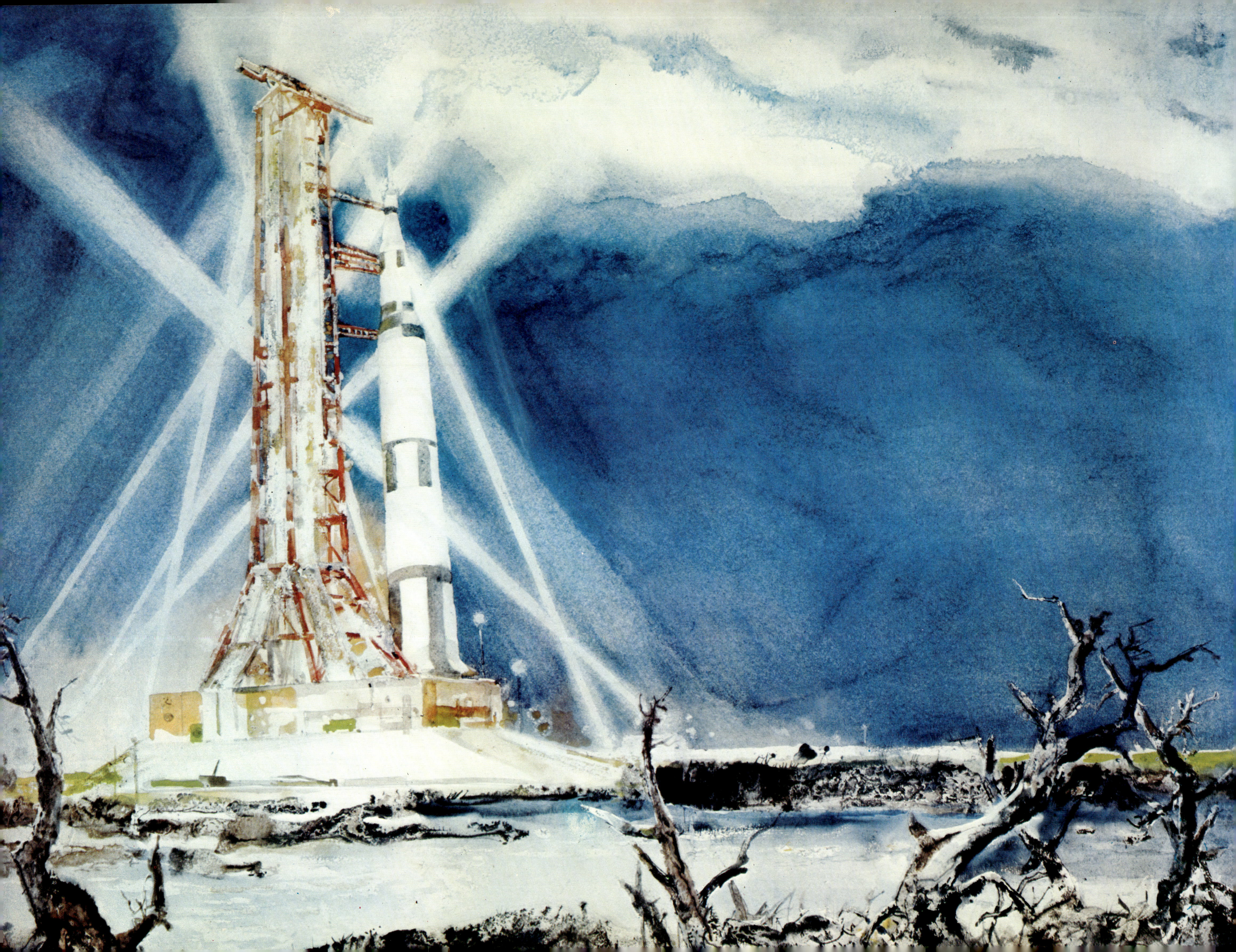

182.
Henry C. Pitz
LAST CHECK
Watercolor on paper, 22×30″

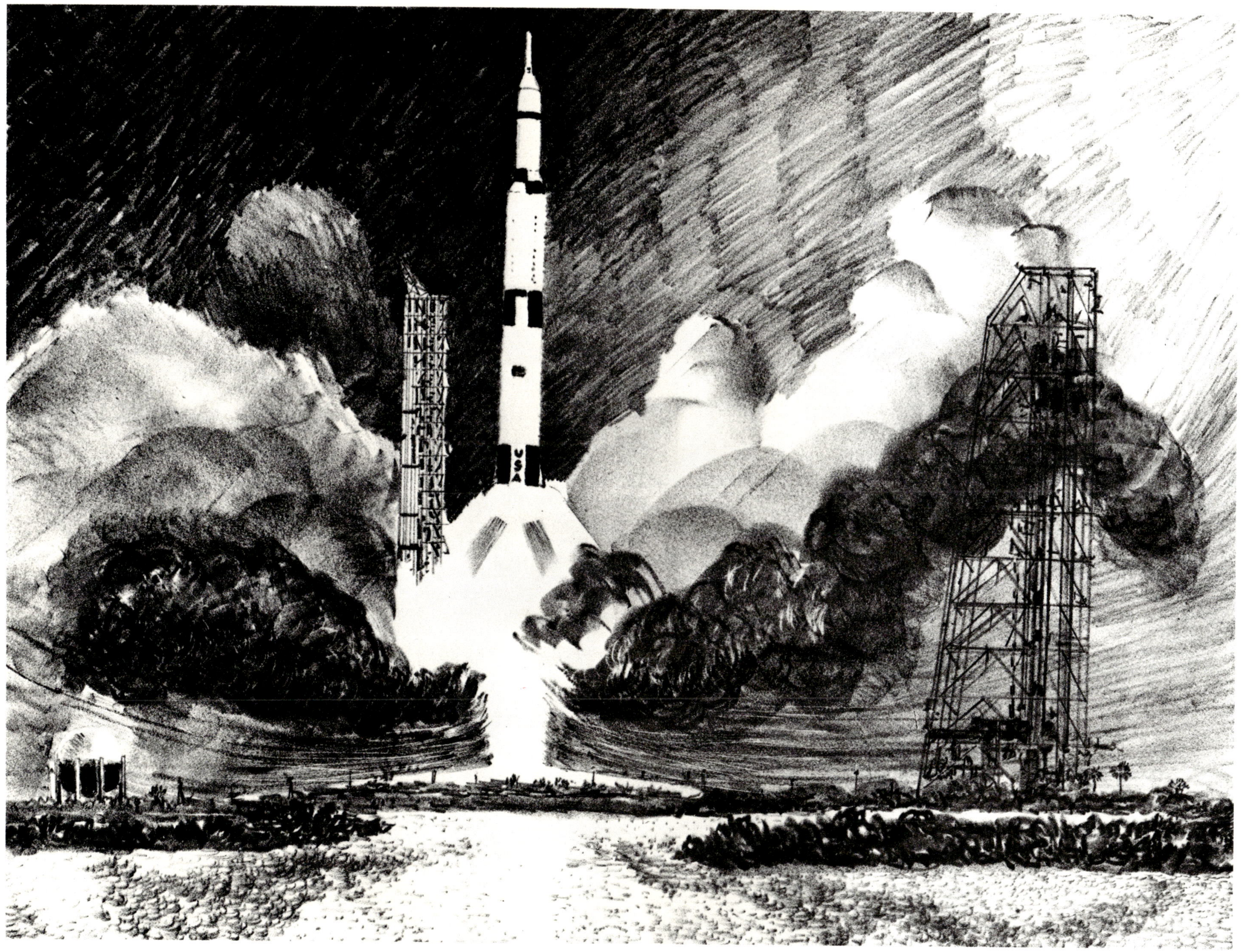

183. *John Meigs.* TO THE MOON. Lithograph on paper, 9×15″

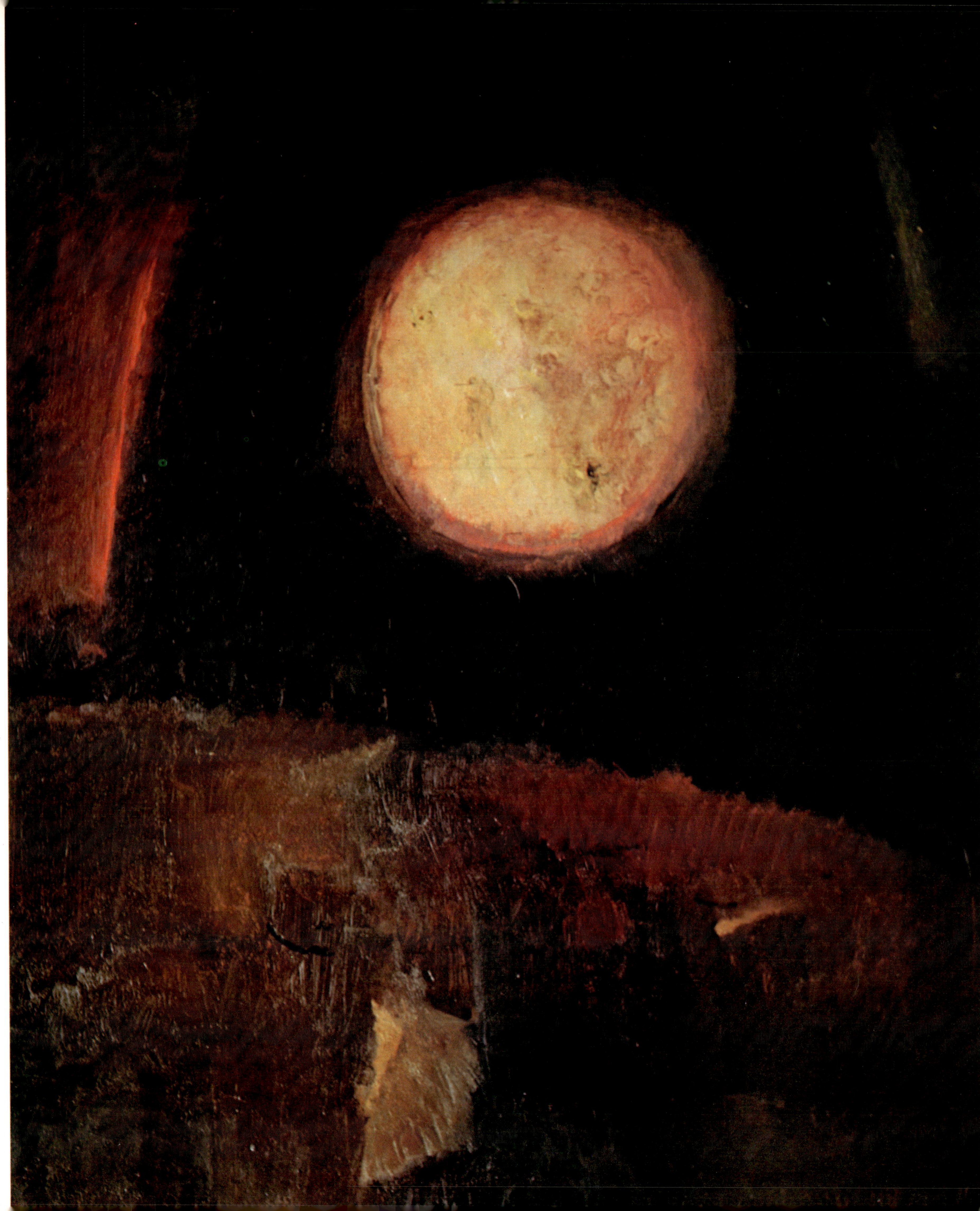

184.
Lamar Dodd
CELESTIAL JOURNEY
Oil on canvas, 50 × 30″

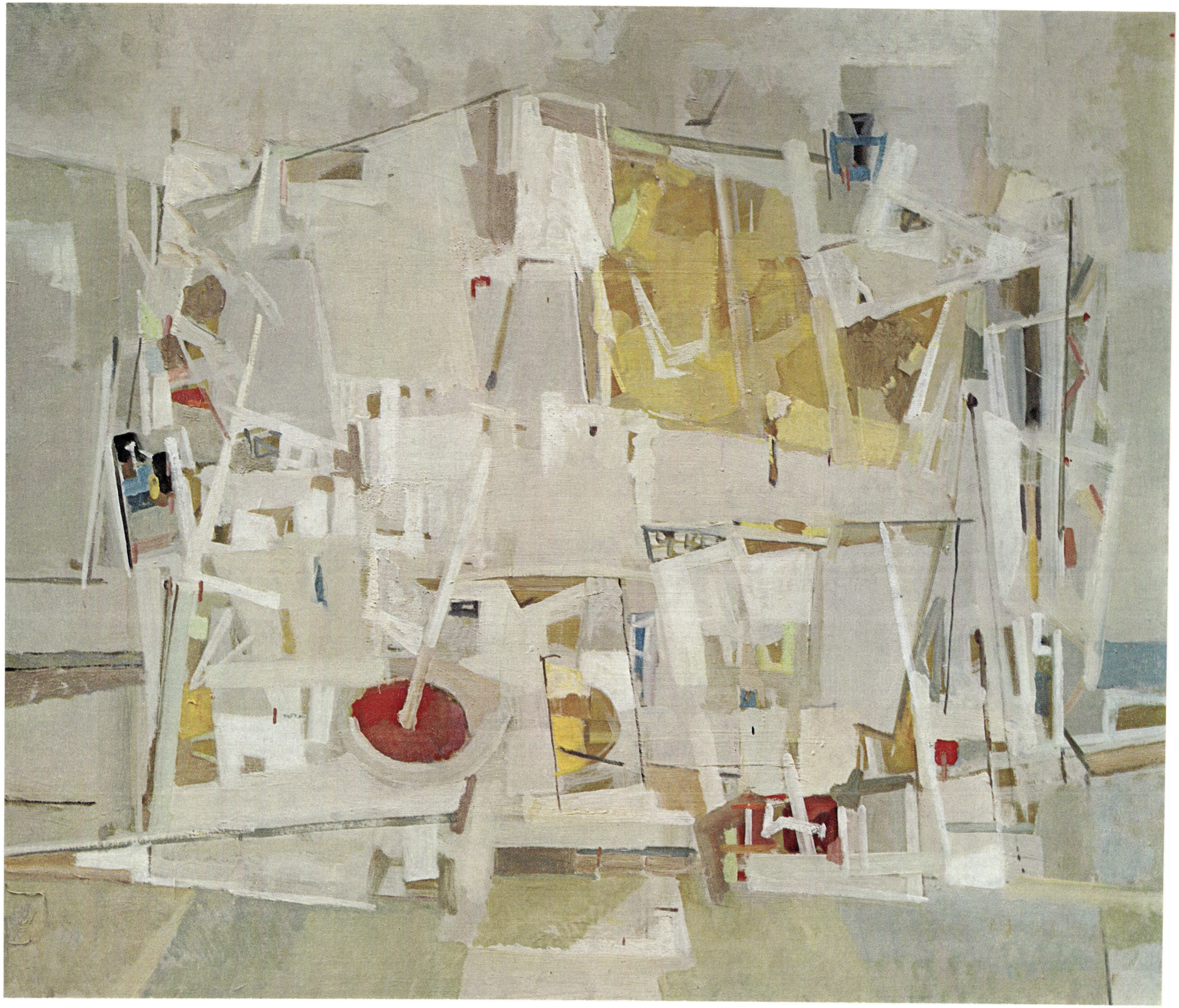

185.
Lamar Dodd
EAGLE
Oil on canvas, 3×4′

Lamar Dodd, like several other artists commissioned by NASA, was invited into the White Room where the Apollo 11 Lunar Module was being prepared for its historic mission. Like a chrysalis wrapped in a cocoon, the folded LM seems to be a mass of plastic coverings, red inspection tags, and gold sheathings. To the imagination of the artist, it appears to be a thing of great beauty of incredible perfection and subtlety.

186.
Paul Calle
LM ASCENT STAGE
Pen and ink on paper, 14×11″

The precious LM arrives from its factory on Long Island.

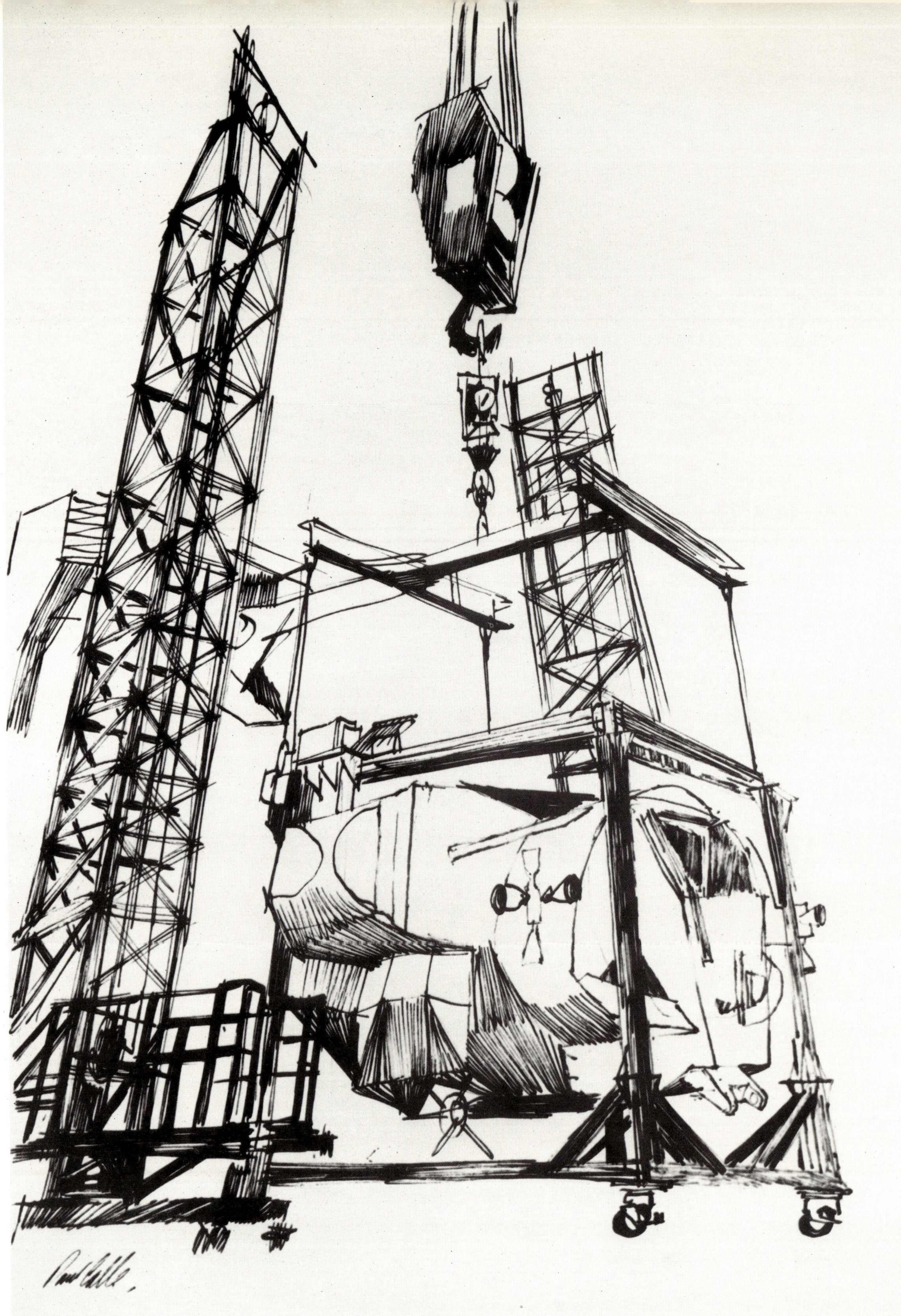

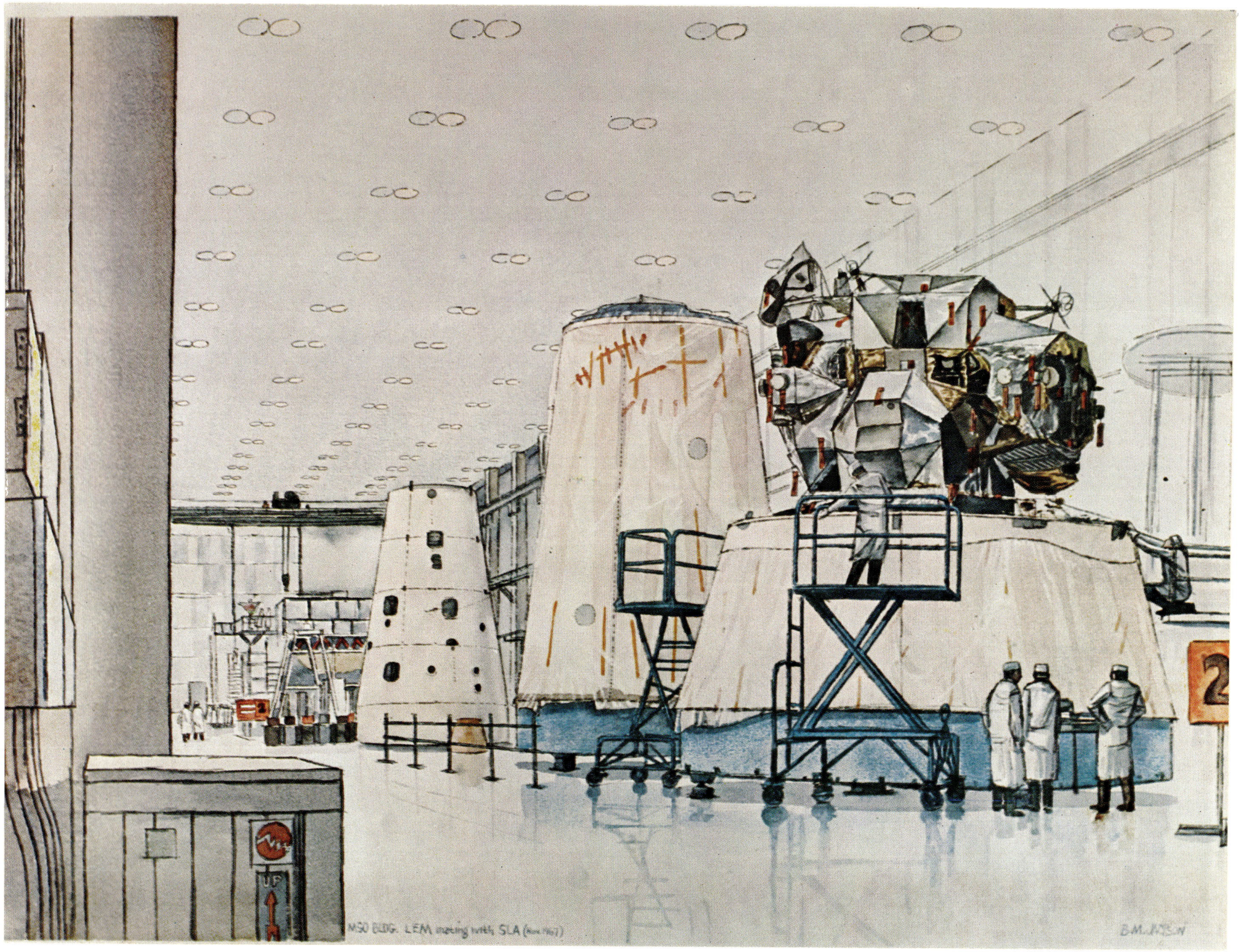

187.
Billy Morrow Jackson
LUNAR MODULE WHITE ROOM
Watercolor on paper, 22 × 30″

The Lunar Module, folded and ready for its epic mission, undergoes final checking before it is encased in its outer covering and taken to the Vehicle Assembly Building. The White Room, which nine years earlier was a canvas-shielded working area on top of a Redstone gantry, for the Apollo program has evolved into a vast hall, white, silent, and surgically clean.

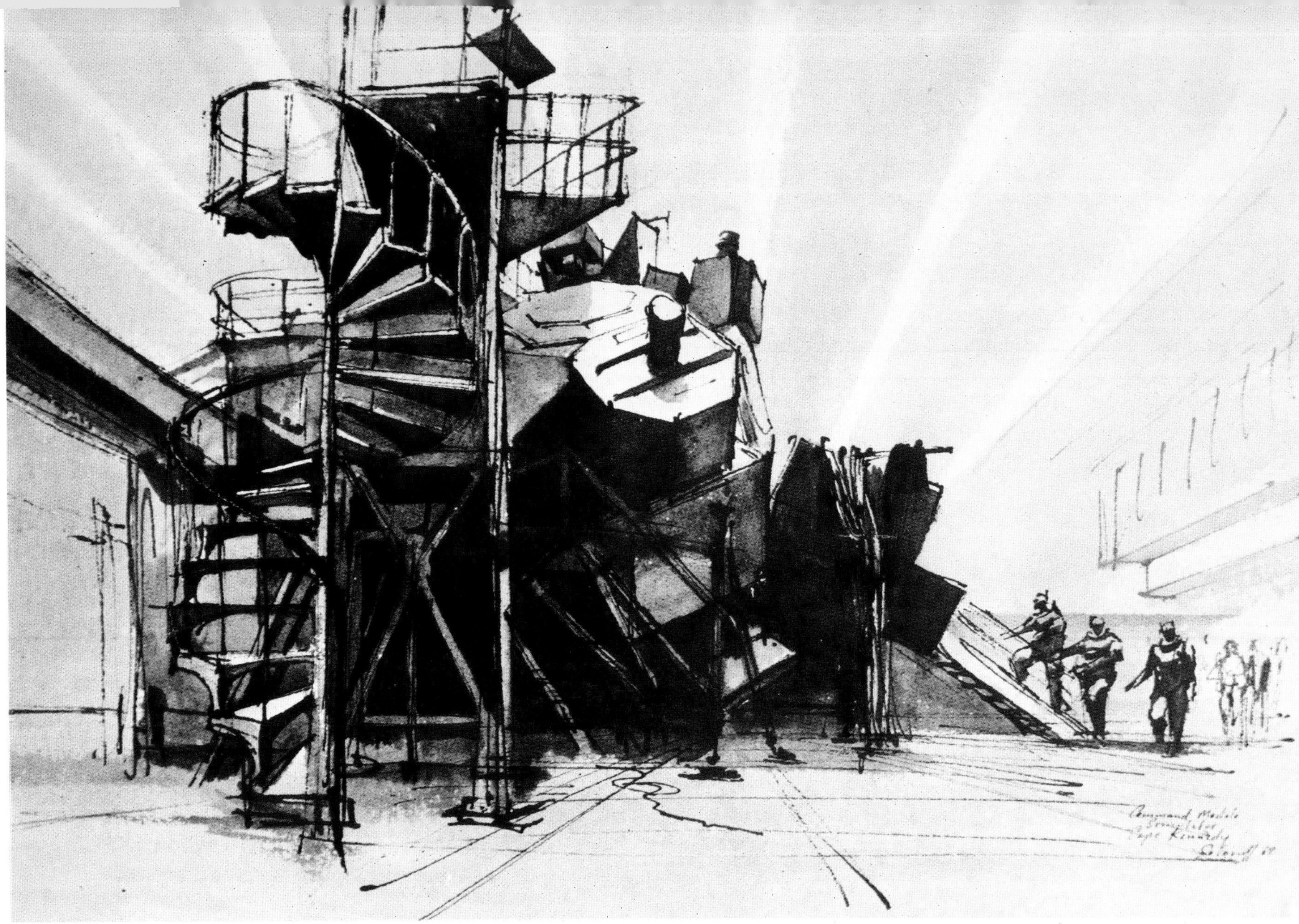

188. *Nicholas Solovioff.* TRAINING SIMULATOR. Watercolor on paper, 14×19″

189.
Nicholas Solovioff
COMMAND MODULE SIMULATOR
Watercolor on paper, 14×19″

No Baroque stage designer ever conceived a more fanciful array of shapes than the creators of the training facility for the lunar landing. Each set of boxlike containers houses a separate TV projector designed to give the astronaut encased in the heart of the machine a facsimile of his destination in space.

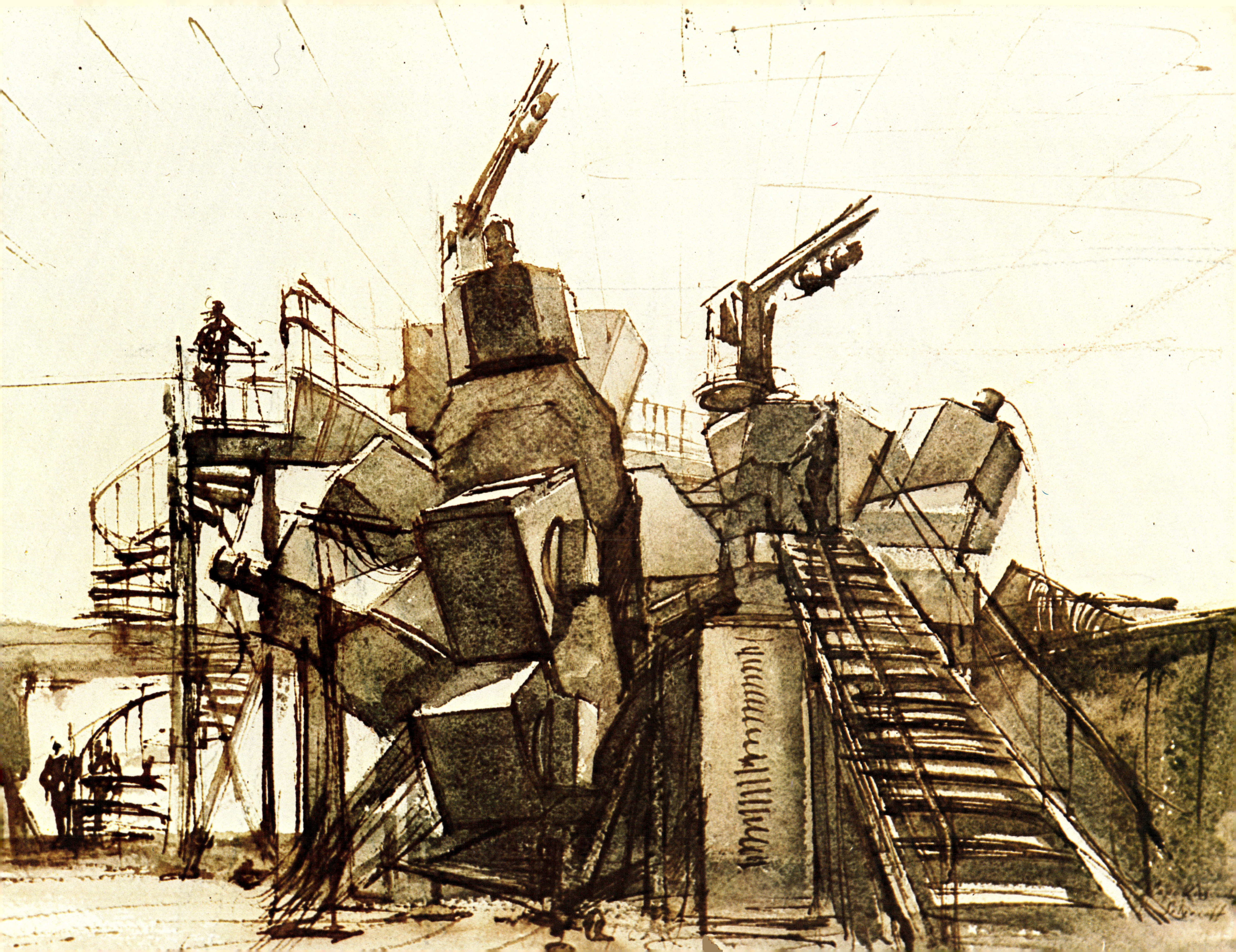

190.
Nicholas Solovioff
GOLDSTONE
Watercolor on paper, 22×16″

Out in California's Mojave Desert, in a valley surrounded by mountains like jagged amethysts, the Goldstone tracking station reaches out into space and picks up the faintest electronic whisperings. The impulses that send back TV pictures from the moon are about as powerful as the flutterings of a butterfly's wings, and the huge dish antenna moves noiselessly, swinging on its base, and must not deviate a millionth of an inch from its target, or the astronaut a quarter of a million miles away will lose his one link with earth.

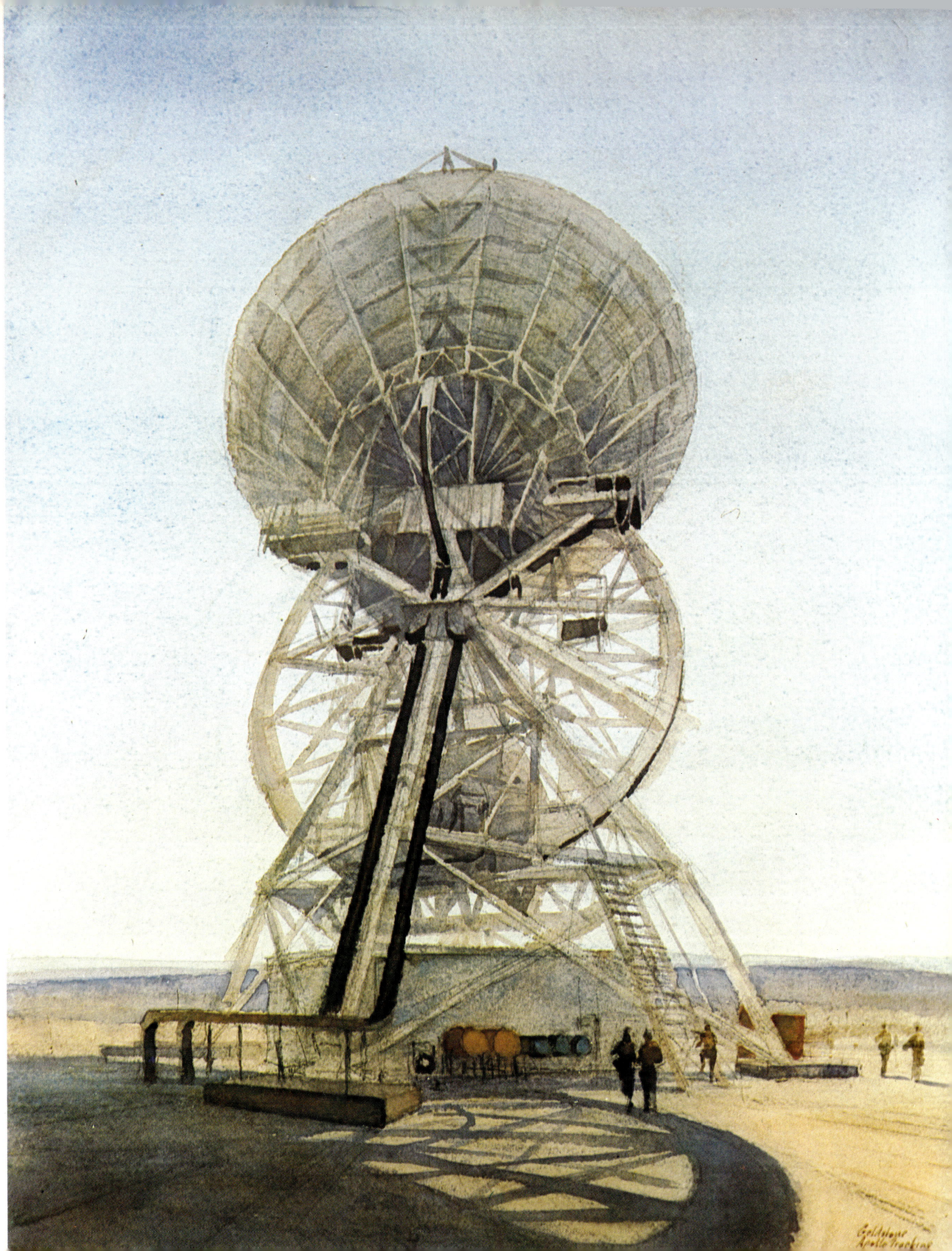

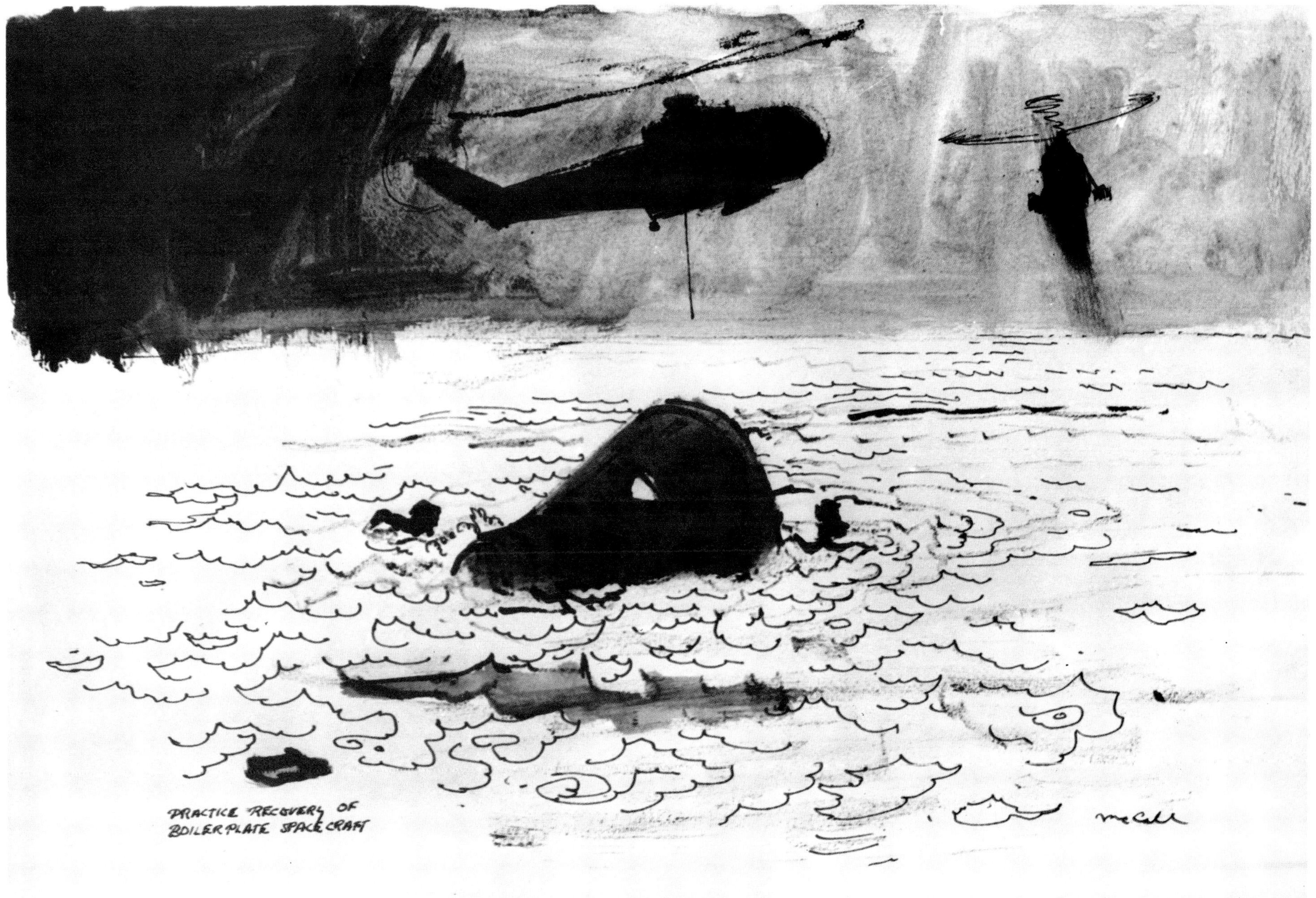

191. *Robert McCall.* PRACTICE RECOVERY OF BOILERPLATE SPACECRAFT. Watercolor and ink on paper, 18×12″

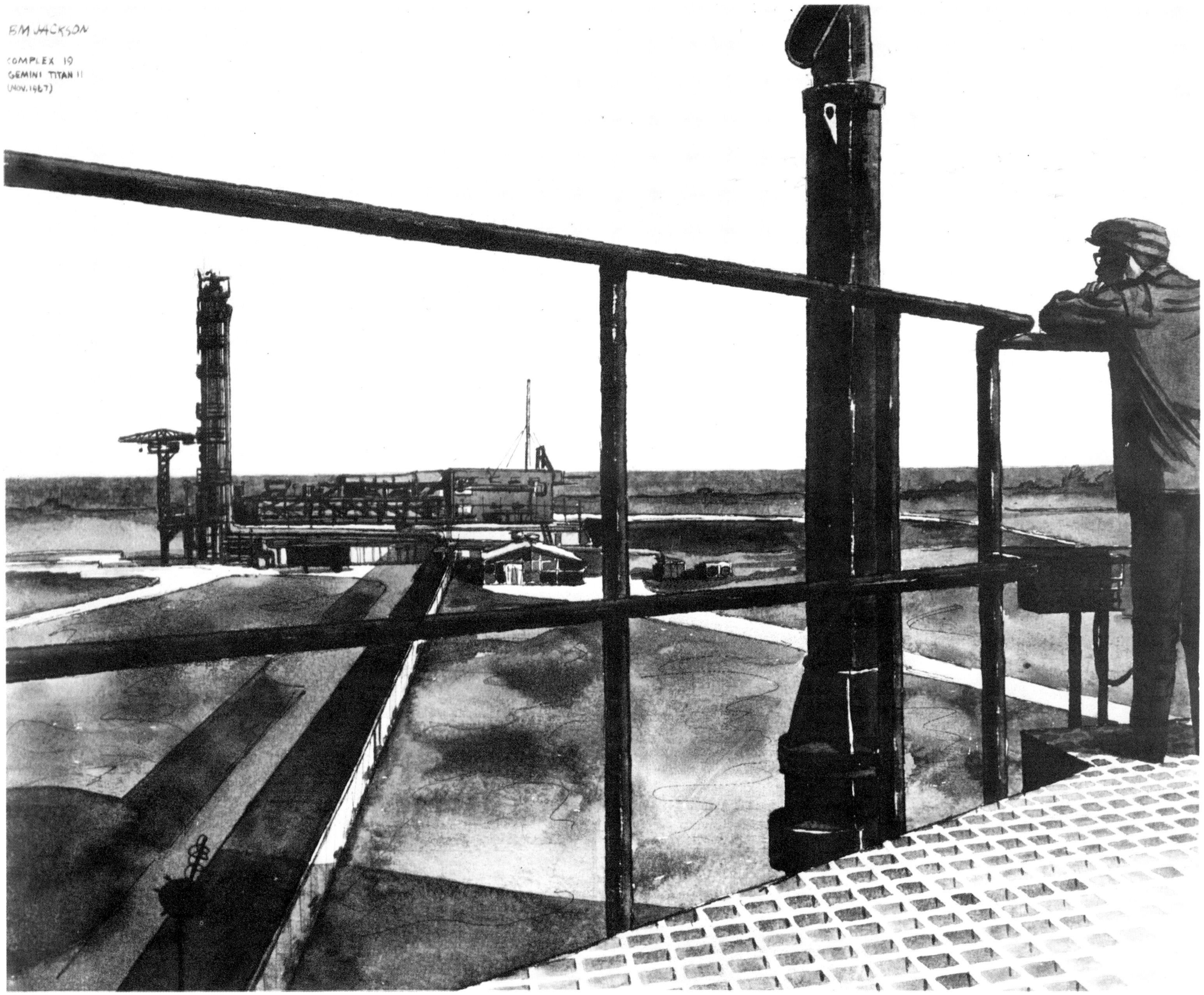

192. *Billy Morrow Jackson*. GEMINI PAD. Watercolor on paper, 22×30″

193.
Nicholas Solovioff
APOLLO LAUNCH
Watercolor on paper, 12×22″

TV coverage has accustomed the world to think of a launch as a spectacle seen at close range. In fact, the nearest spectator is about three miles away from the launch pad. One of the VIP viewing stands was built on a causeway across the Banana River. Here, the brilliant fireball is seen across the placid waters of the lagoon. The artist, while awed by the event, has put the action into perspective and has not forgotten the beauty of the pearly gray light of dawn.

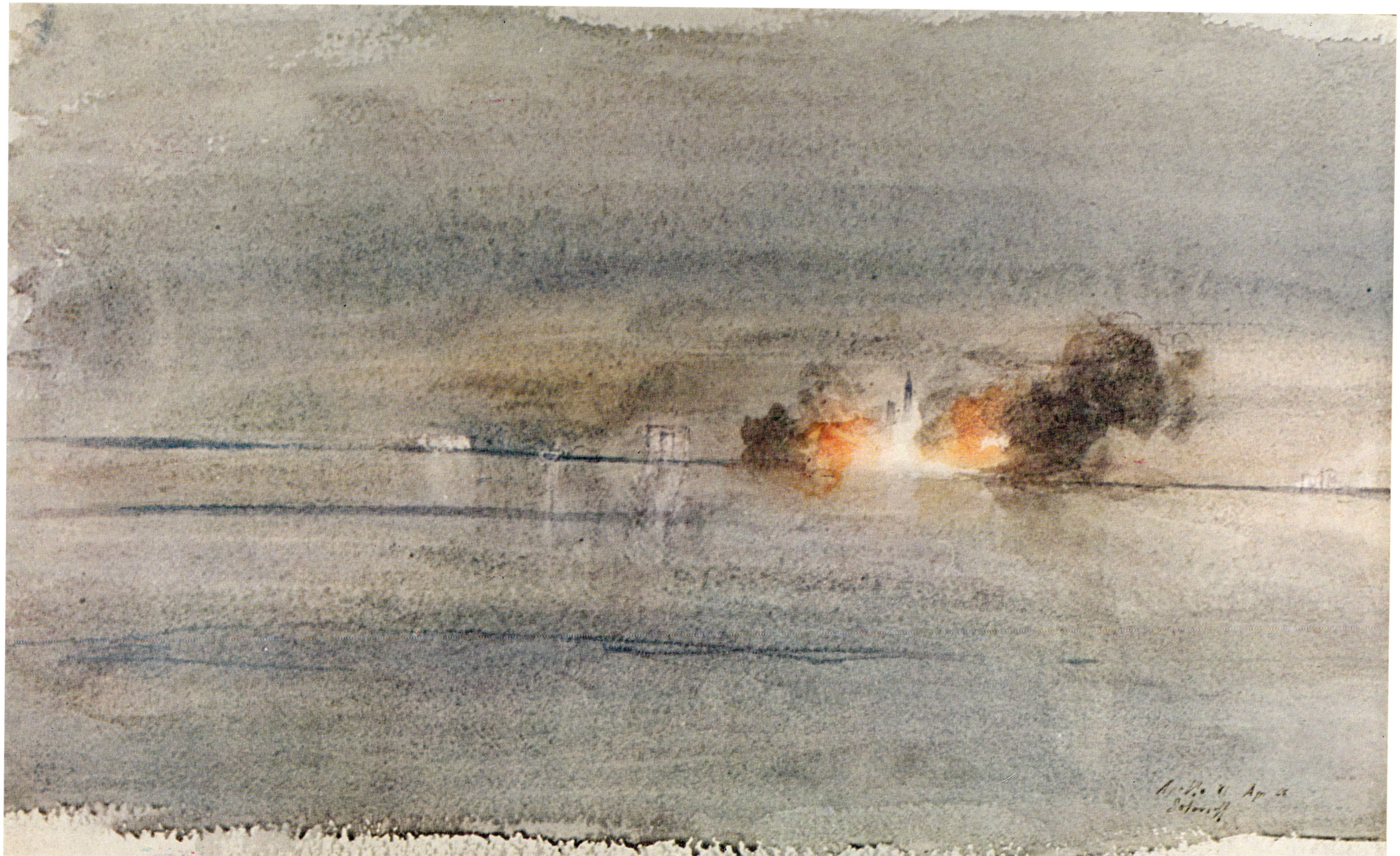

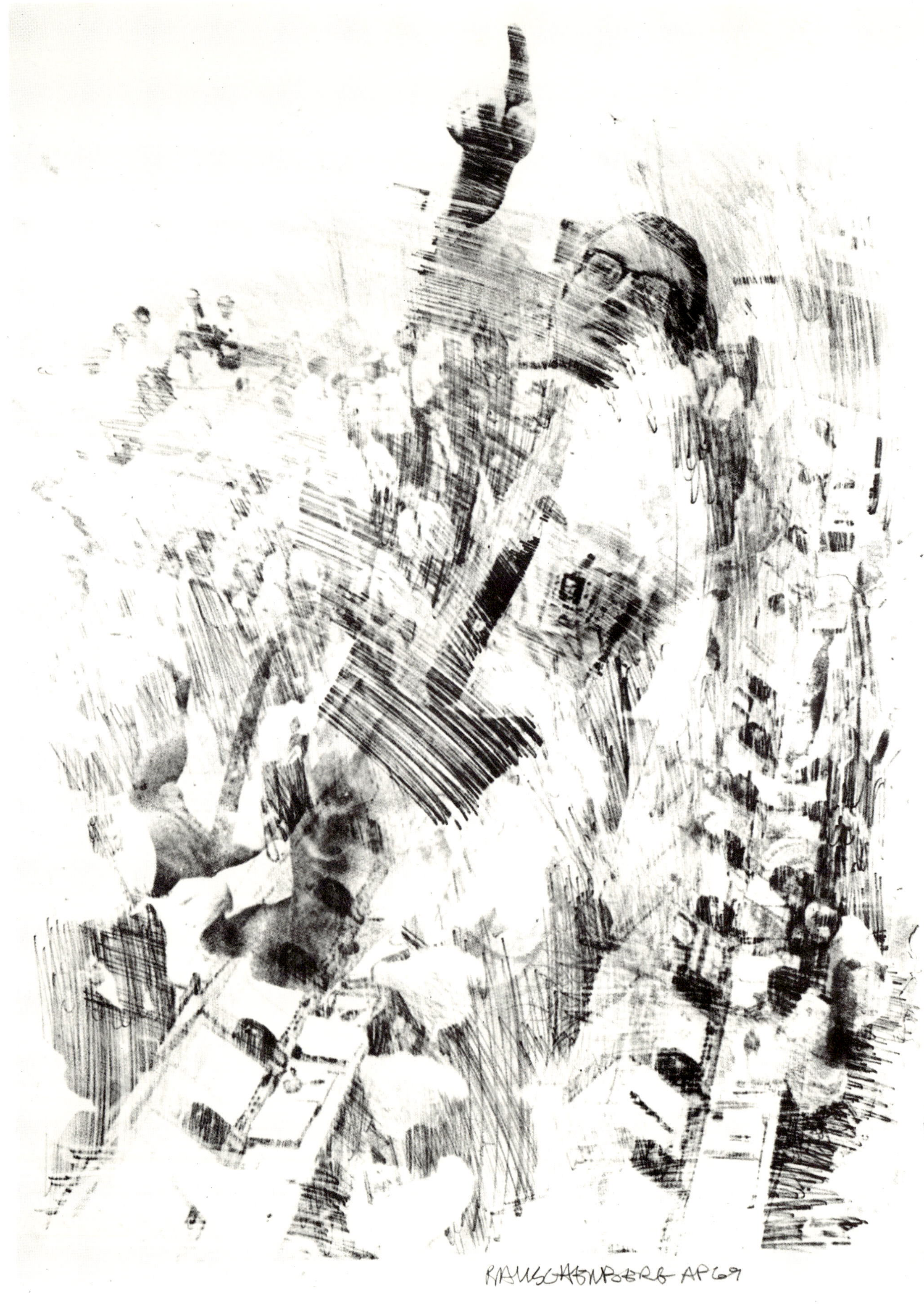

194.
Robert Rauschenberg
SKY-RITE
Color lithograph on paper, 33×23″
From the Stoned Moon series
reproduced courtesy of Gemini G.E.L.,
Los Angeles, Calif.

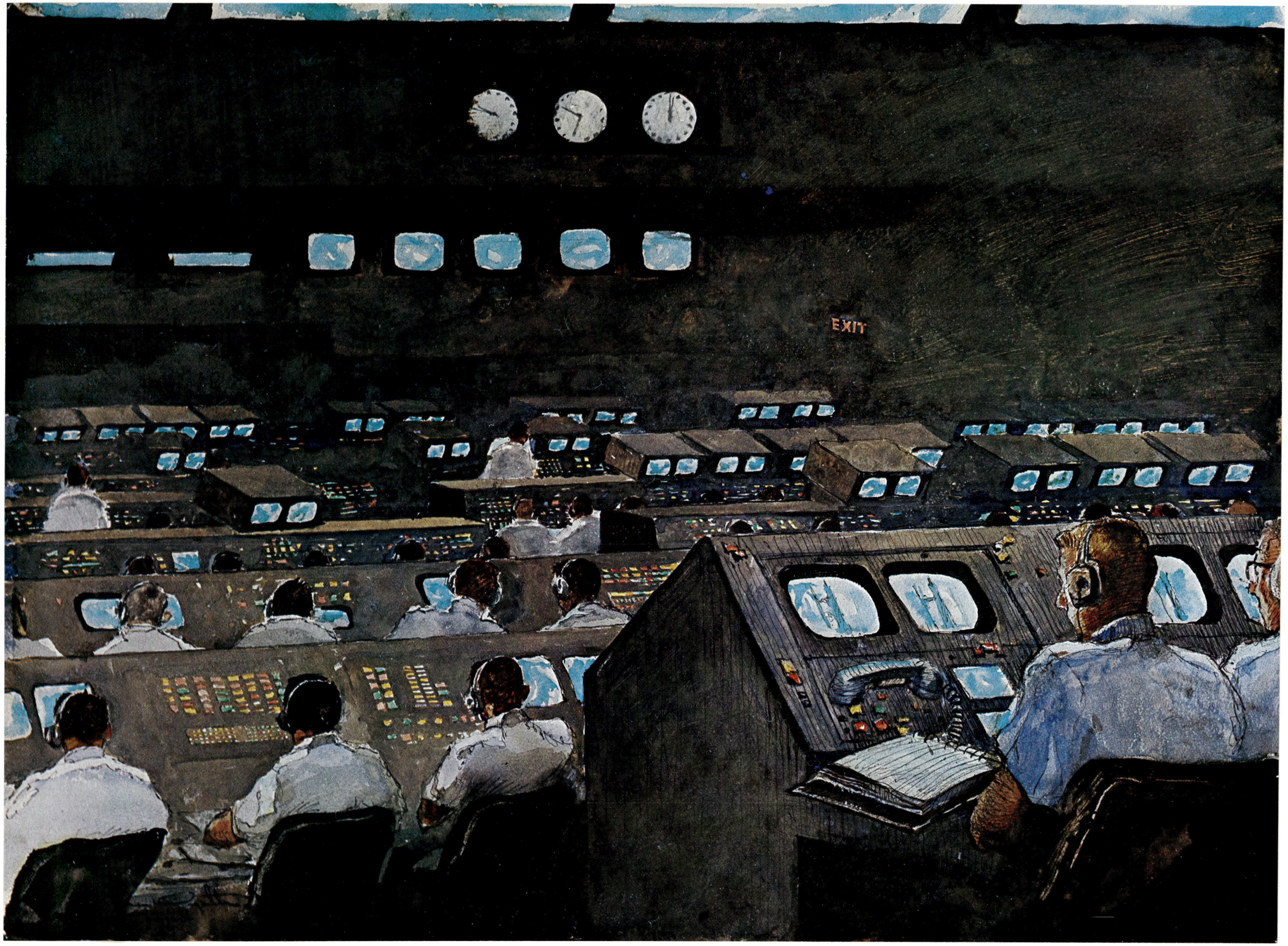

195. *James Wyeth.* FIRING ROOM Watercolor on paper, 18×25″

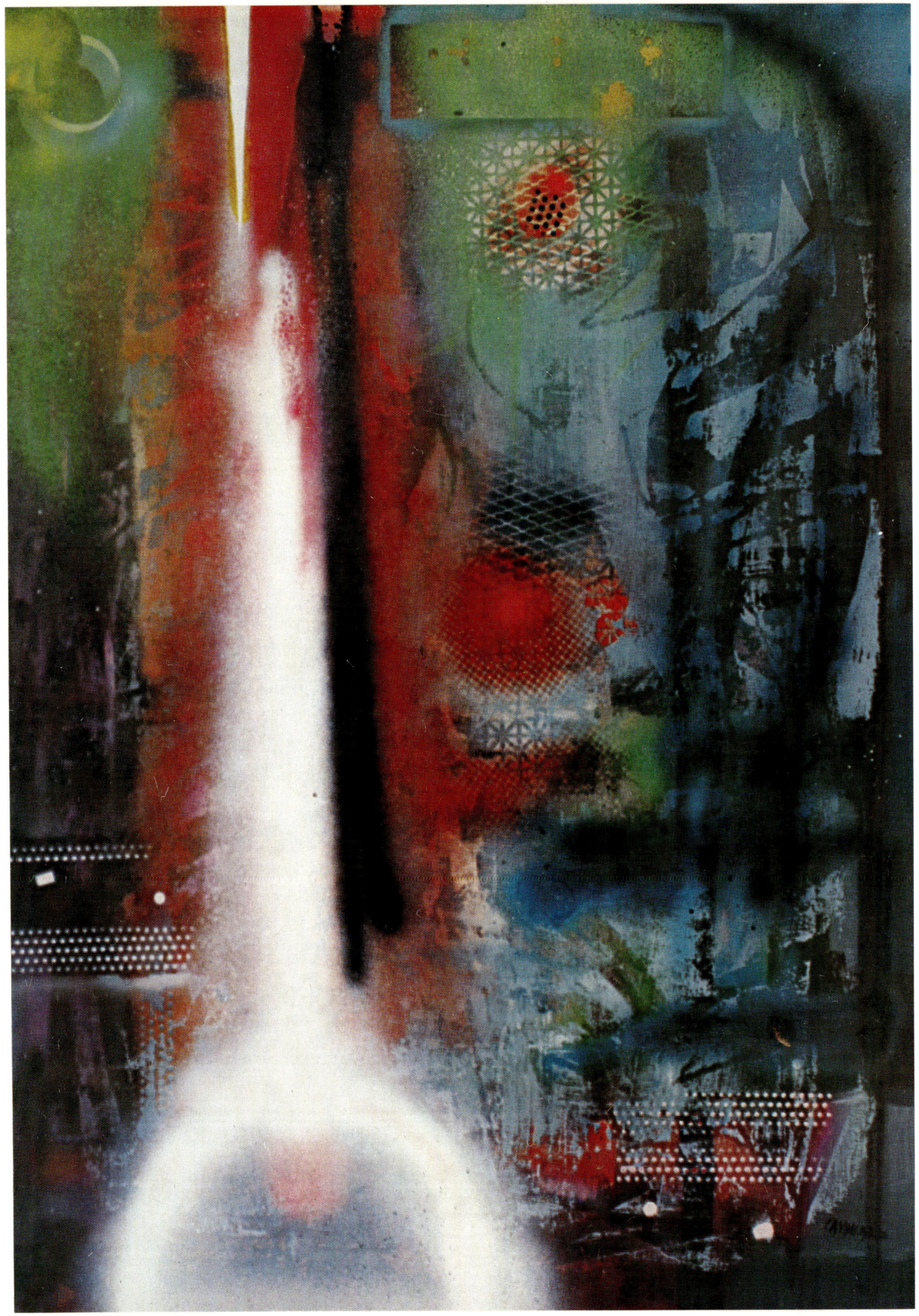

190.
Vincent Cavallaro
IGNITION SEQUENCE
Oil on canvas, 50×36″

Those who watch a major launch carry away with them images of brilliant color, shattering sound, and immense power. Cavallaro set himself the task of transposing these inner images onto canvas.

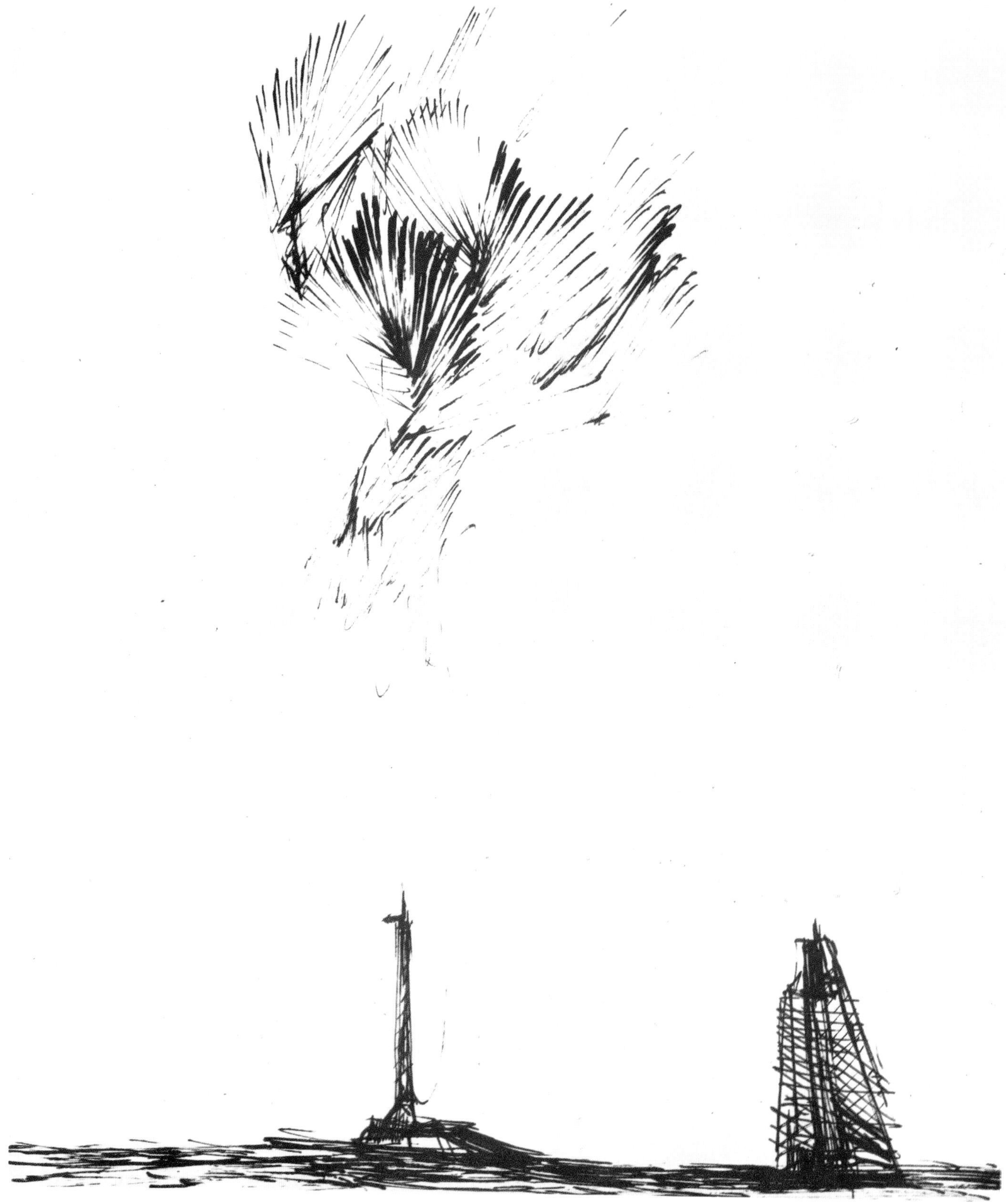

197.
Paul Calle
T PLUS 50 SECONDS
Ink on paper, 22 × 18″

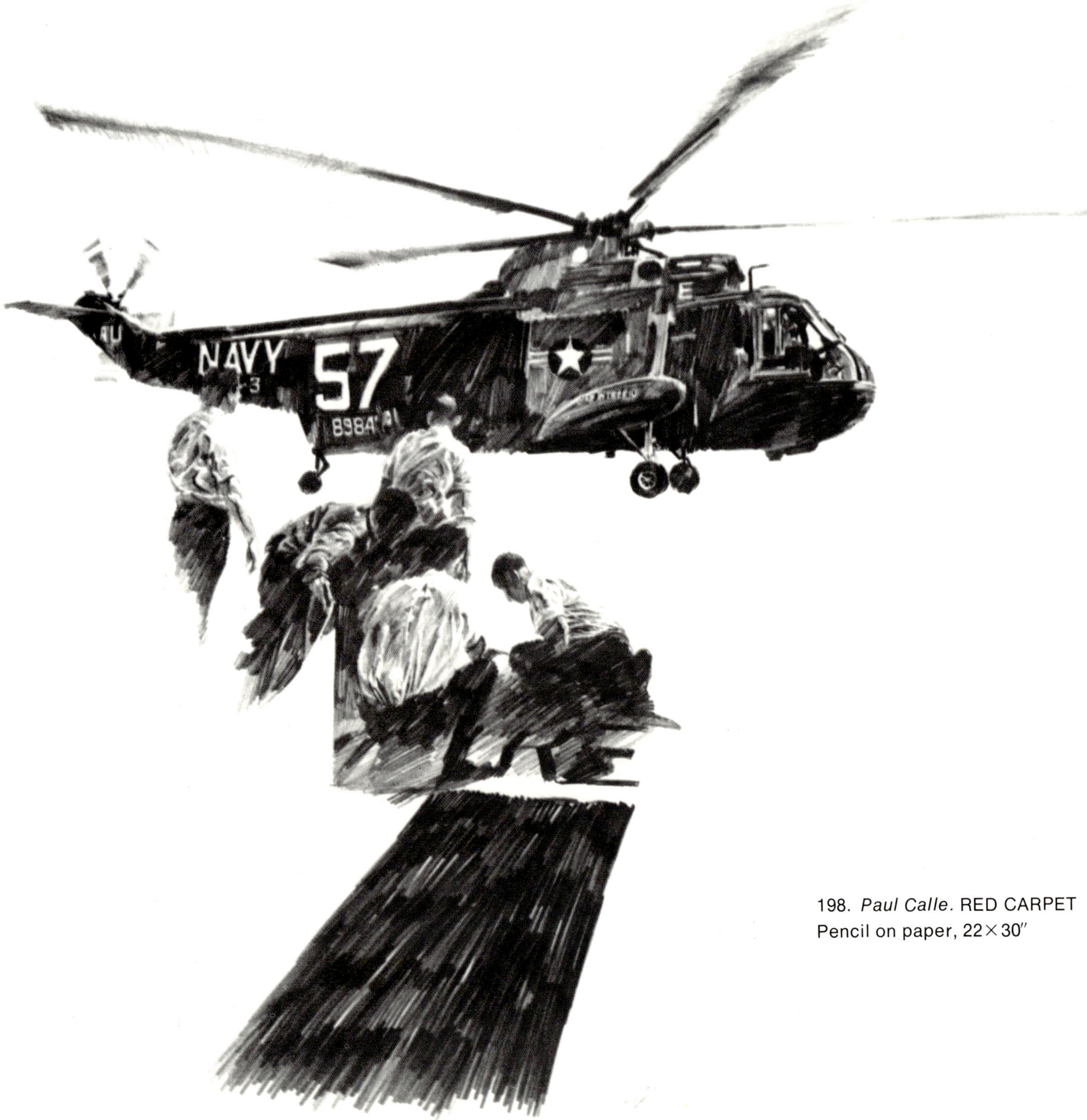

198. *Paul Calle*. RED CARPET
Pencil on paper, 22×30″

199.
Tom O'Hara
APOLLO 10 IS HAULED ABOARD
Acrylic on paper, 22×30″

The Apollo spacecraft, which started its journey with seven and a half million horsepower, is pulled aboard the carrier by nine sailors.

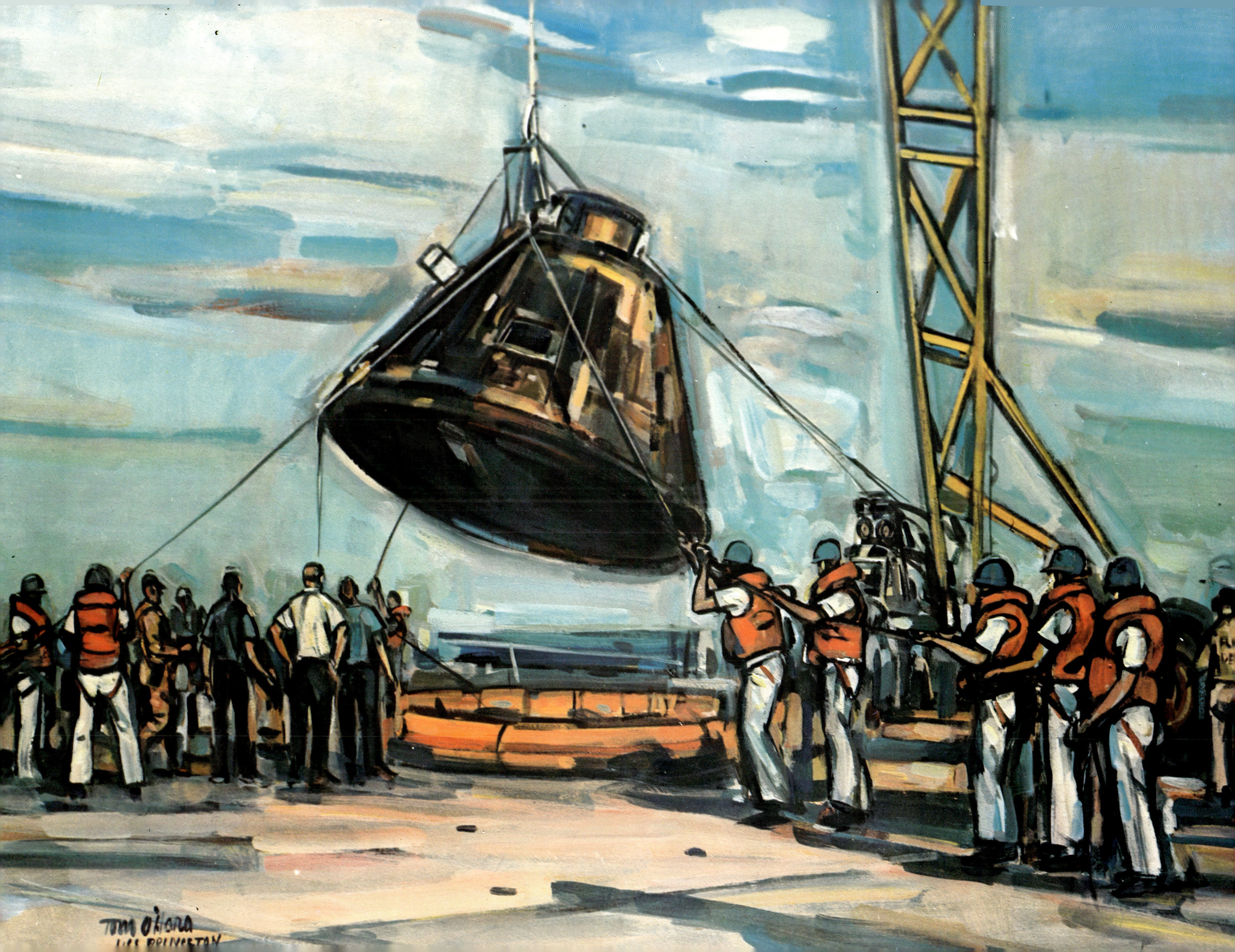
Tom O'Hara

200.
Leonard Dermott
END OF THE VOYAGE
Acrylic on canvas, 28¾ × 19″

"When the sun came up, even the sea looked like gold leaf," Dermott noted when Apollo 9 was rescued from the Atlantic. He also noted with regret that when it was being picked up some of the flotation gear broke loose, giving the spacecraft a disheveled, woebegone appearance, unworthy of such a successful mission.

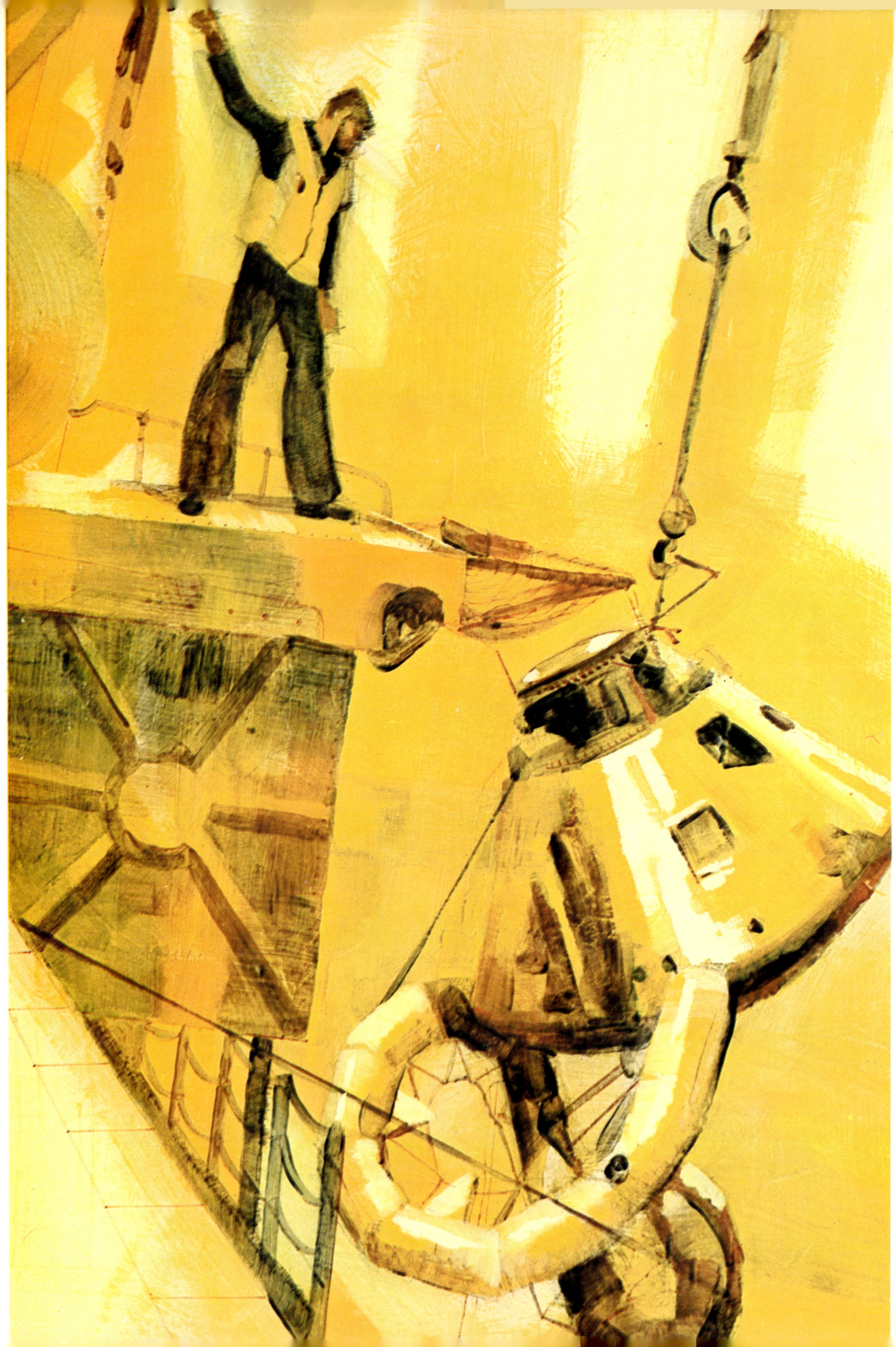

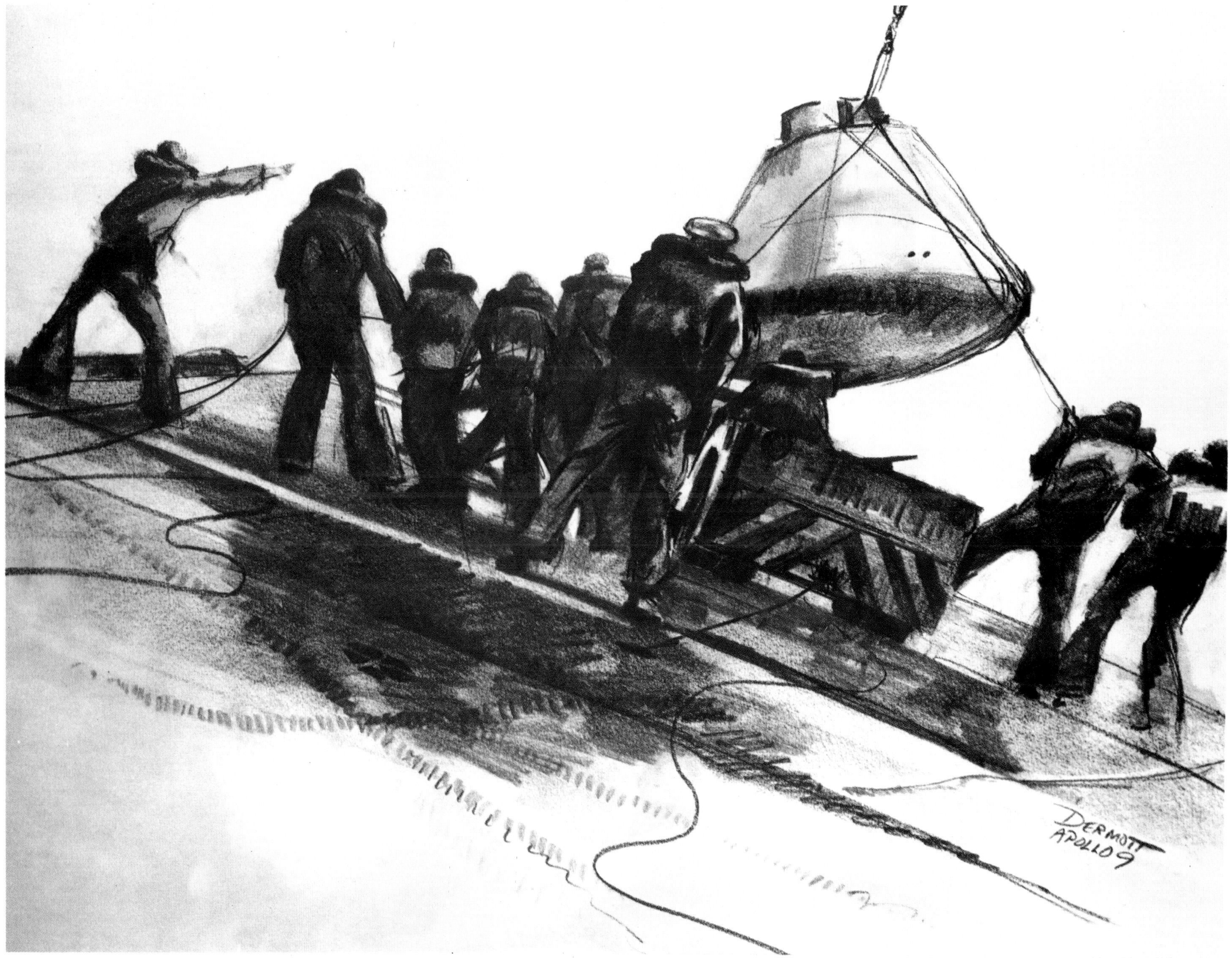

201. *Leonard Dermott.* COMING ABOARD. Charcoal on paper, 18×24″

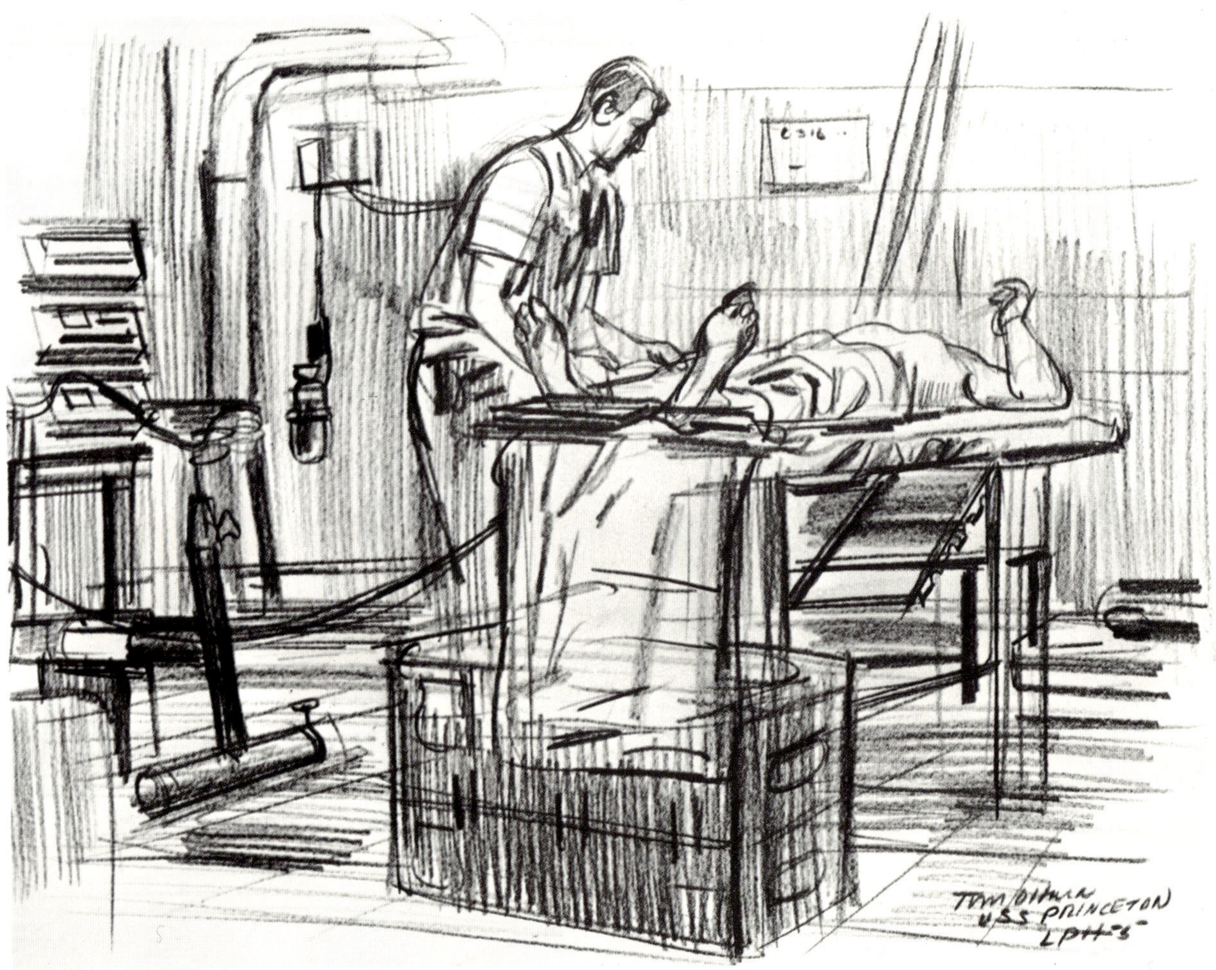

202. *Tom O'Hara.* MEDICAL EXAMINATION
Crayon on paper, 10×13½″

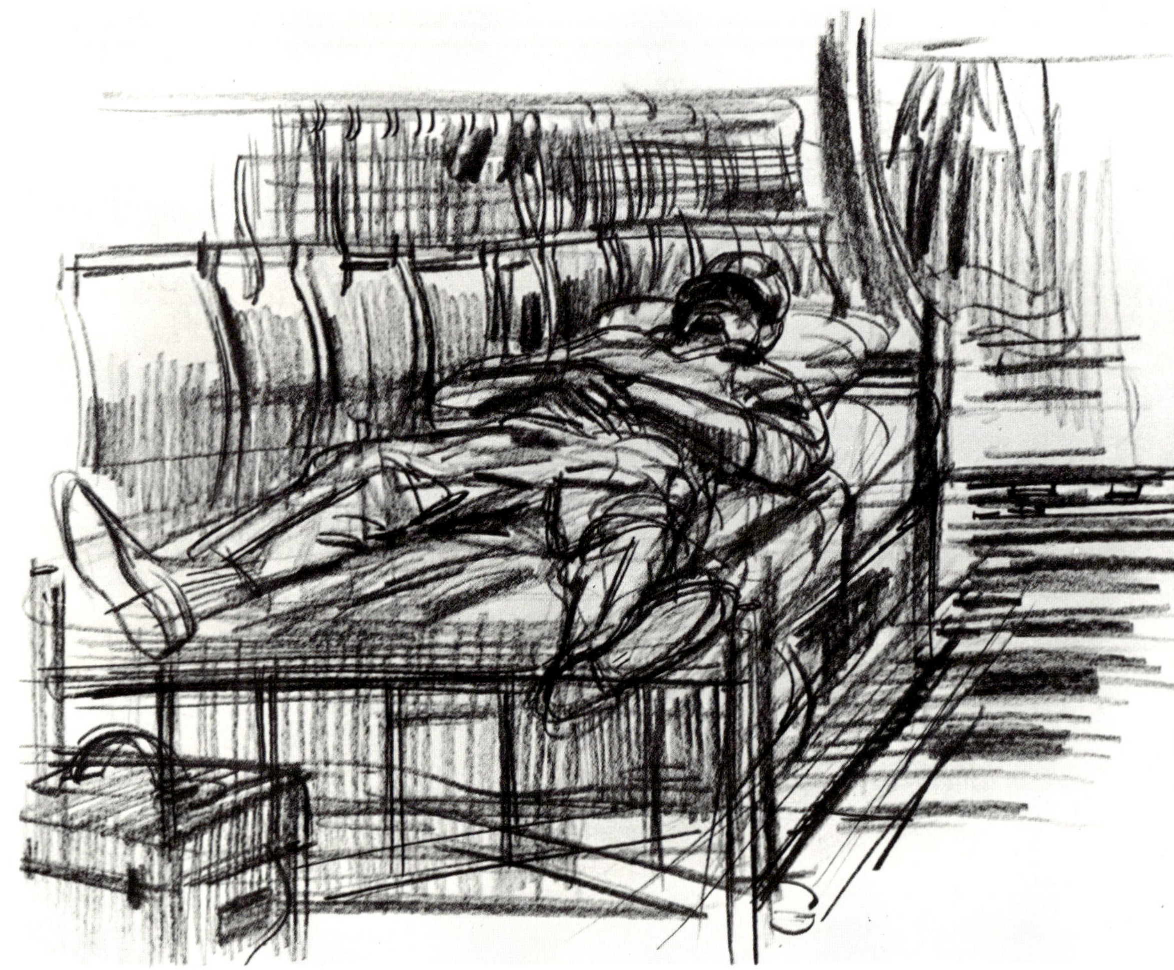

203. *Tom O'Hara.* ASTRONAUT RETURNS TO HOUSTON
Crayon on paper, 10×13½″

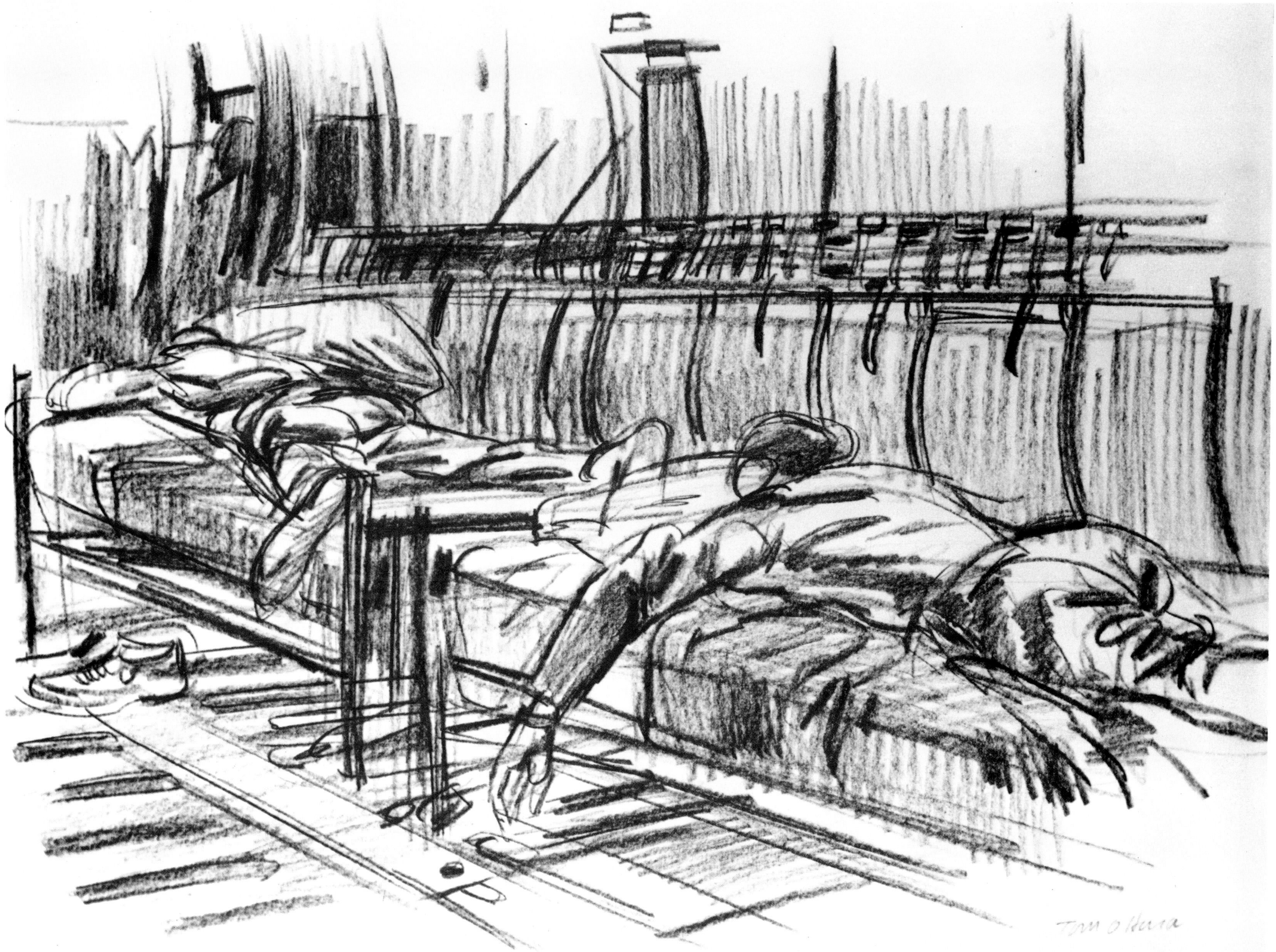

204. *Tom O'Hara.* APOLLO 10 ASTRONAUTS SLEEP ON PLANE BACK TO HOUSTON. Black crayon on paper, 10×13″

205.
Mitchell Jamieson
FIRST LOOK
Ink on paper, 18 × 11″

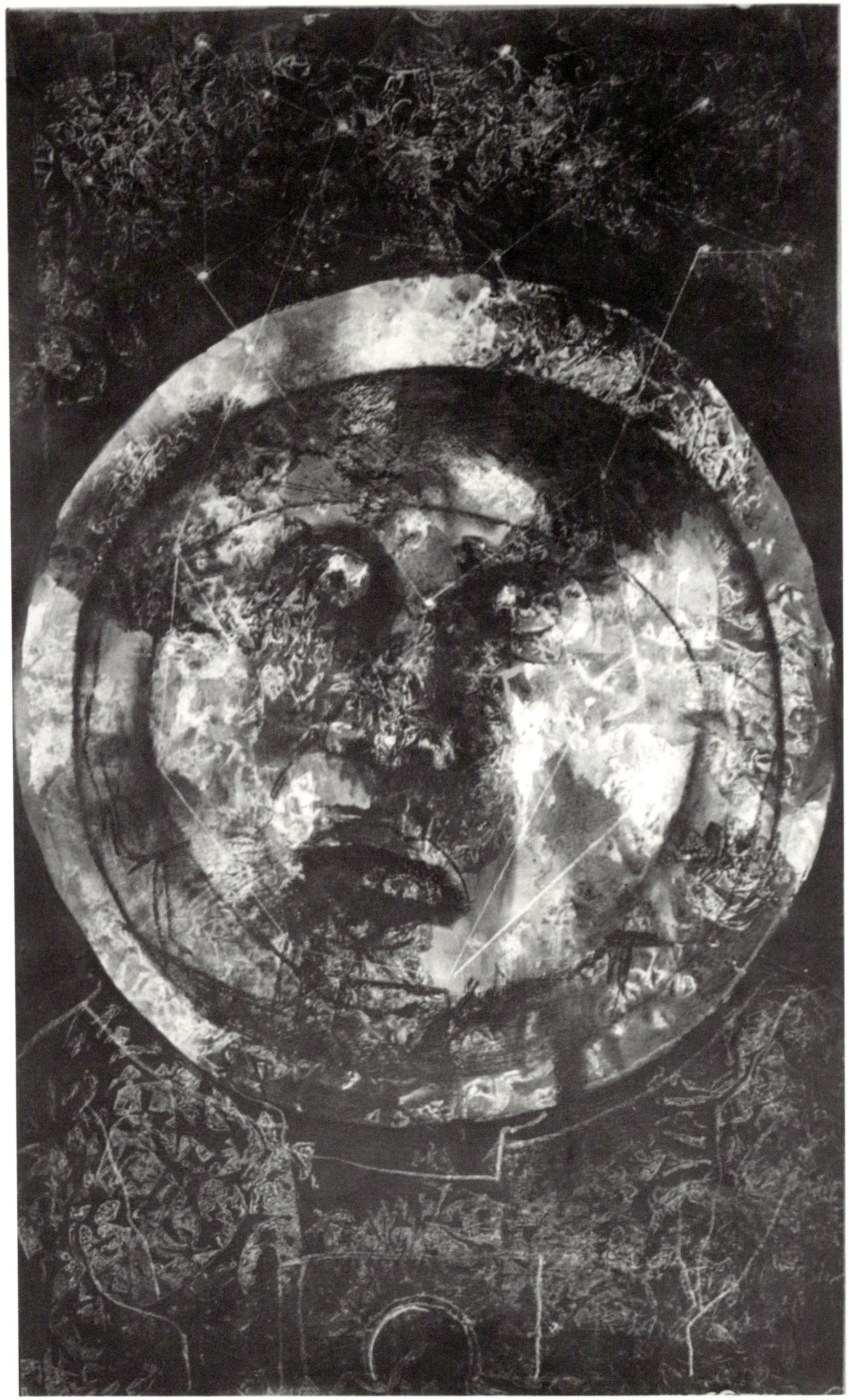

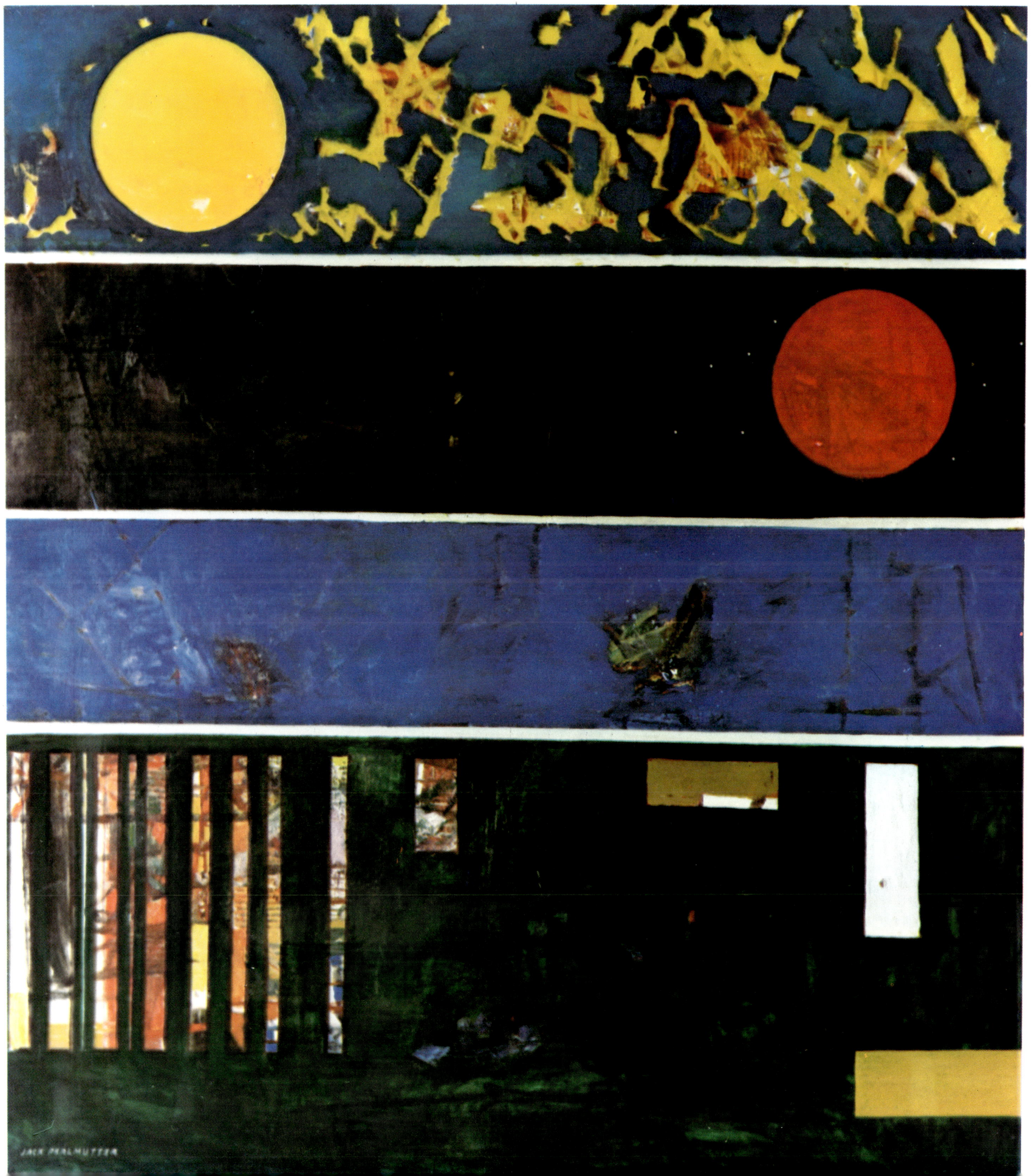

206.
Jack Perlmutter
SUN AND MOON
Oil on canvas, 46×42"

Millions of factors contribute toward a successful mission, not least of which is the position in the heavens of the sun and the moon and the "windows" in space through which the craft must pass.

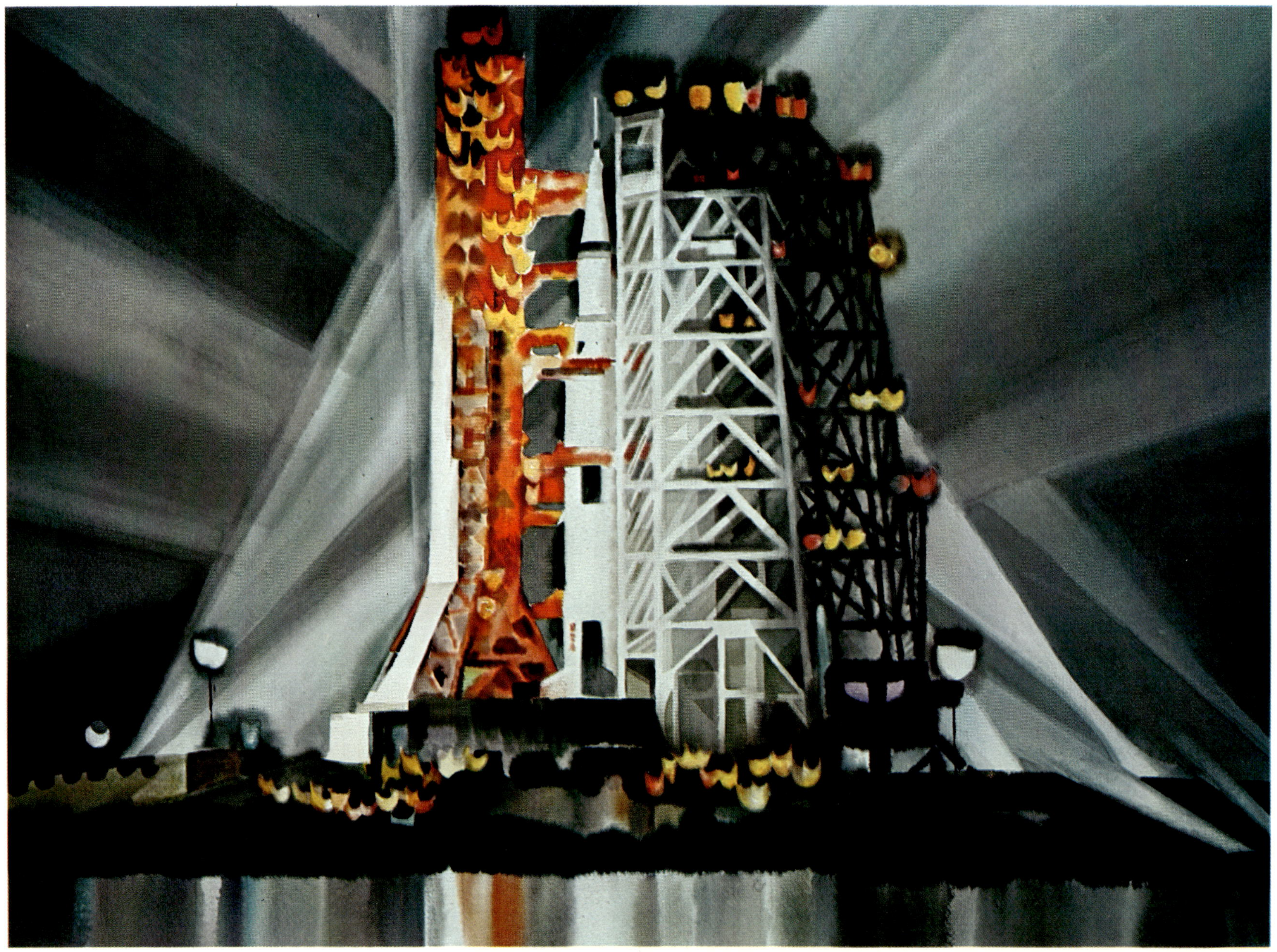

207. *Mario Cooper*. APOLLO 11. Watercolor on paper, 20×30″

208.
Dale Meyers
APOLLO 11—FIREFLY TO THE MOON
Watercolor on paper, 21×28″

One of the most beautiful sights of the Space Age is the cathedral-like structure of Saturn-Apollo during the first light of dawn. The encircling searchlights, the myriad floodlights within the structure, and the opalescent light of dawn for a brief period are balanced to produce an ethereal effect in silvery blue.

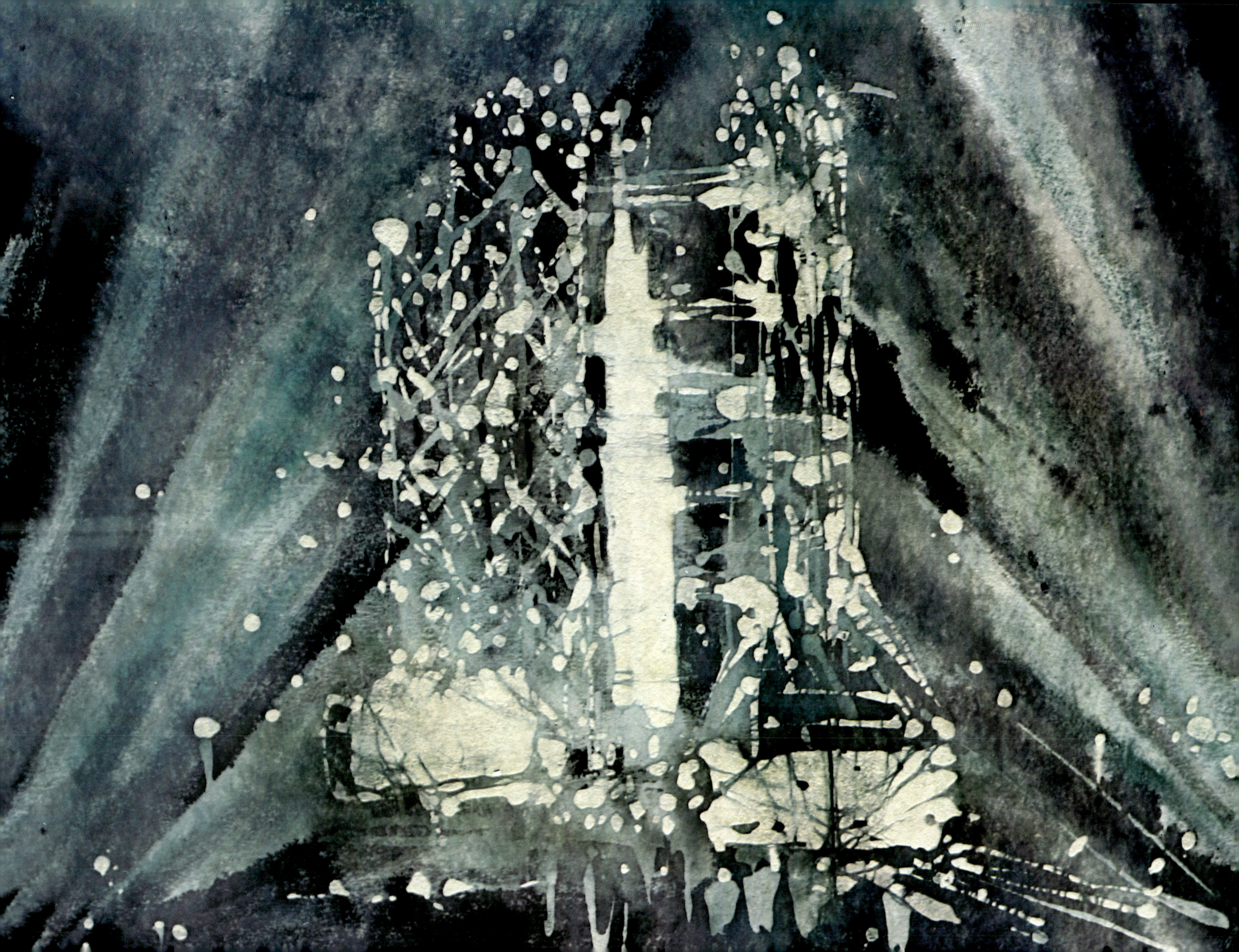

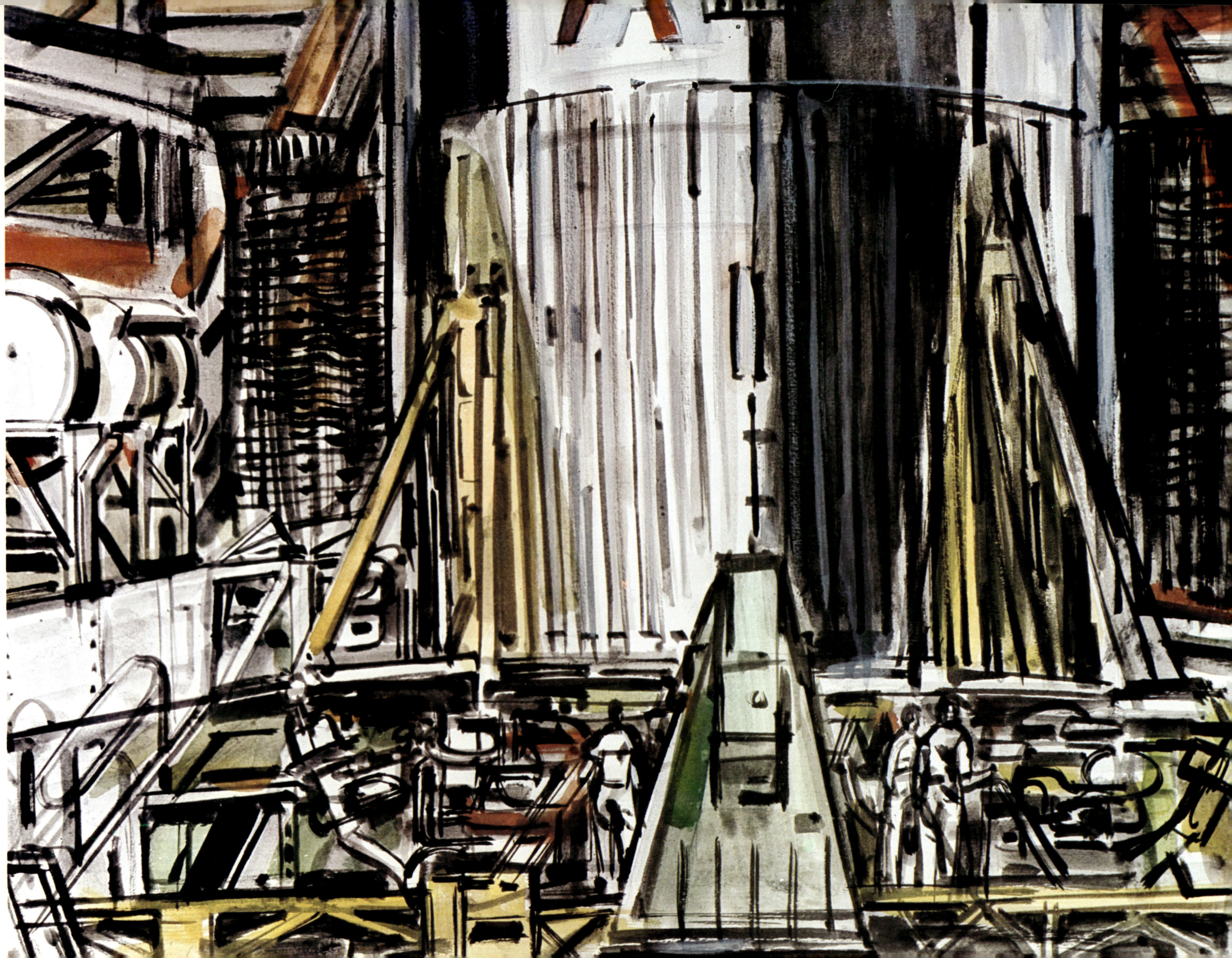

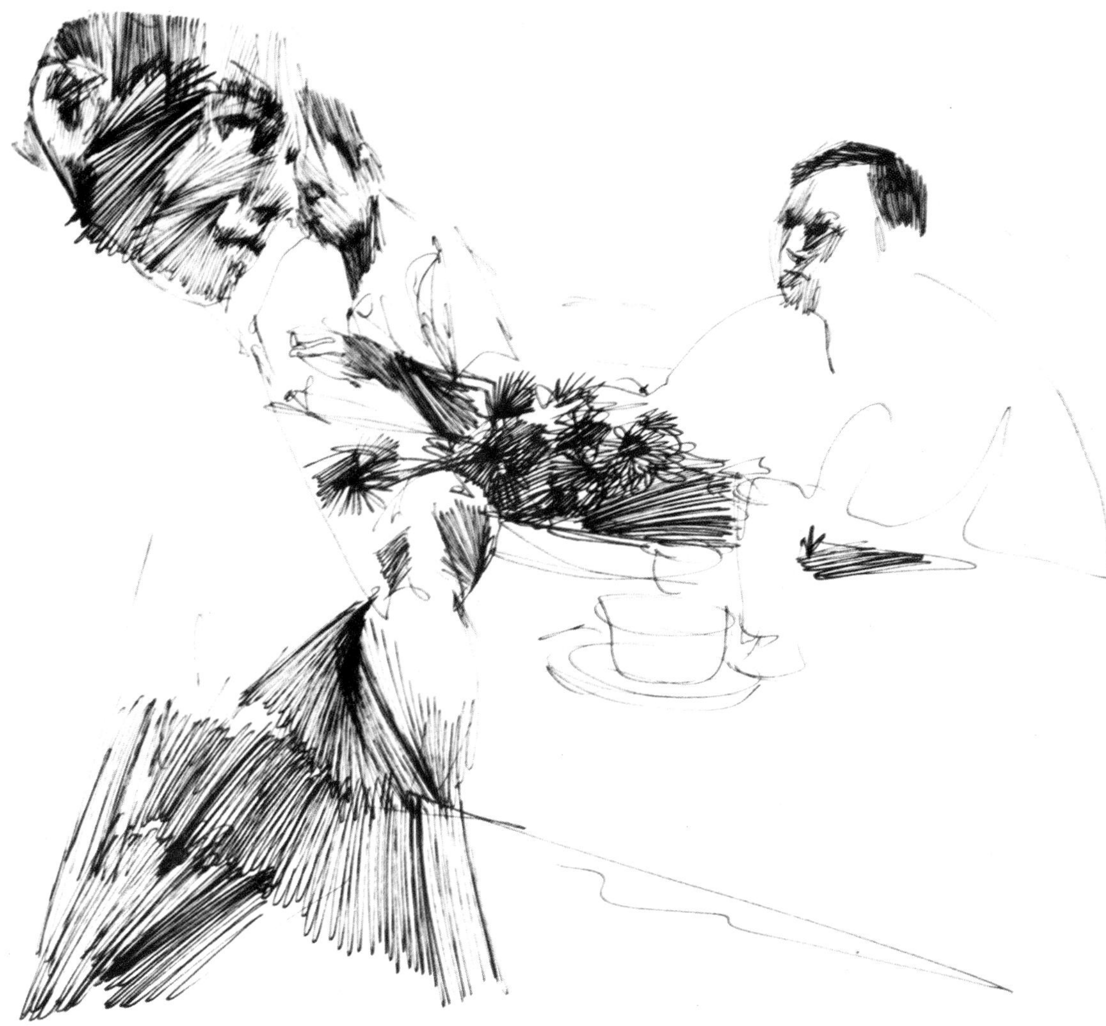

210.
Paul Calle
BREAKFAST FOR THE APOLLO 11 CREW
Pen and ink on paper, 11×14″

211.
Paul Calle
ARMSTRONG BEING HELPED WITH HIS SPACESUIT
Pen and ink on paper, 11×14″

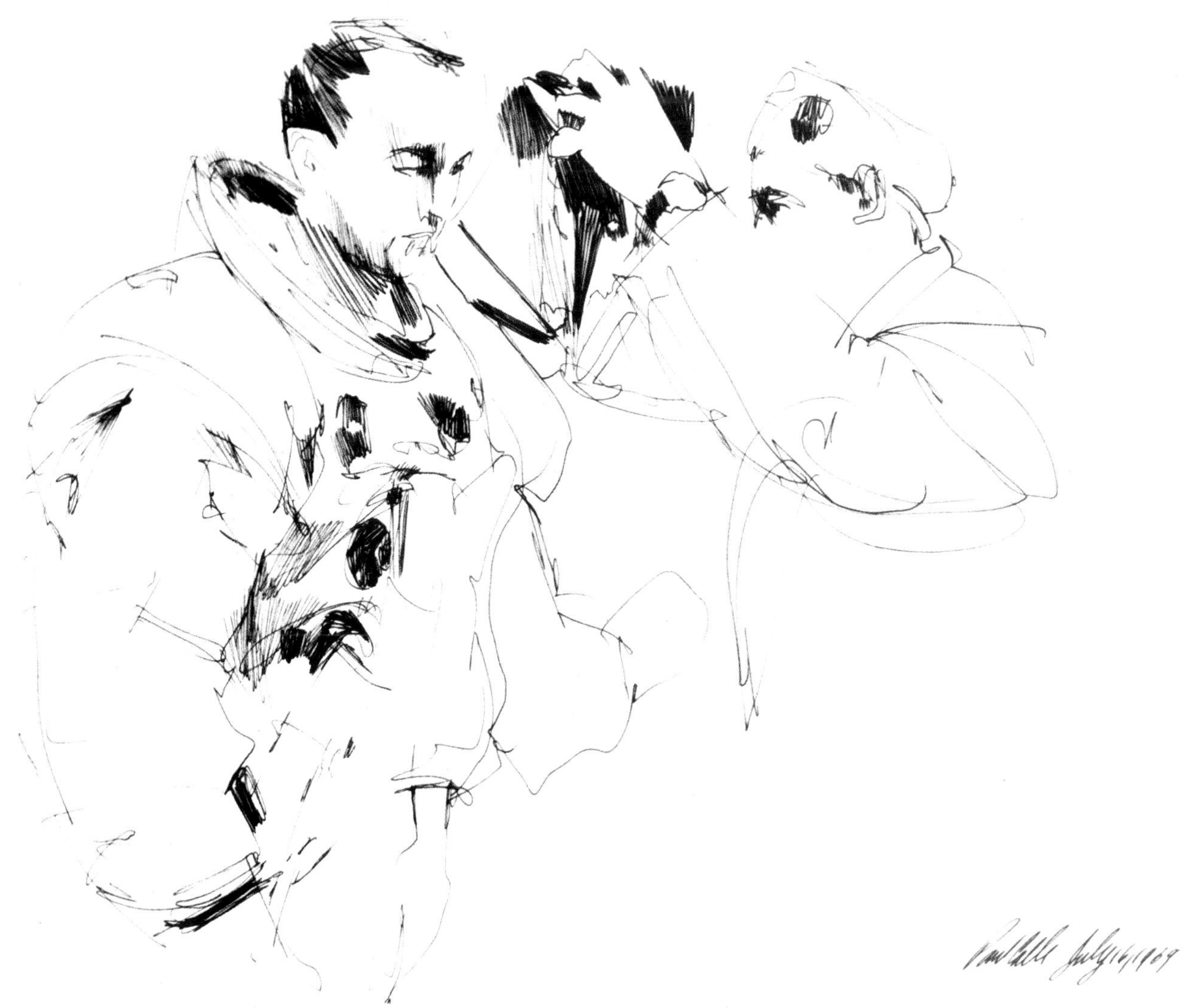

209.
Tom O'Hara
CHECKING THE ENGINES
Watercolor on paper, 11×14″

213.
Paul Calle
SUITING UP FOR APOLLO 11
Pen and ink on paper, 18 × 23″

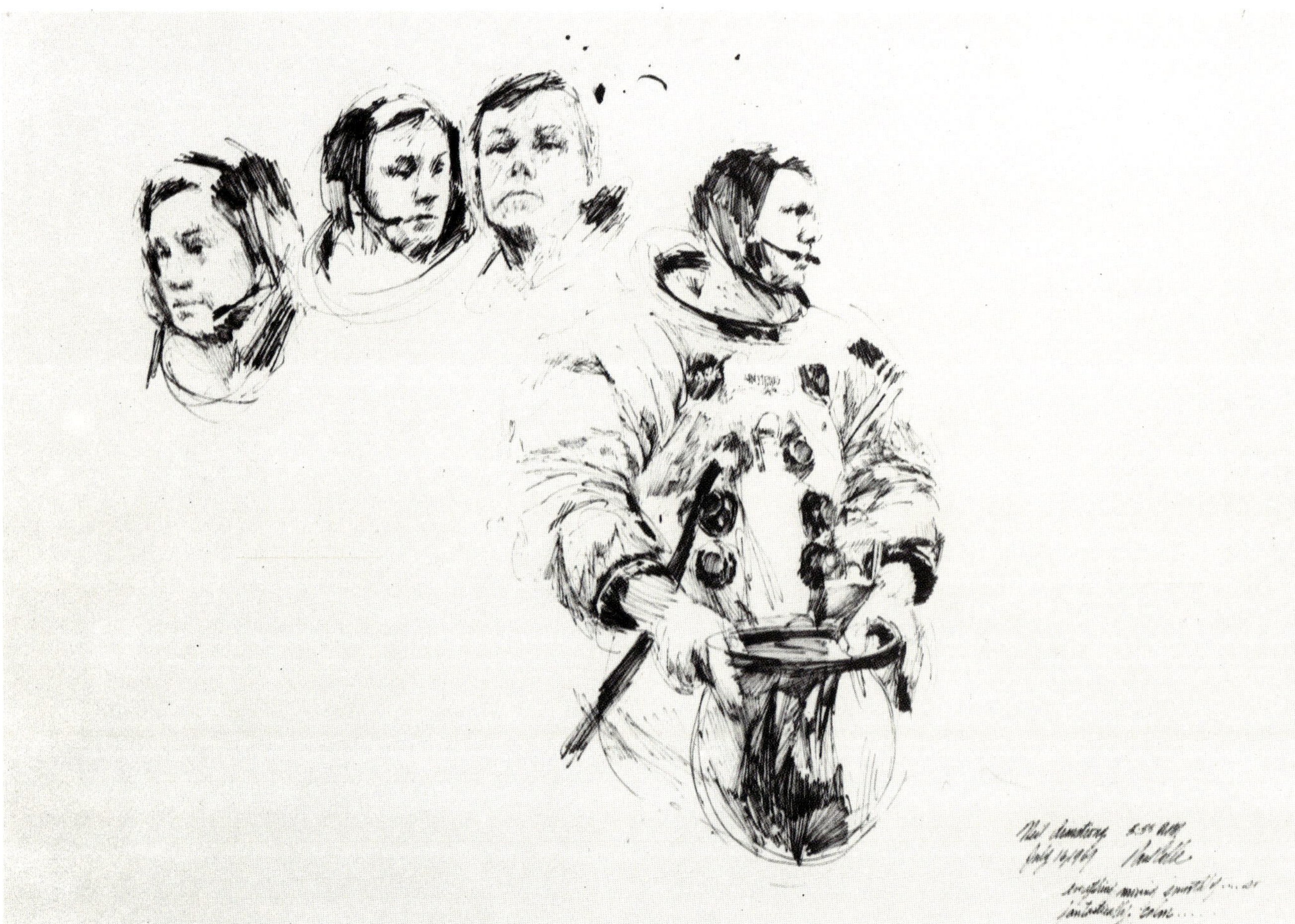

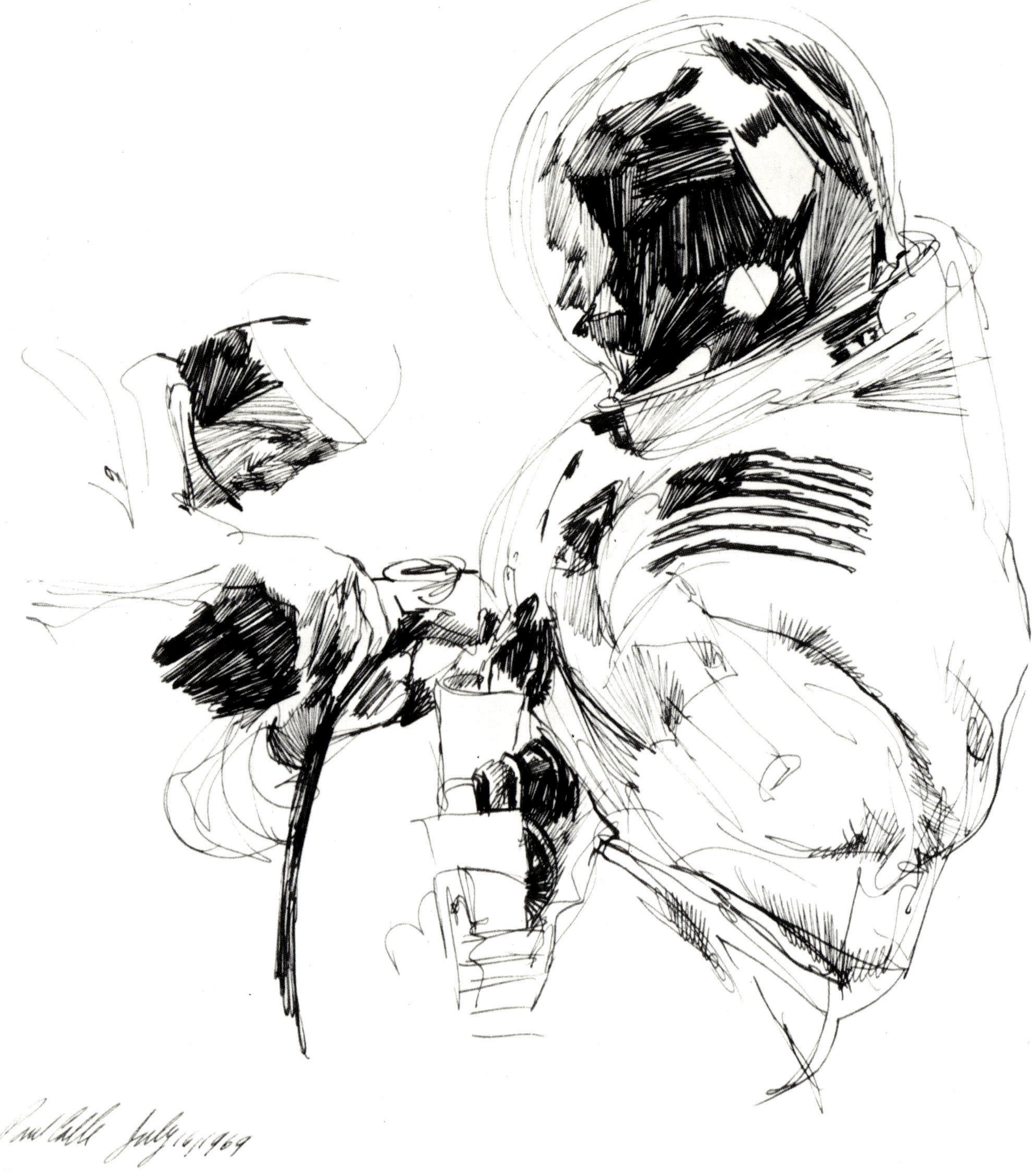

212.
Paul Calle
LAUNCH MORNING
Pen and ink on paper, 16½ × 22″

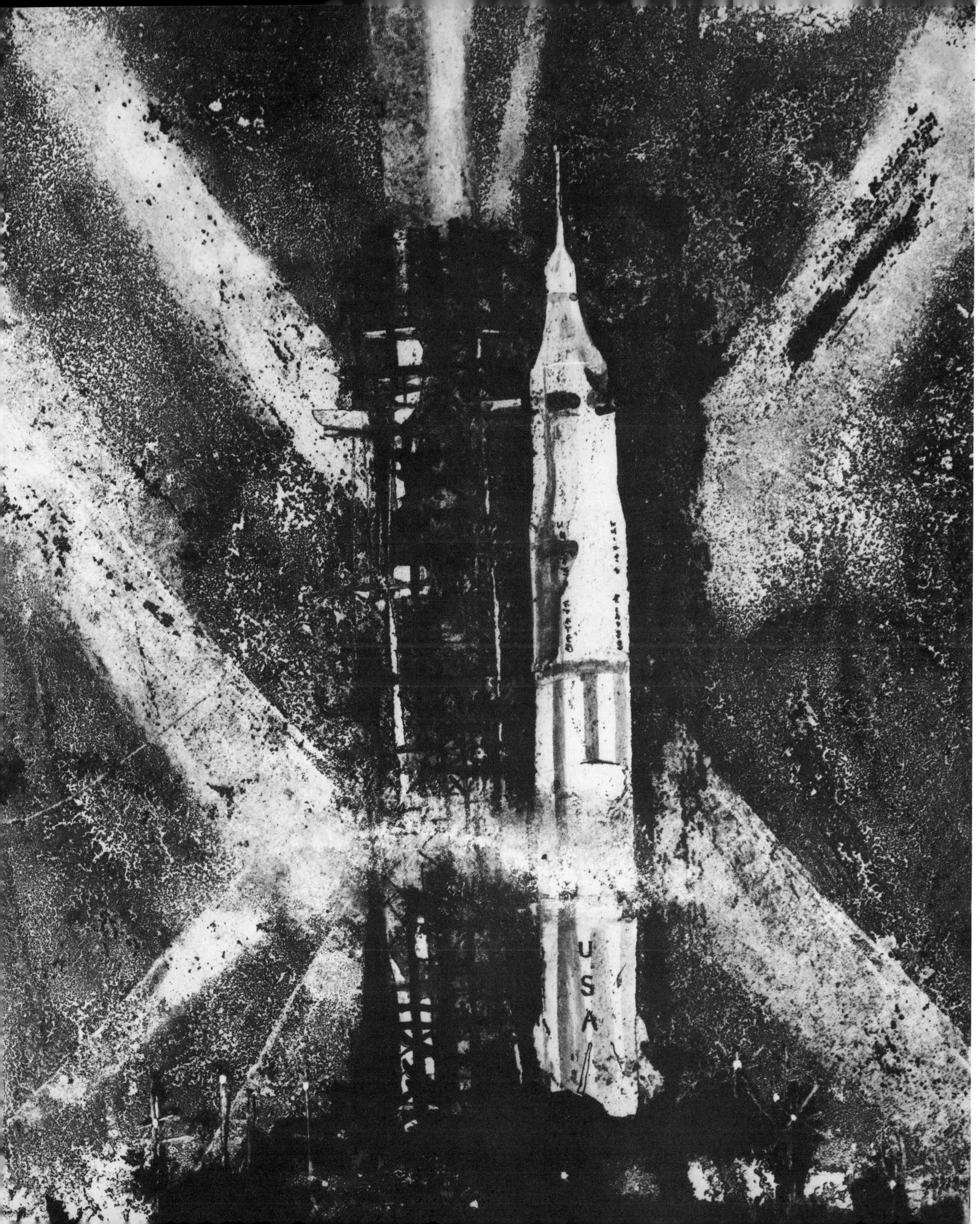

214.
William Thon
DAWN FOR APOLLO 11
Watercolor on paper, 23 × 17½″

History, NASA, and art fortunately converged to meet on the morning of July 16, 1969, when the astronauts were suiting up and the giant rocket was waiting for them at Cape Kennedy.

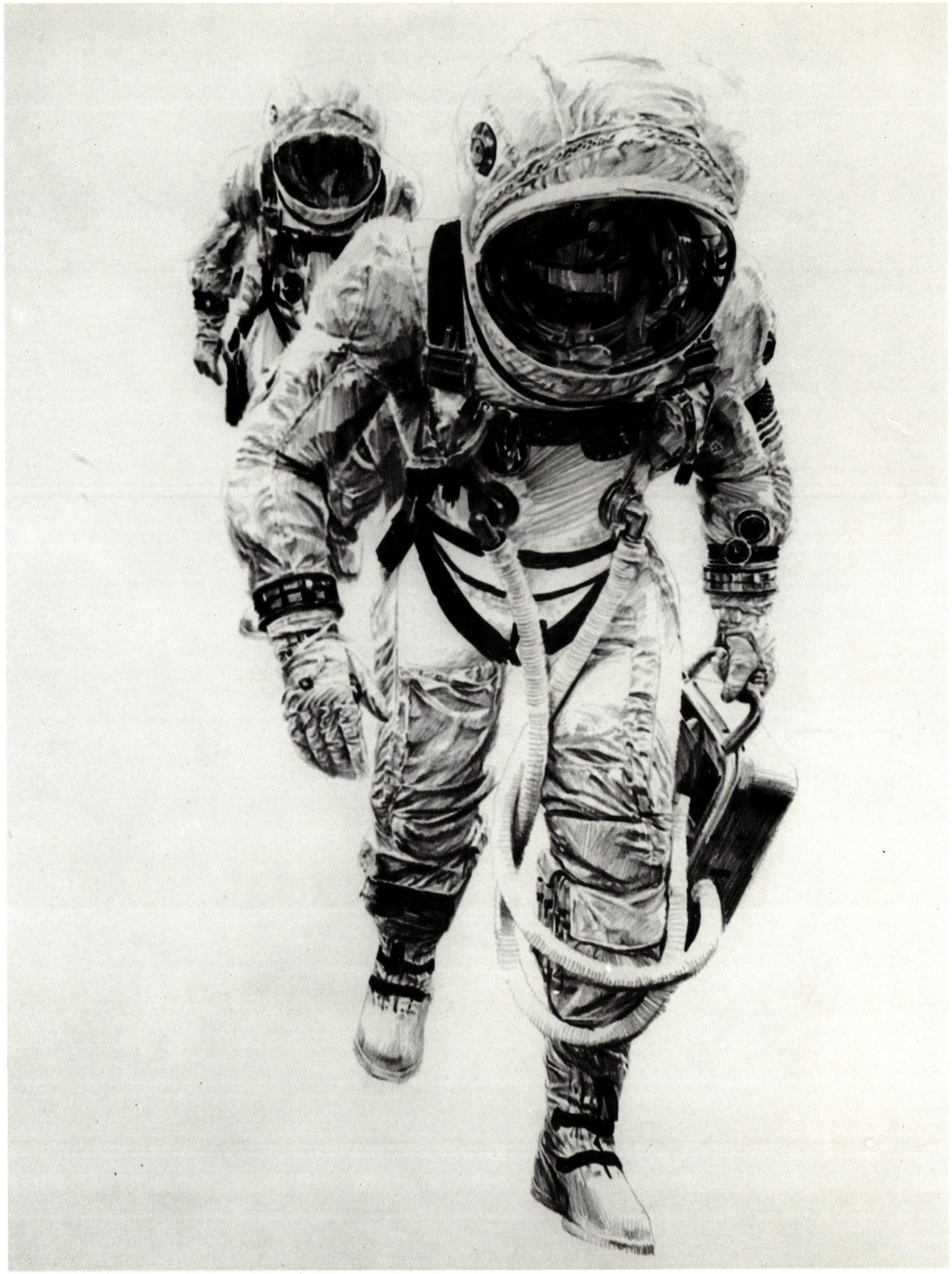

215. *Paul Calle*. GEMINI ASTRONAUTS. Pencil on paper, 22×30″

216.
James Wyeth
T MINUS 3 HOURS 30 MINUTES AND COUNTING
Watercolor on paper, 24×31″

Before dawn, on July 16, 1969, hundreds of thousands of spectators and journalists from every quarter of the world converged on Cape Kennedy. Interspersed among the batteries of cameras and looking like survivors from the Middle Ages were a group of artists. Their impressions of the great event varied enormously. This watercolor was finished the morning man first went to the moon.

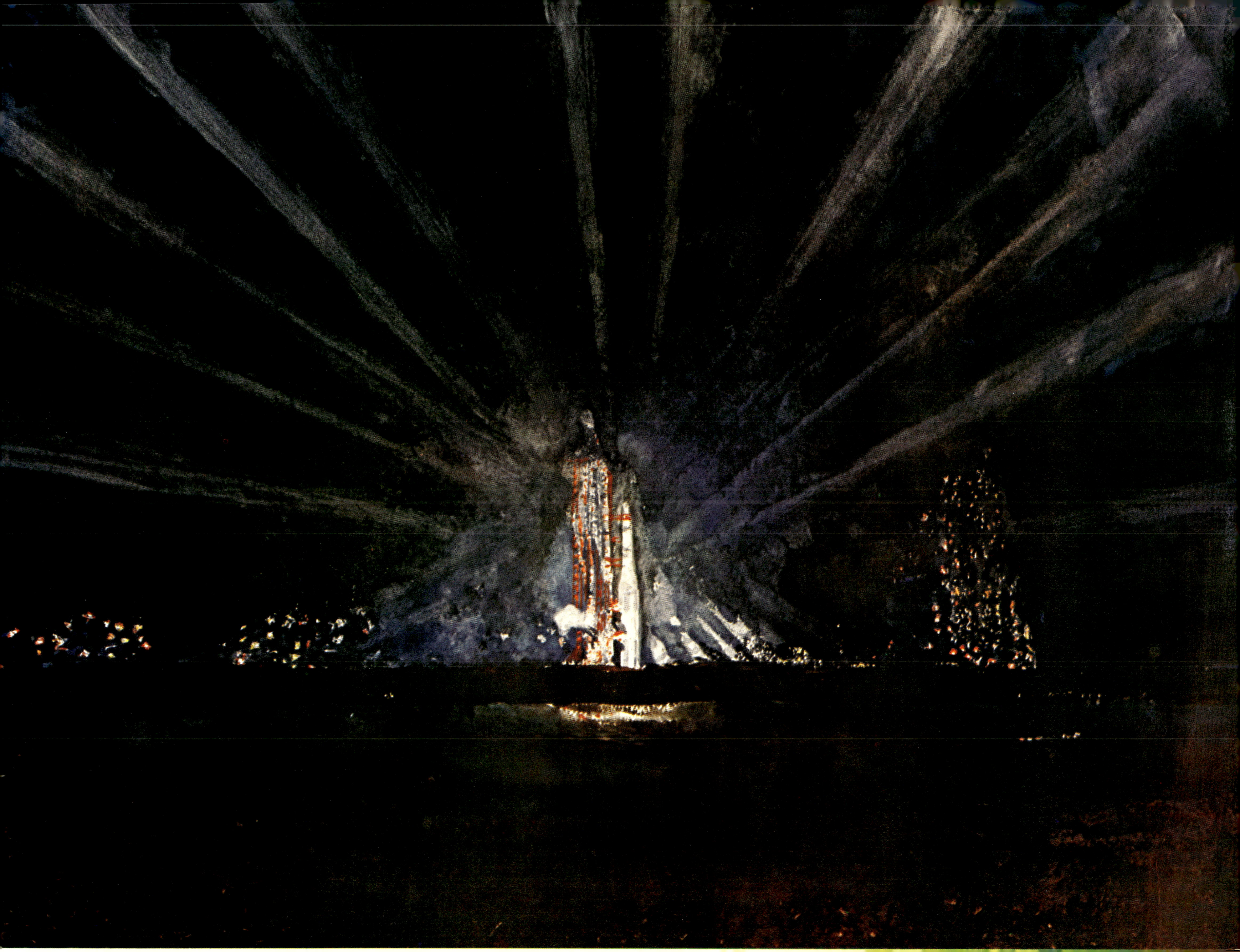

217.
Lamar Dodd
DESTINATION
Oil on canvas, 5×4′

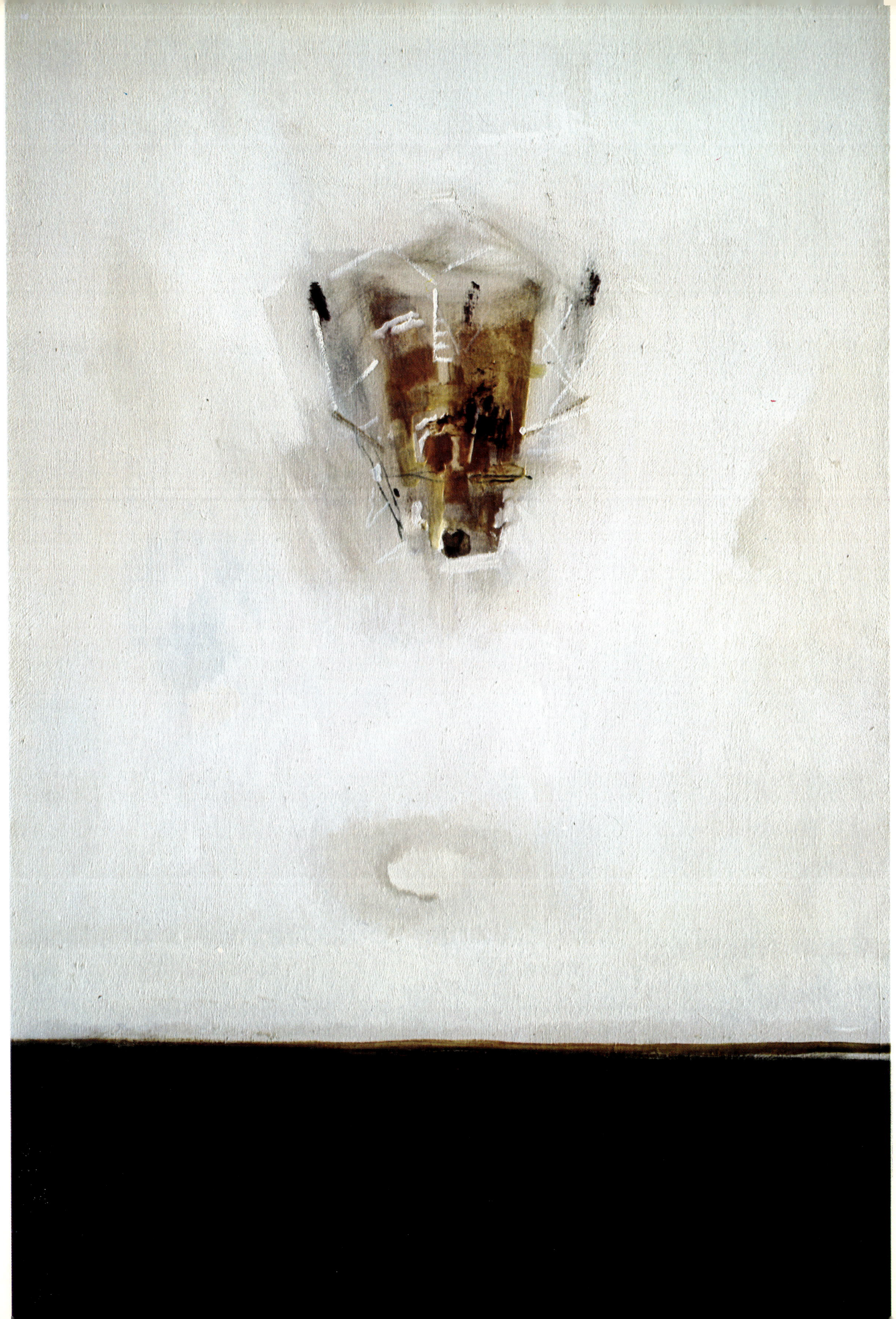

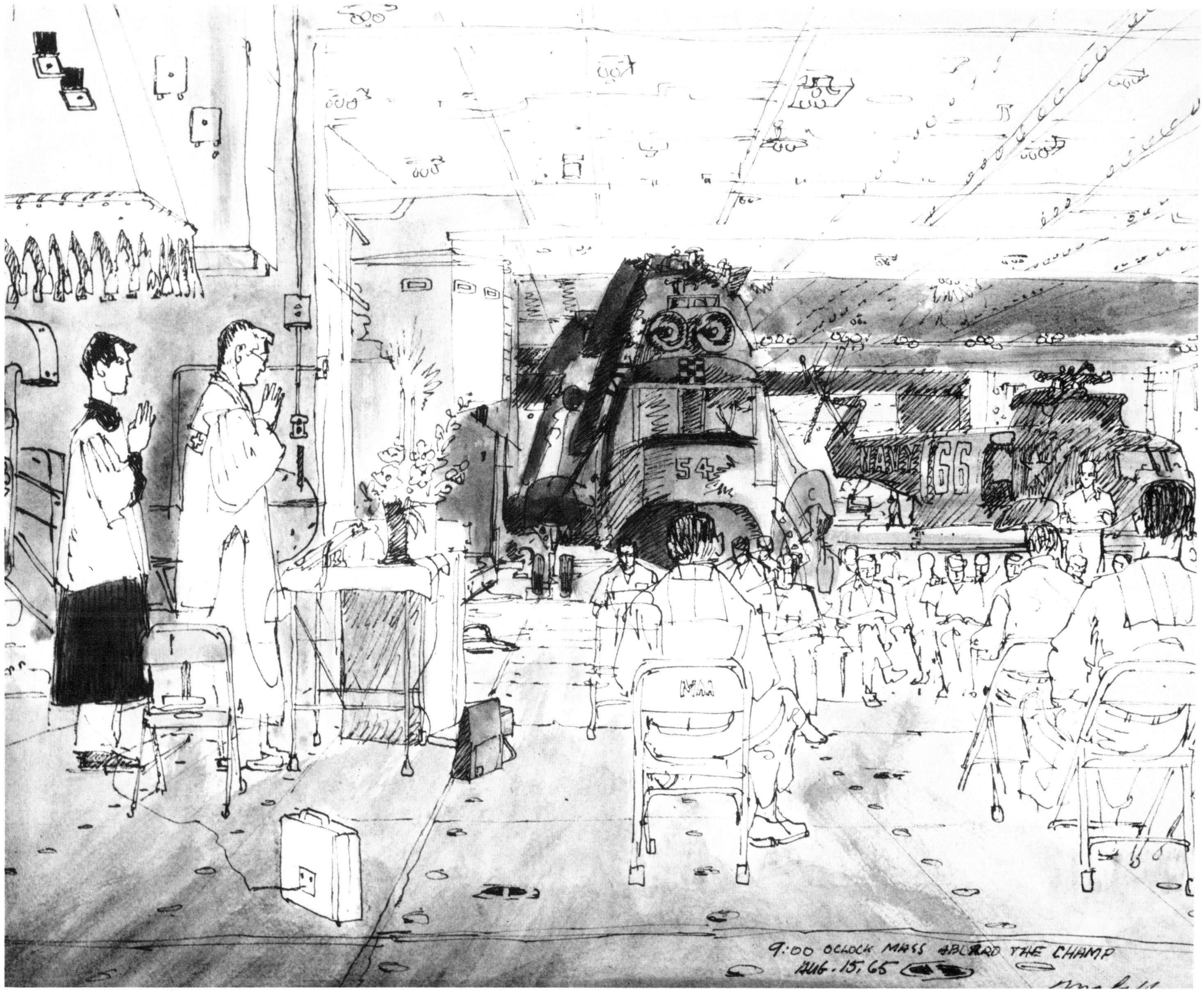

218. *Robert McCall.* SUNDAY SERVICE. Ink and wash on paper, 12×18″

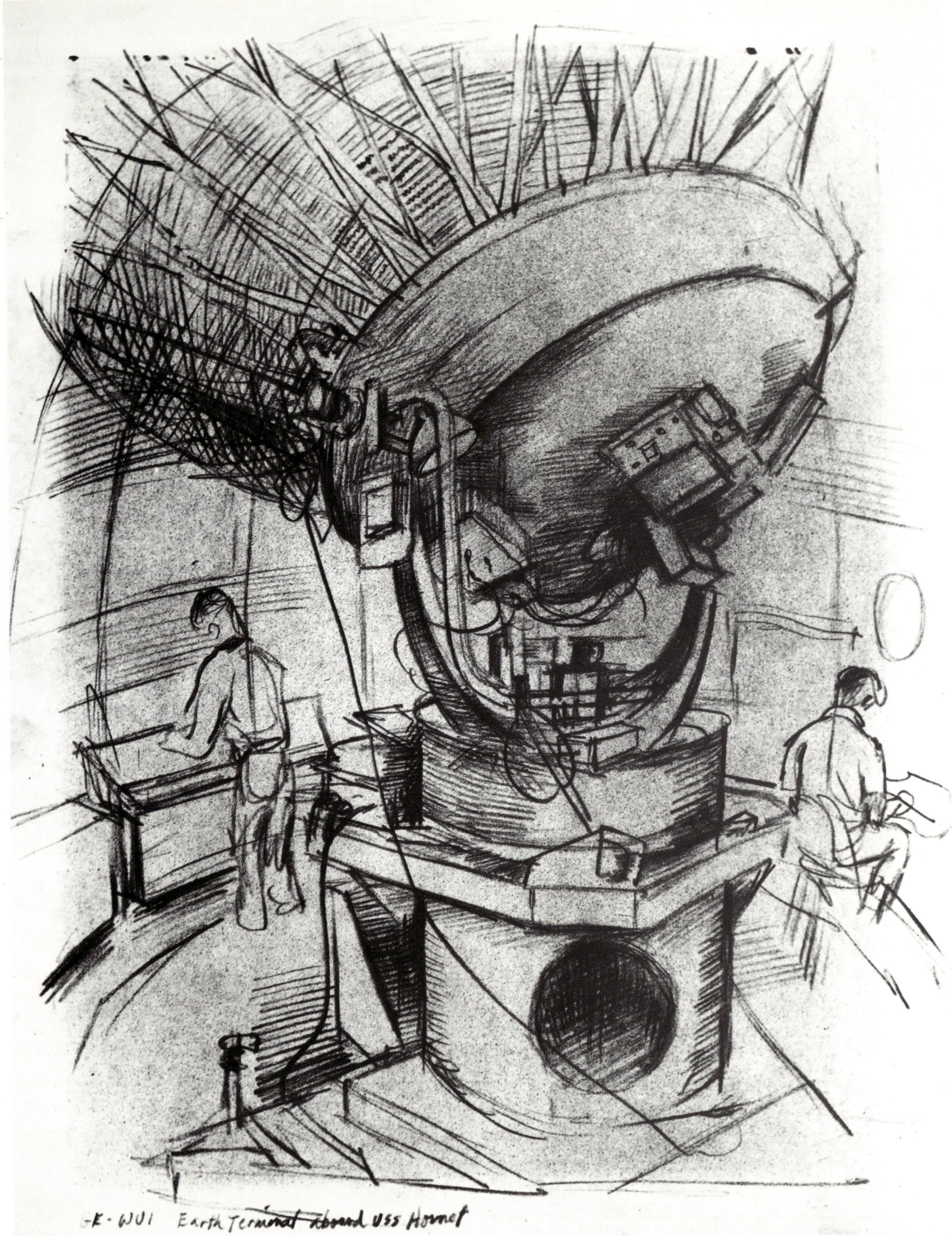

219.
Mitchell Jamieson
EARTH TERMINAL
Pencil on paper, 18×12″

220.
Julio Fernandez
APOLLO 11
Watercolor on paper, 13½ × 16″

A launch appears differently to different people. Julio Fernandez saw it as a delicate and graceful leap from the earth's surface, as the audience along the bank of the Banana River, luminescent in the morning sunlight, stands in reverent silence.

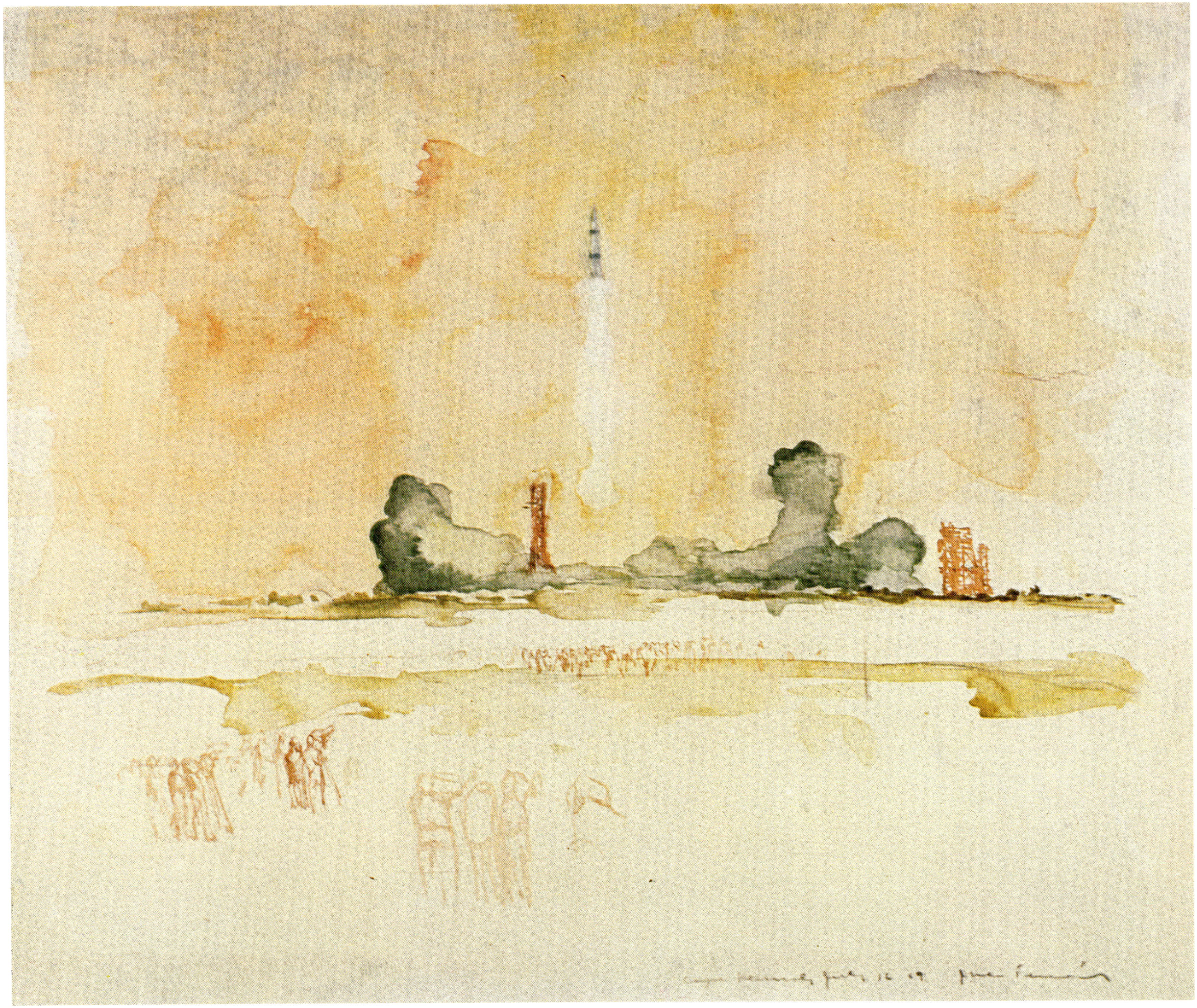

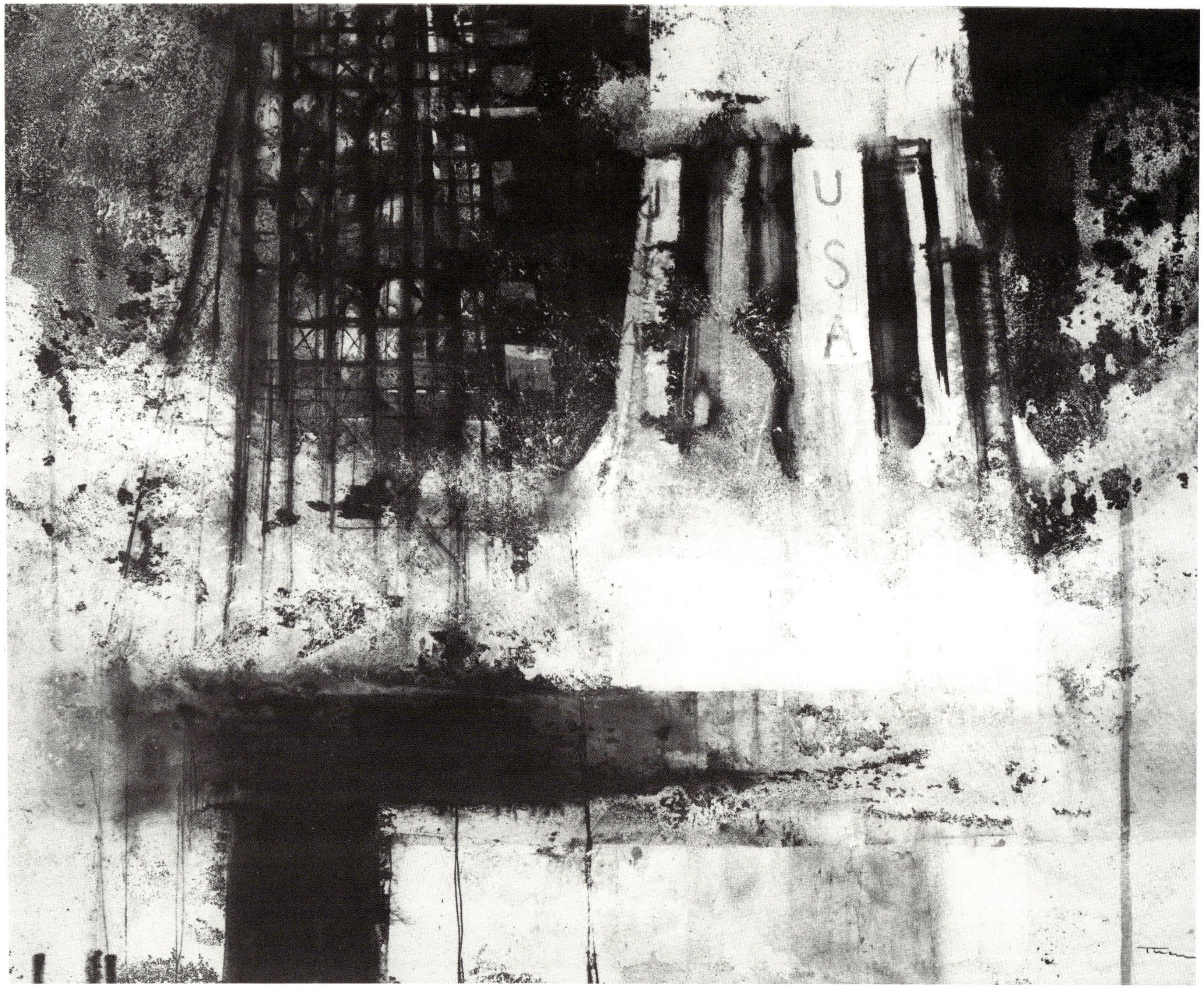

221.
William Thon
"WE HAVE LIFTOFF"
India ink heightened with white on paper, 18×23½″

William Thon captured not only the stark beauty of Apollo 11, but also the breath-holding emotions that everyone felt as the rocket, tied by invisible bonds of gravity, tore itself free from earth and inched its way upward into history.

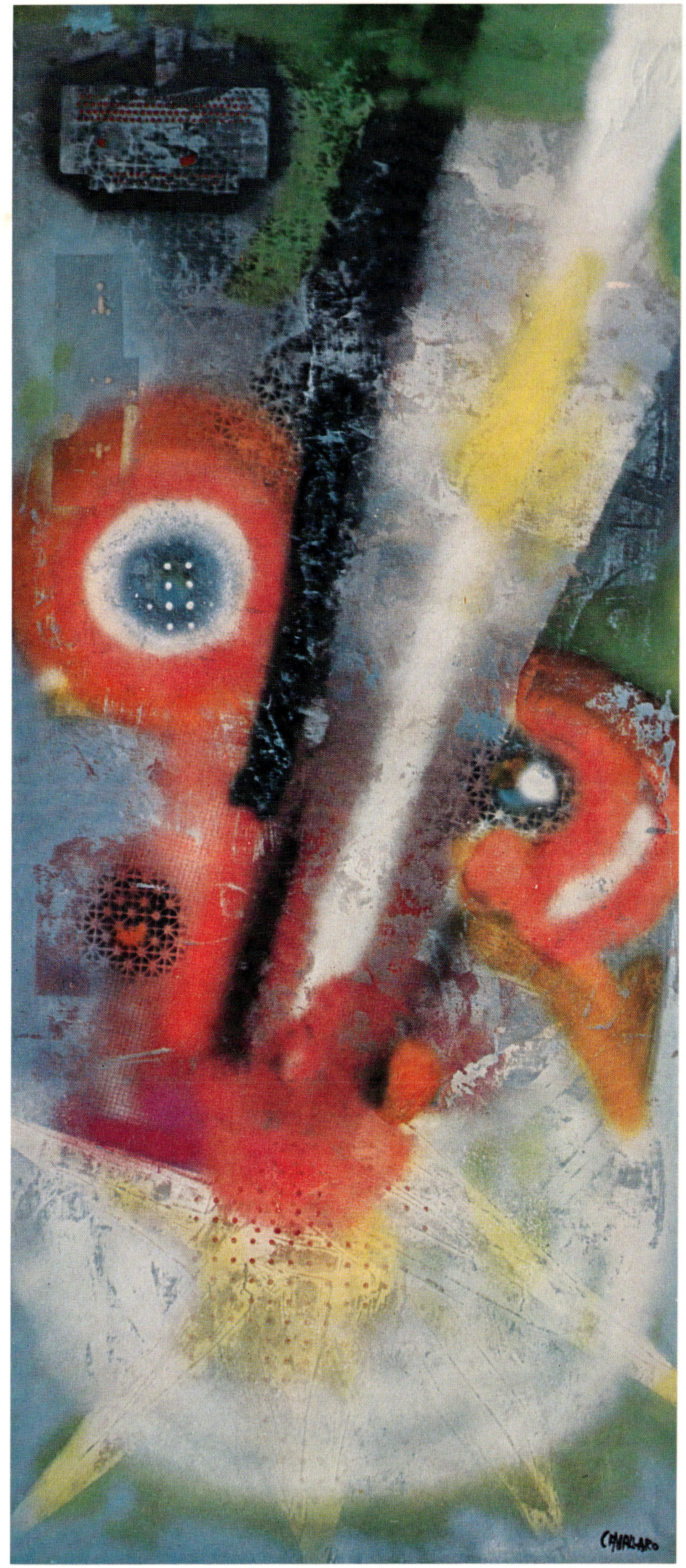

222.
Vincent Cavallaro
POWER TO GO
Oil on canvas, 80×36½″

For Cavallaro, the launch was a pyrotechnic display of majestic beauty.

224.
Franklin McMahon
LM ON THE MOON
Pencil on paper, 22×30″

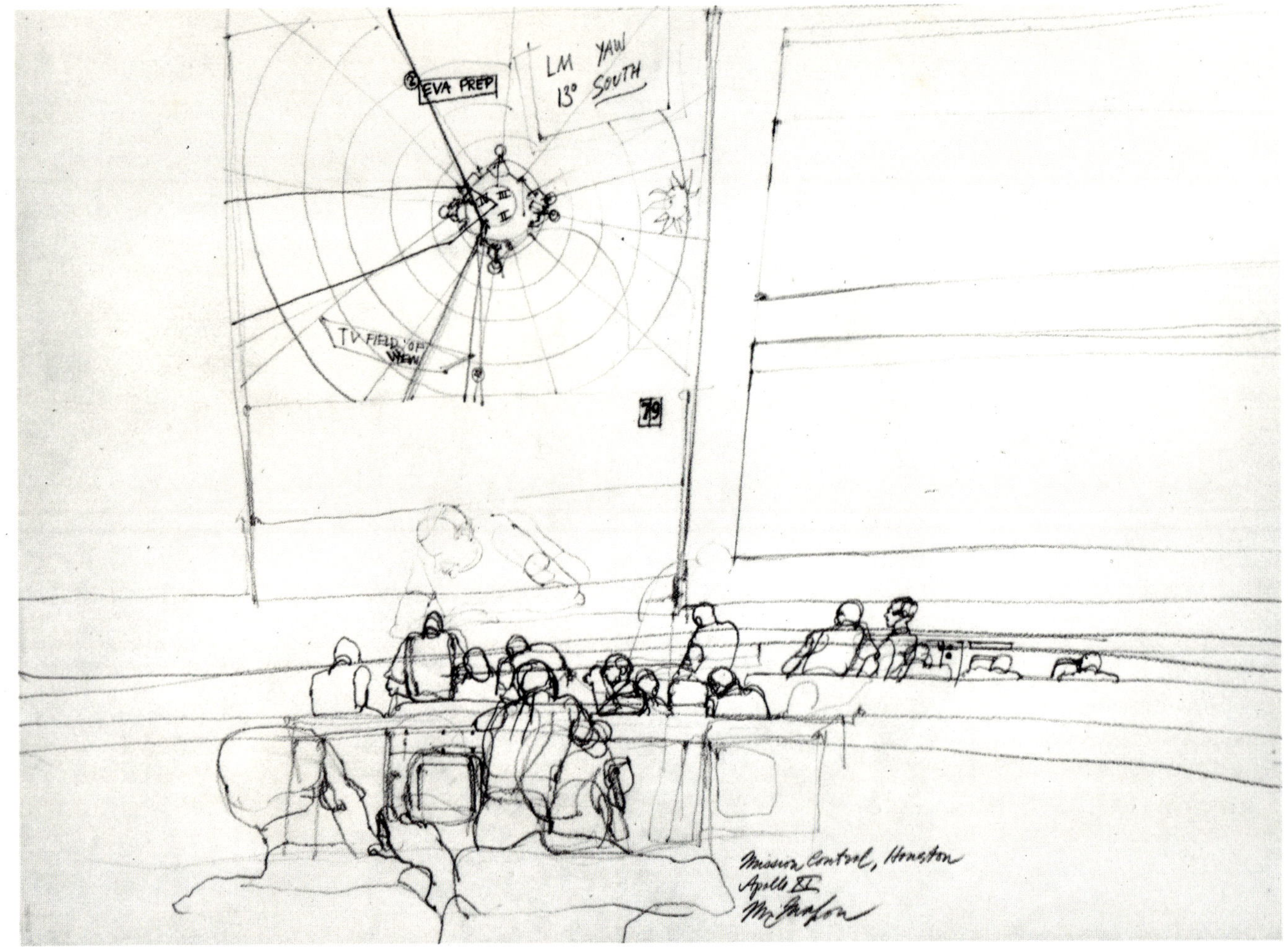

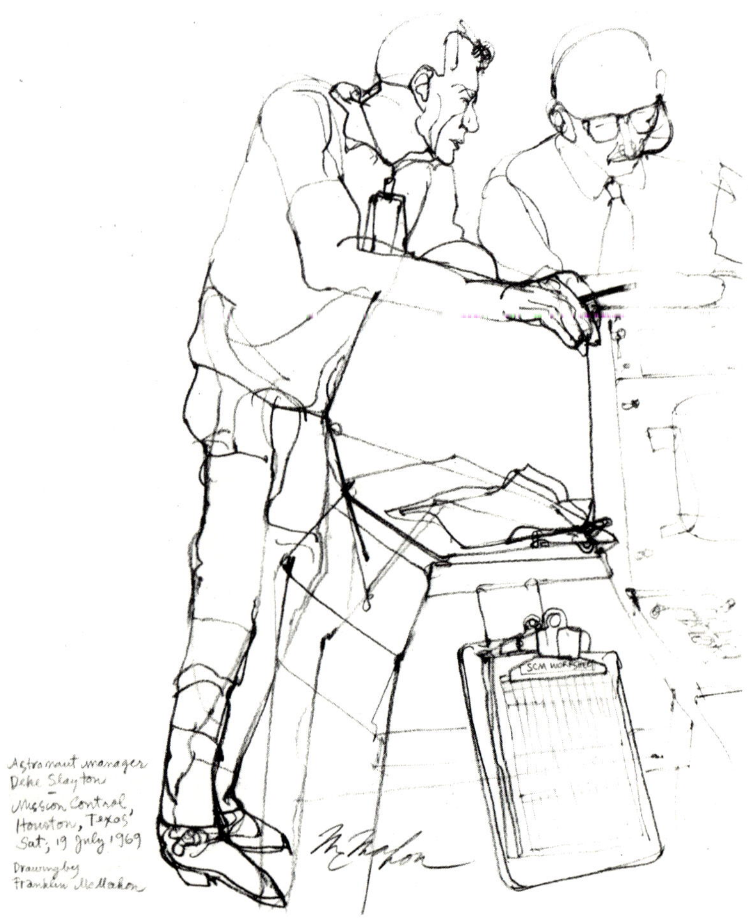

223.
Franklin McMahon
ASTRONAUT MANAGER
Pencil on paper, 17×14″

225.
Franklin McMahon
HOUSTON CONTROL
Watercolor on paper, 22½ ×30″

The space Pilgrim's Progress to the moon was registered in Houston, Texas, mostly in constantly moving graphs, columns of figures, and flashing lights. In a sense, the mission was being converted into a form of nonobjective art. The controlling crew, working in four-hour shifts, watched every heartbeat, pressure gauge, and navigation instrument on the distant LM.

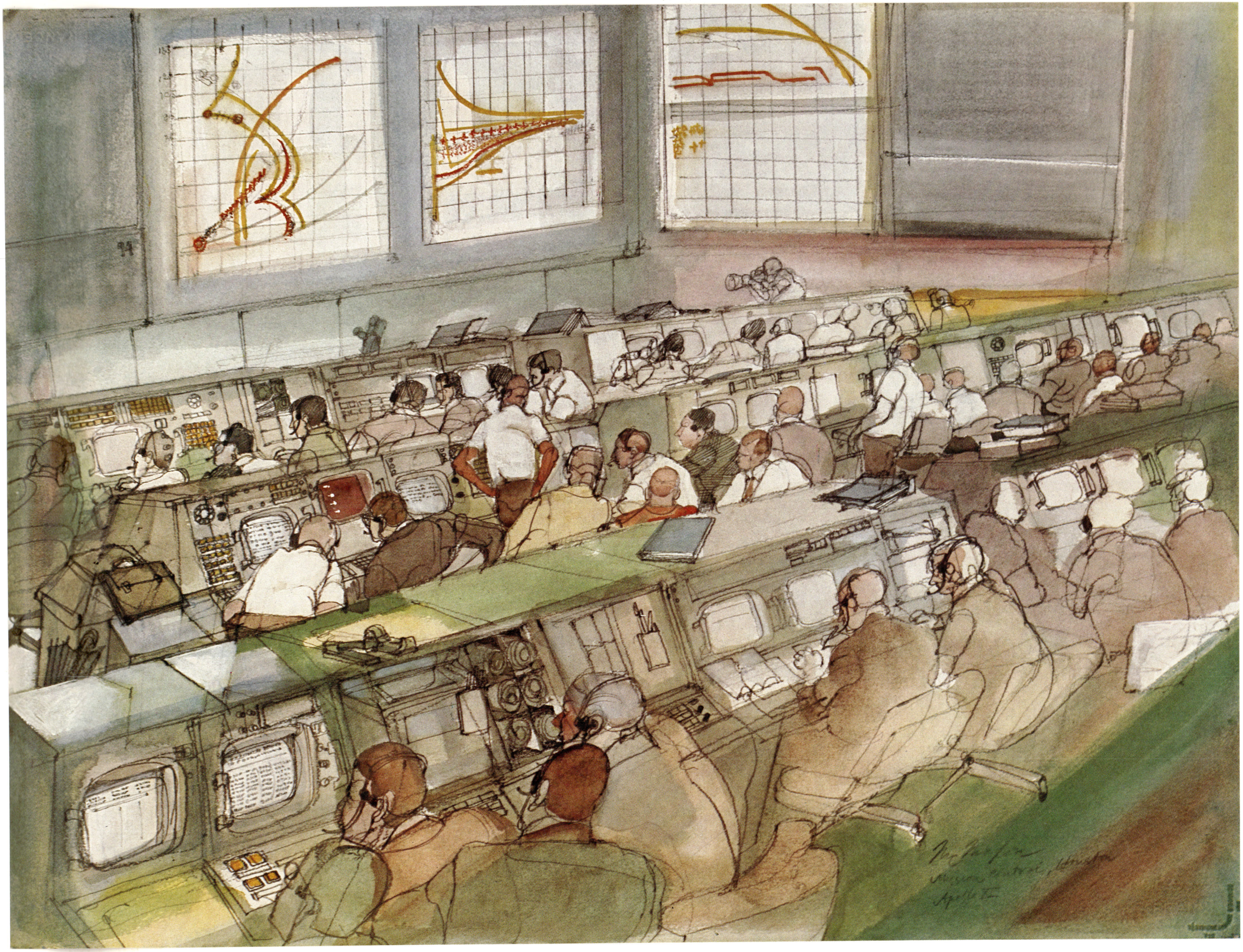

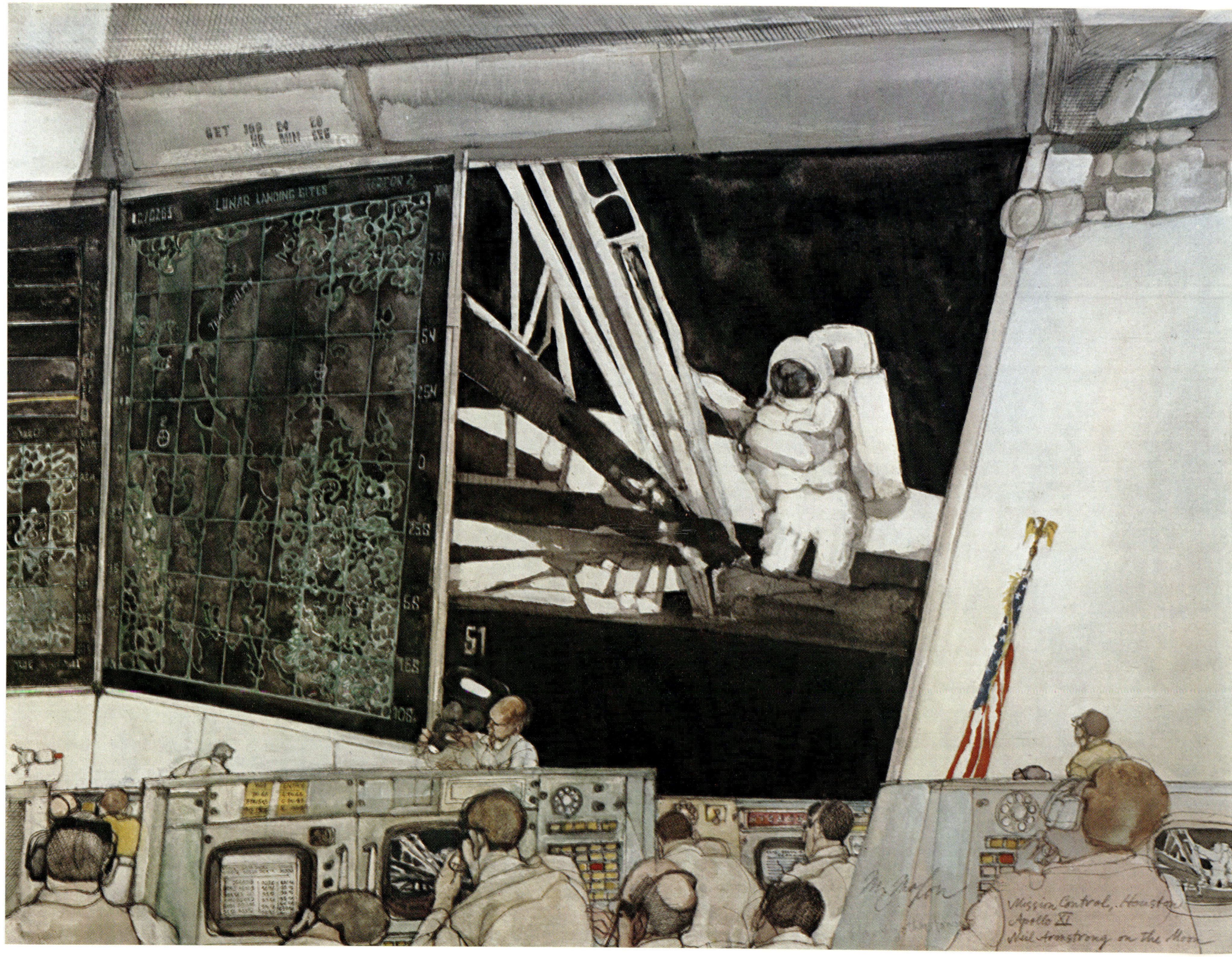

226.
Franklin McMahon
"THE EAGLE HAS LANDED"
Watercolor on paper, 22½ × 30″

At the moment when the first TV pictures of Armstrong on the moon were received in the control room at Houston, there was a ripple of applause, and then all concerned immediately returned to their tasks of monitoring, checking, and advising the crew more than a quarter of a million miles away. Next to the giant TV image were schematic diagrams of the moon, showing the presumed position of the Lunar Module on its surface and the Command Module orbiting above.

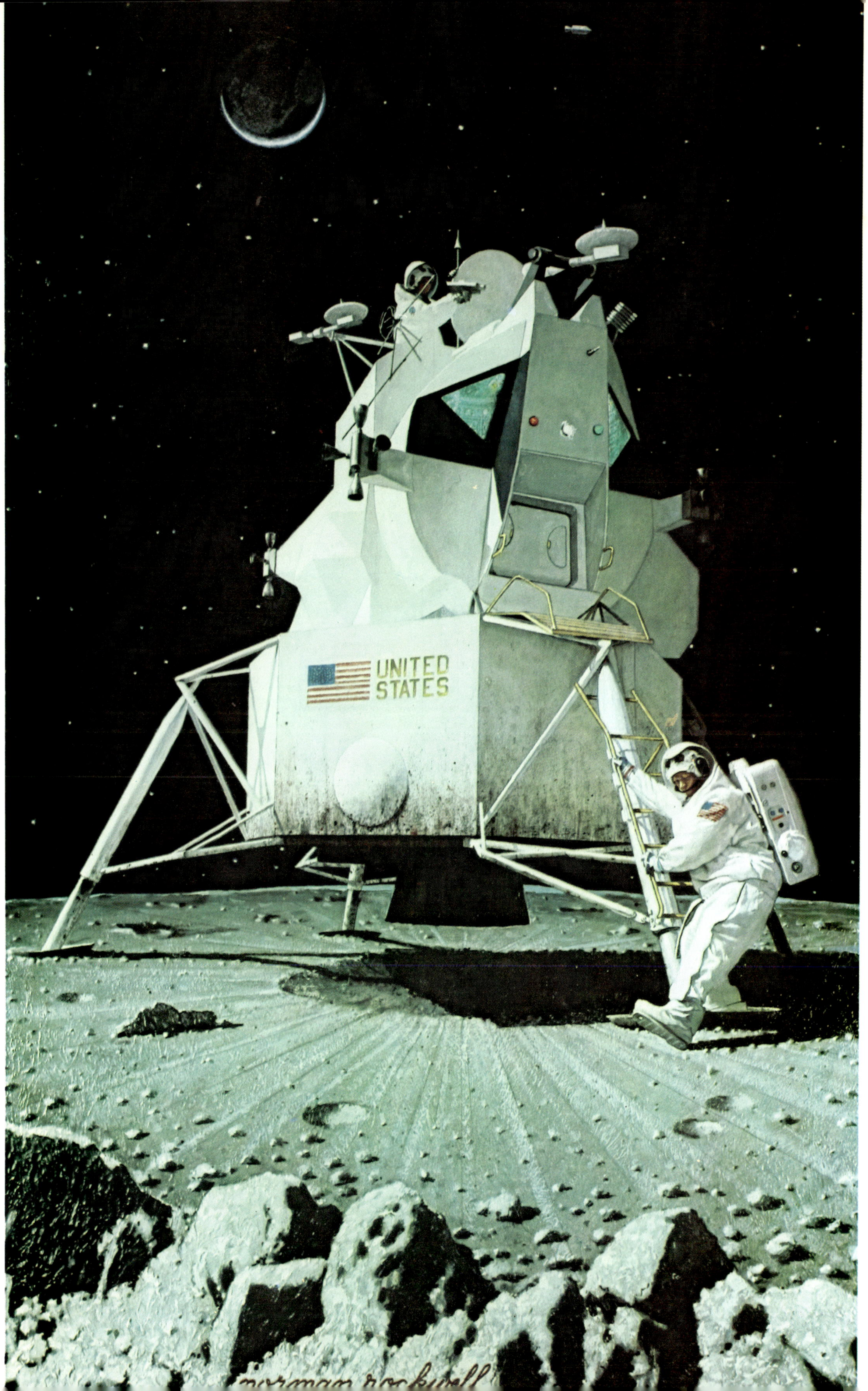

227.
Norman Rockwell
ASTRONAUTS ON THE MOON
Mixed media on canvas, 64×40½″
Reproduced courtesy of *Look* magazine,
© 1966 Cowles Communications, Inc.,
and the Air and Space Museum,
Smithsonian Institution

Many months before the first manned landing on the moon, *Look* magazine commissioned Norman Rockwell to paint the subject. His prophetic picture turned out to be 90 per cent accurate.

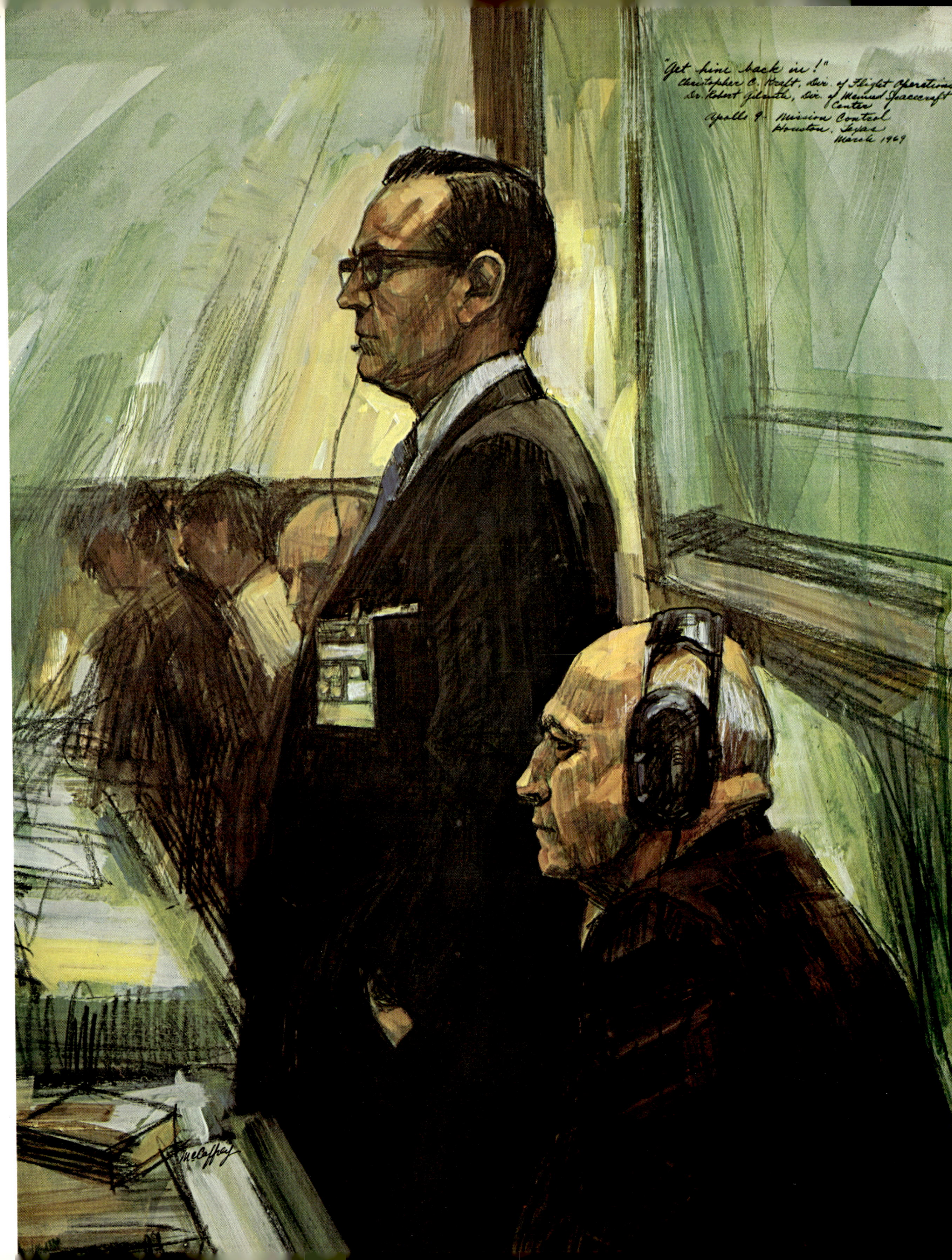

228.
M. McCaffrey
CHRIS KRAFT AND DR. GILRUTH
Mixed media on paper, 35×28″

Although rarely in the public eye, two of the most important men in the manned space flight program have been Christopher Columbus Kraft, standing, and Dr. Robert Gilruth. These studies were made in the Houston control room during the mission of Apollo 9.

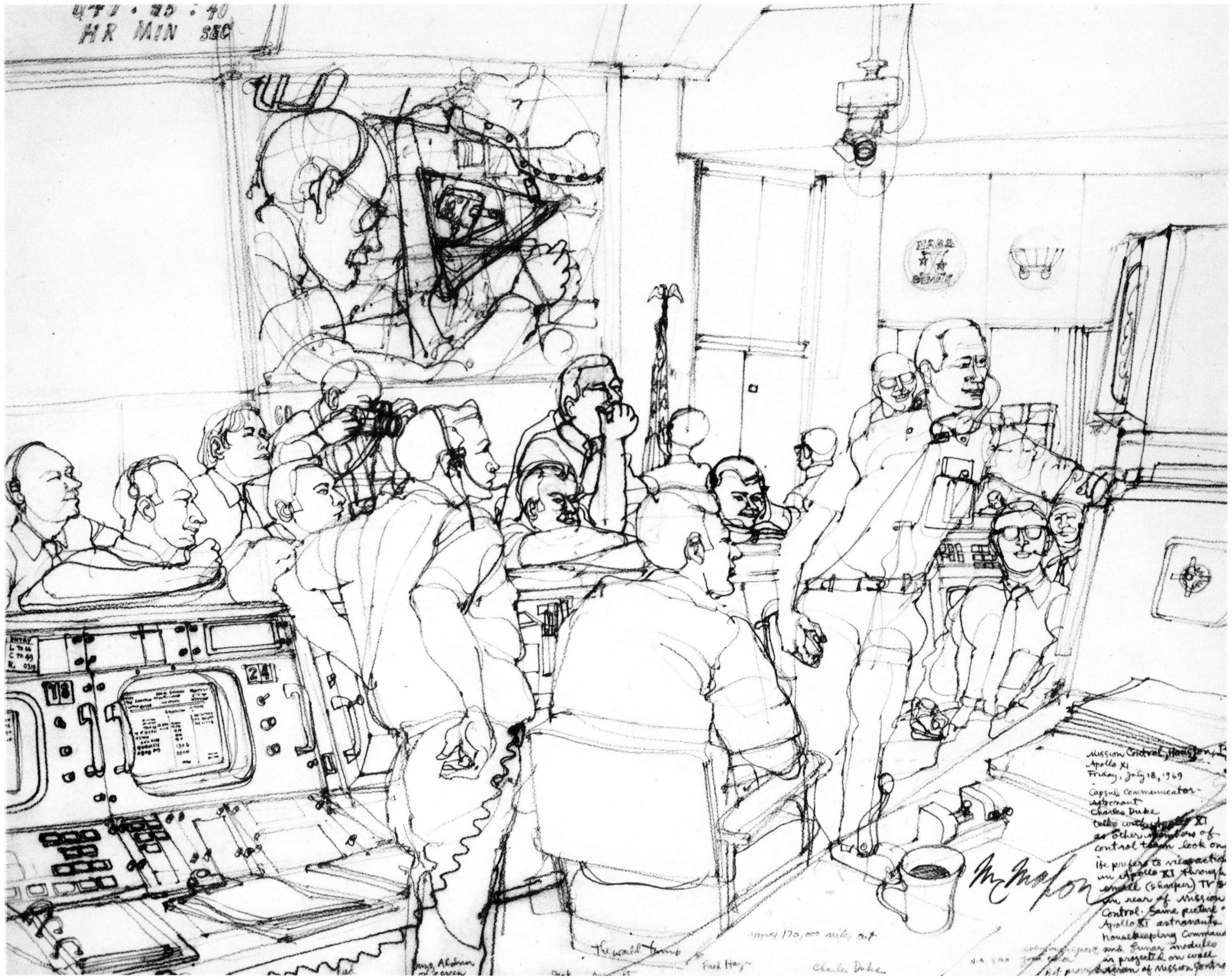

229. *Franklin McMahon.* SPACE CONVERSATION. Pencil on paper, 22×30″

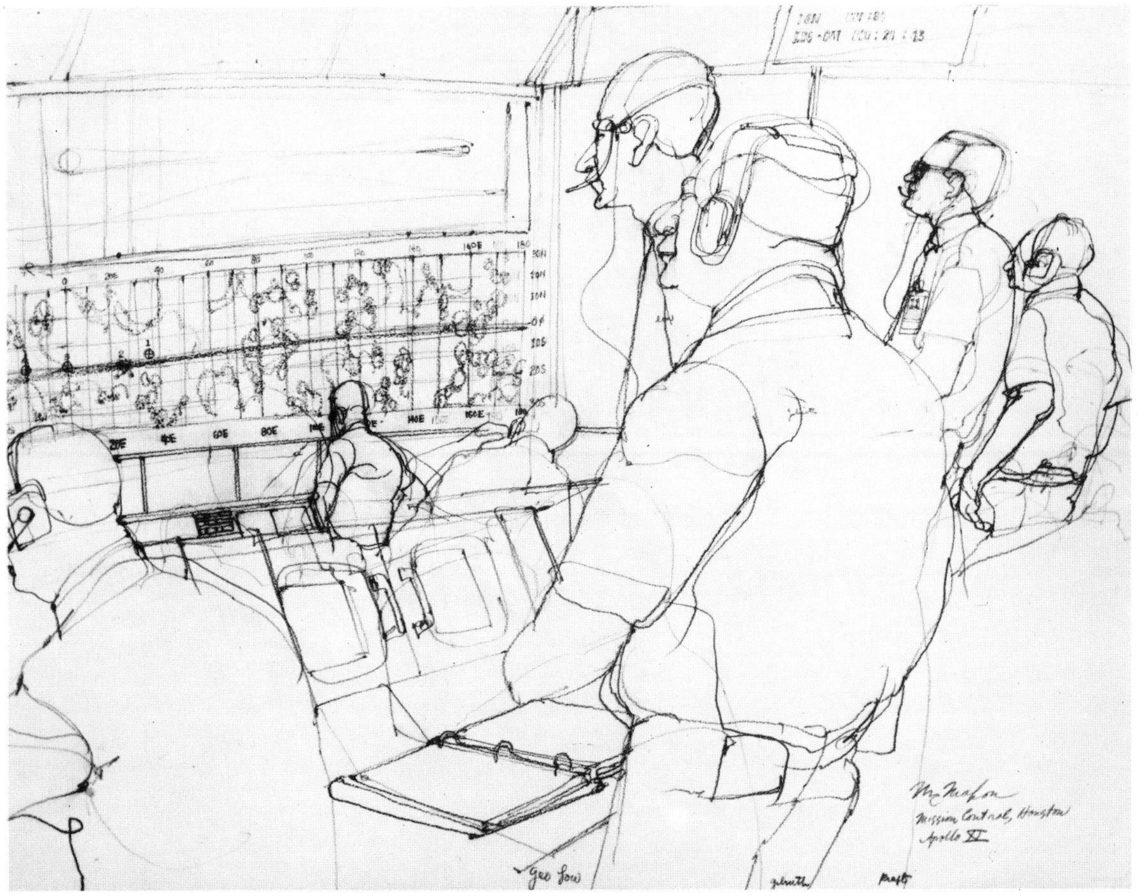

230. *Franklin McMahon.* ADVICE FOR A MAN ON THE MOON. Pencil on paper, 22×30″

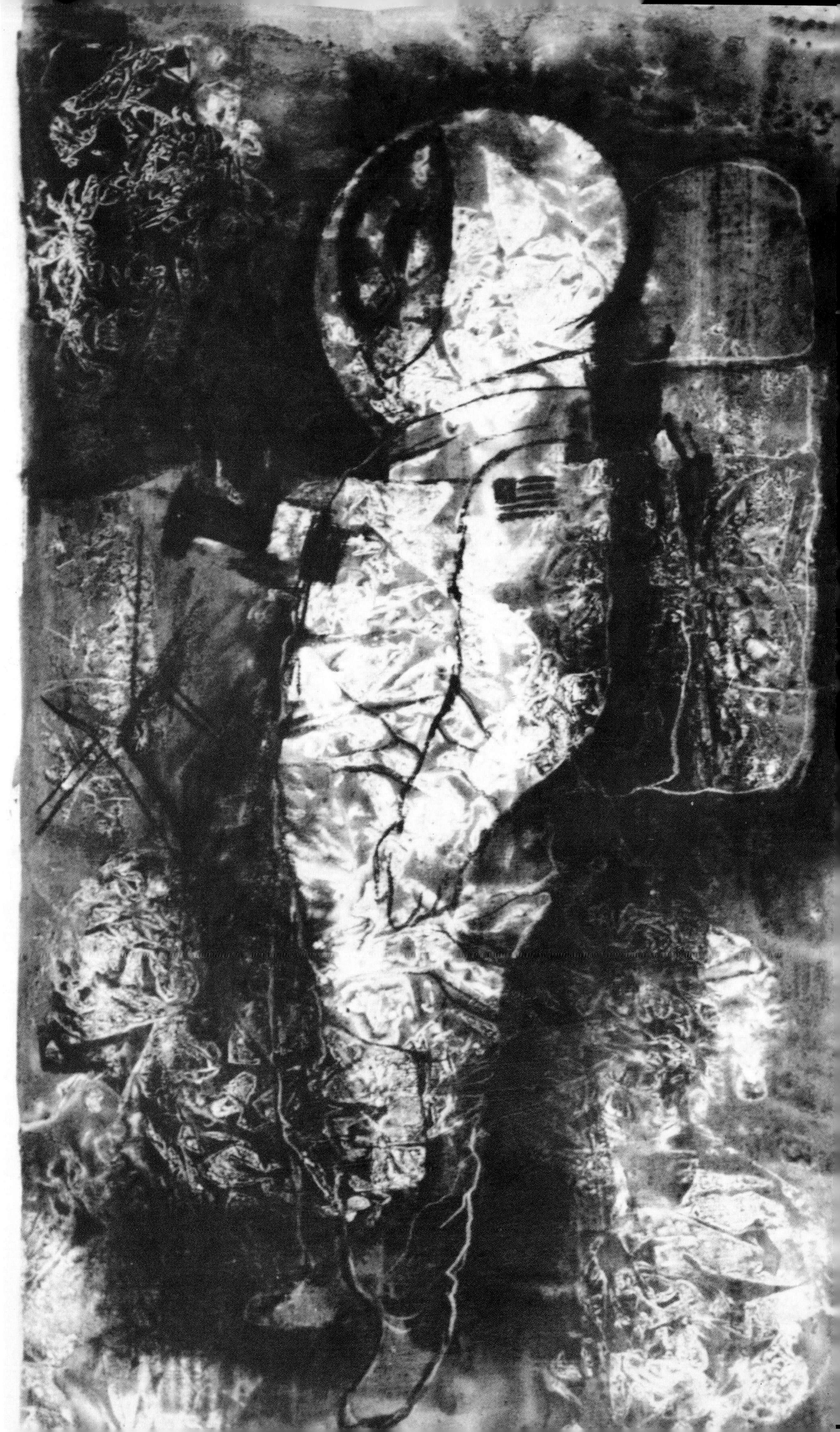

231.
Mitchell Jamieson
ON THE MOON
Ink on paper, 18×12″

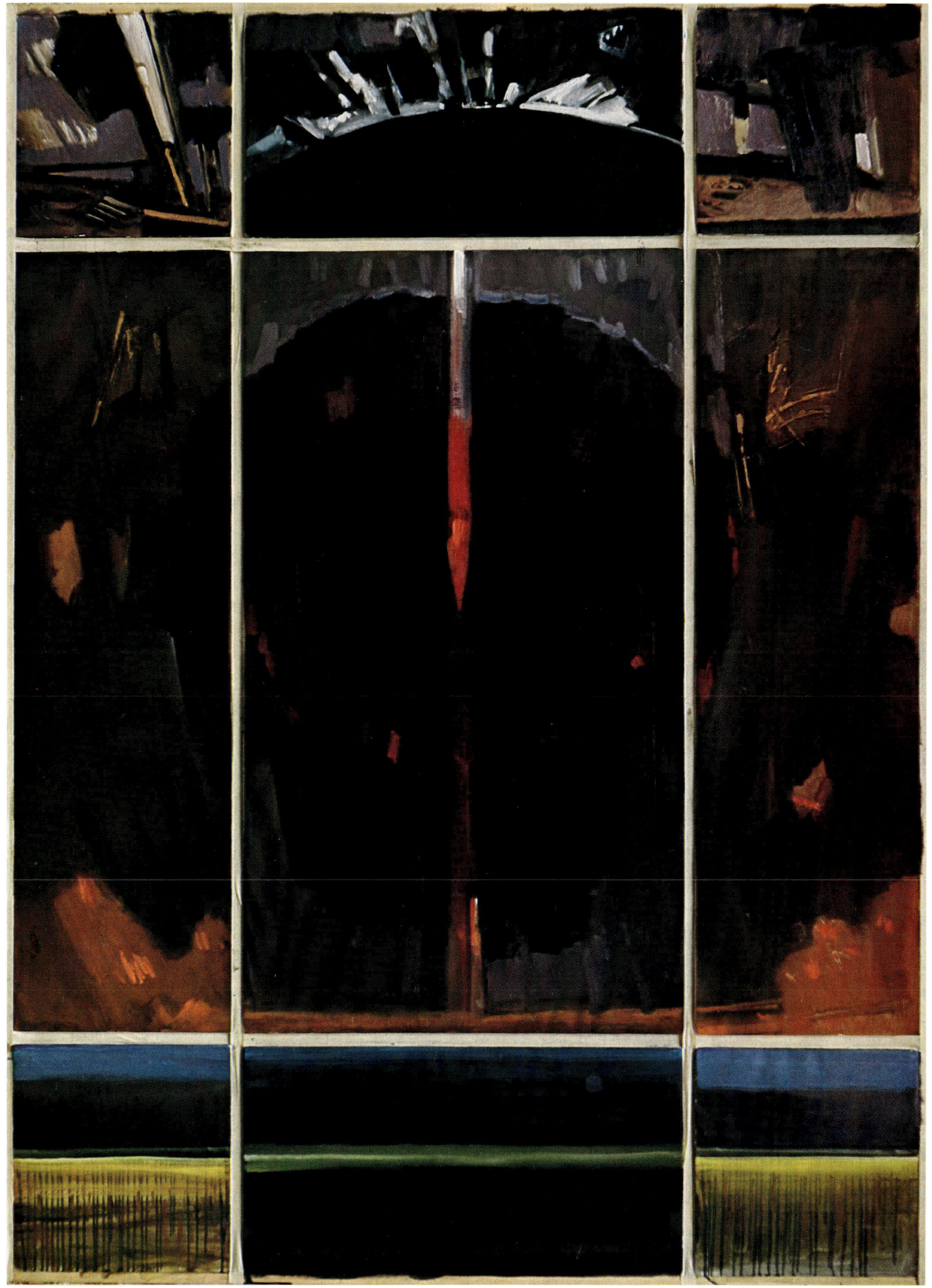

232.
Lamar Dodd
THE COSMOS AWAITS
Oil on canvas, 8×6′

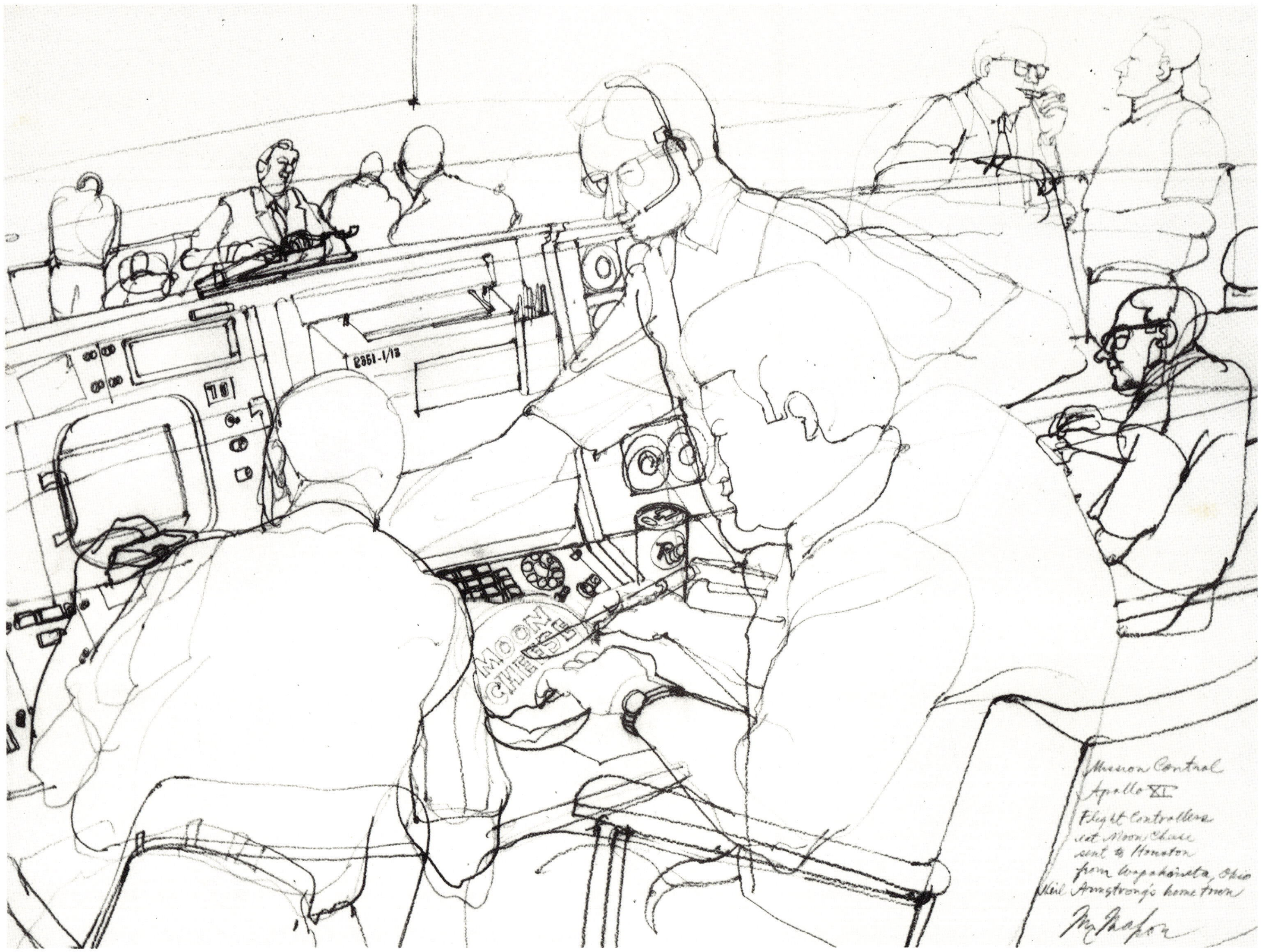

233.
Franklin McMahon
HOME TOWN CHEESE
Pencil on paper, 22×30″

One of the minor moments in history was when a moon cheese, specially made in Wapakoneta, Ohio, Apollo 11 astronaut Neil Armstrong's home town, was ceremoniously eaten by members of the launch crew at the Houston Control Center.

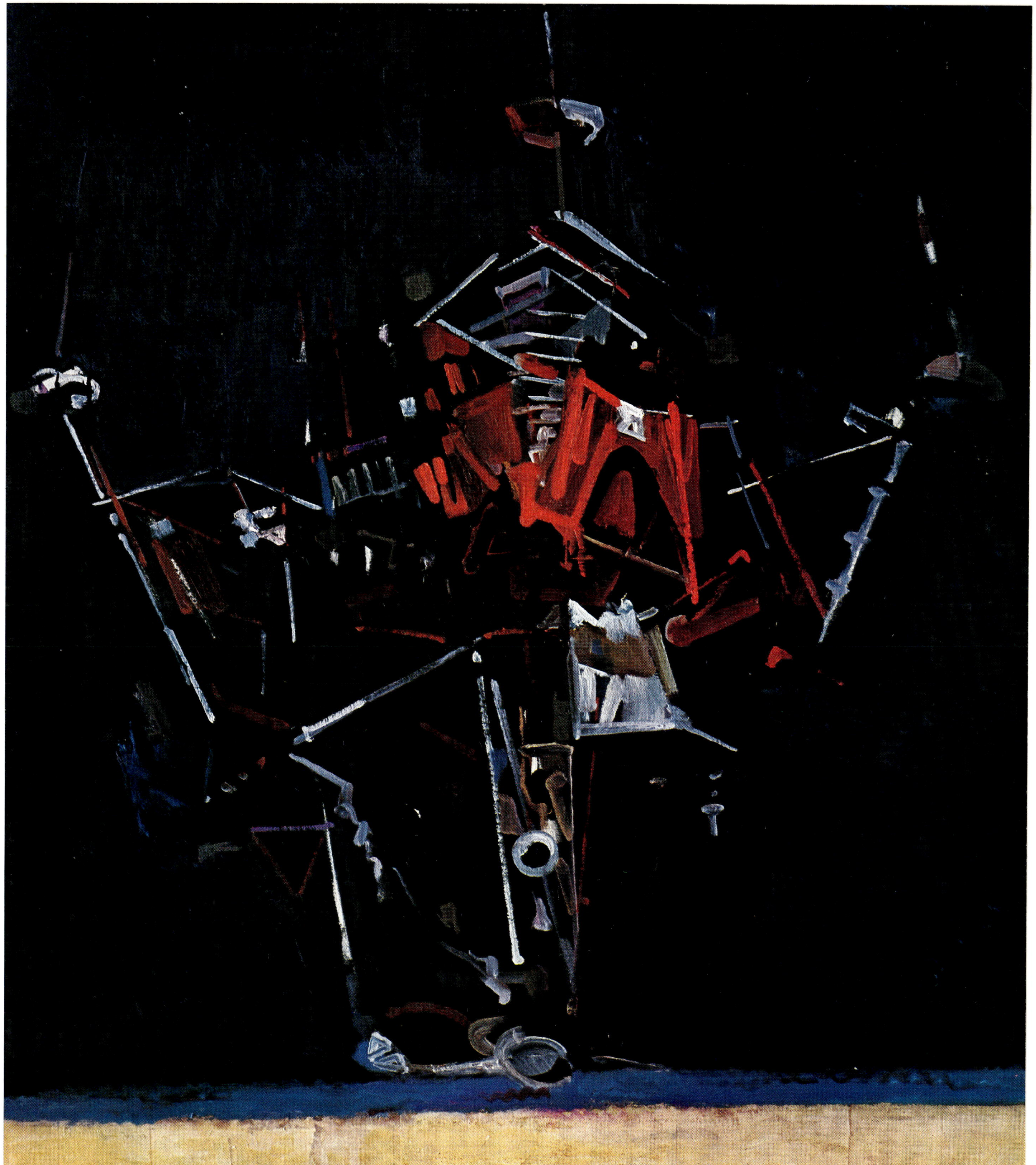

234.
Lamar Dodd
LUNAR MODULE
Oil on canvas, 4×4′

Only a privileged few astronauts have ever seen the LM, or Lunar Module, in flight. It is a spidery, fragile creation resembling nothing that has ever flown, swam, rolled, or walked before. While in space and prepared for landing, it appears to have no top, bottom, front, or back. The LM perhaps more than any other invention represents the most adventurous technical thinking of our times.

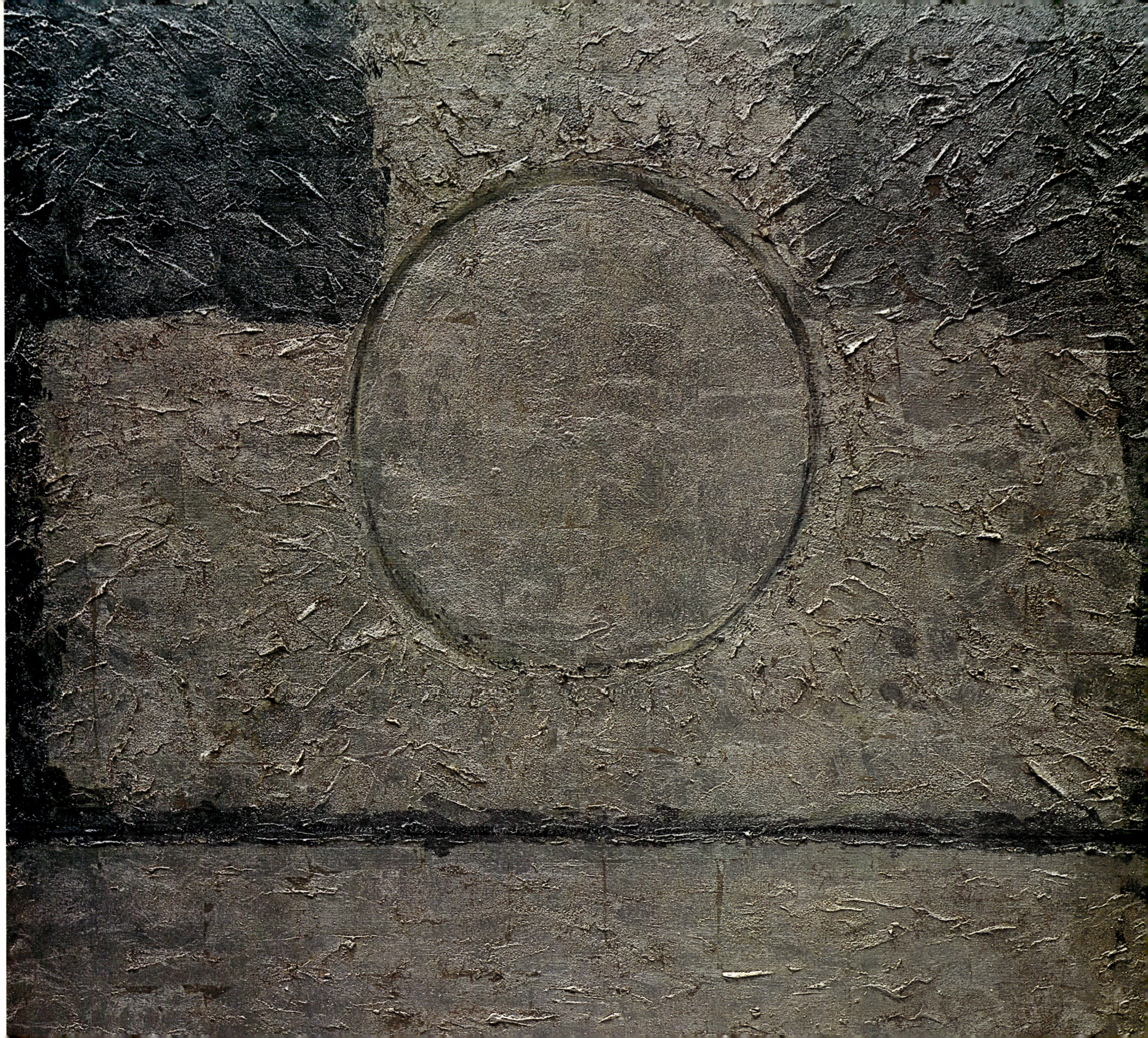

235.
Lamar Dodd
CELESTIAL BODY
Oil on canvas, 4×5′

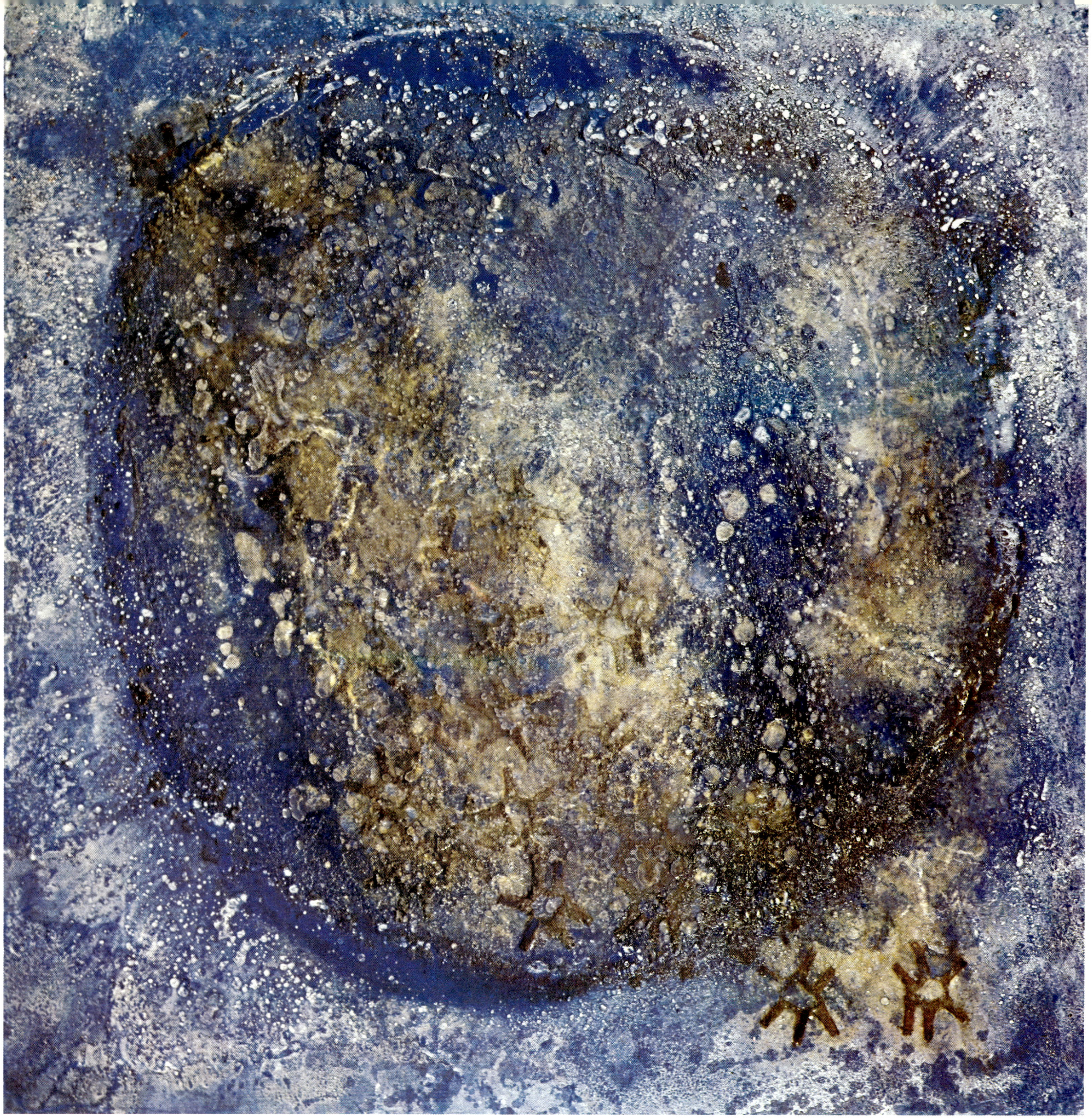

236.
Yeffe Kimball
COPERNICUS
Acrylic on canvas, 5×5′

One of the spin-off advantages of NASA's art program is that the dividing line between so-called realistic art and so-called nonobjective art is blurred and virtually eliminated. By the standards of yesterday, a picture like Yeffe Kimball's *Copernicus* would rank as an abstract work of art. By Space Age standards, it may be a surprisingly accurate rendition of what man may find on the moon. The realm of outer space may have in store for us planetscapes that will be like cosmic revelations to everyone except the artist, who always precedes the scientist.

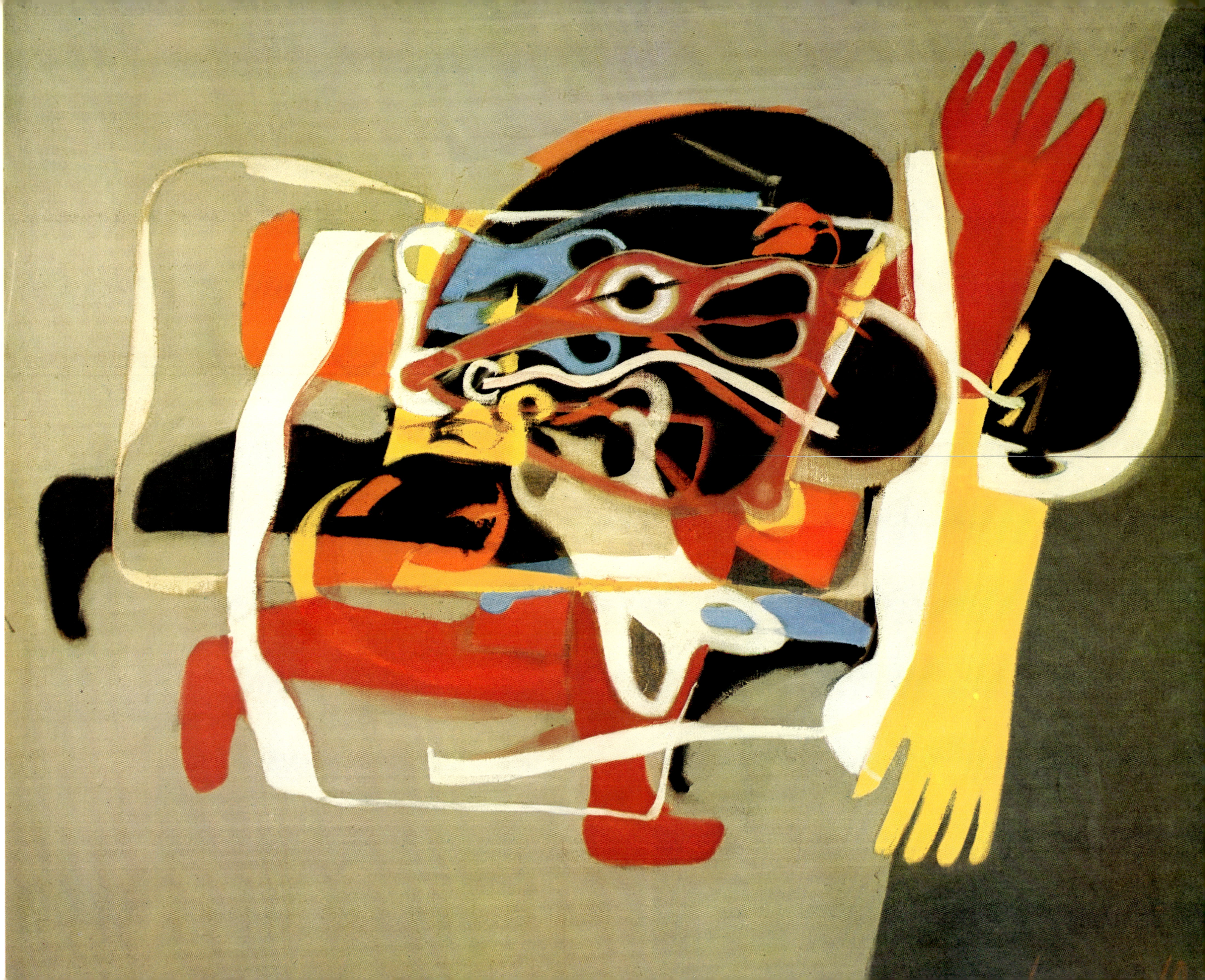

237.
Hans Cremers
METAMORPHOSES
Acrylic on canvas, 39×47″

A theme that several artists have taken is the change in man himself that occurs when he breaks his bonds with earth. In a sense, man in space loses his entity to his monitors on the ground, where he is a pulse, a pair of lungs, eyes, ears, each watched separately and charted by a different specialist at the Control Center. Cremers, from Düsseldorf, Germany, has specialized in space themes and has made the metamorphoses of the human being the subject of a series of paintings.

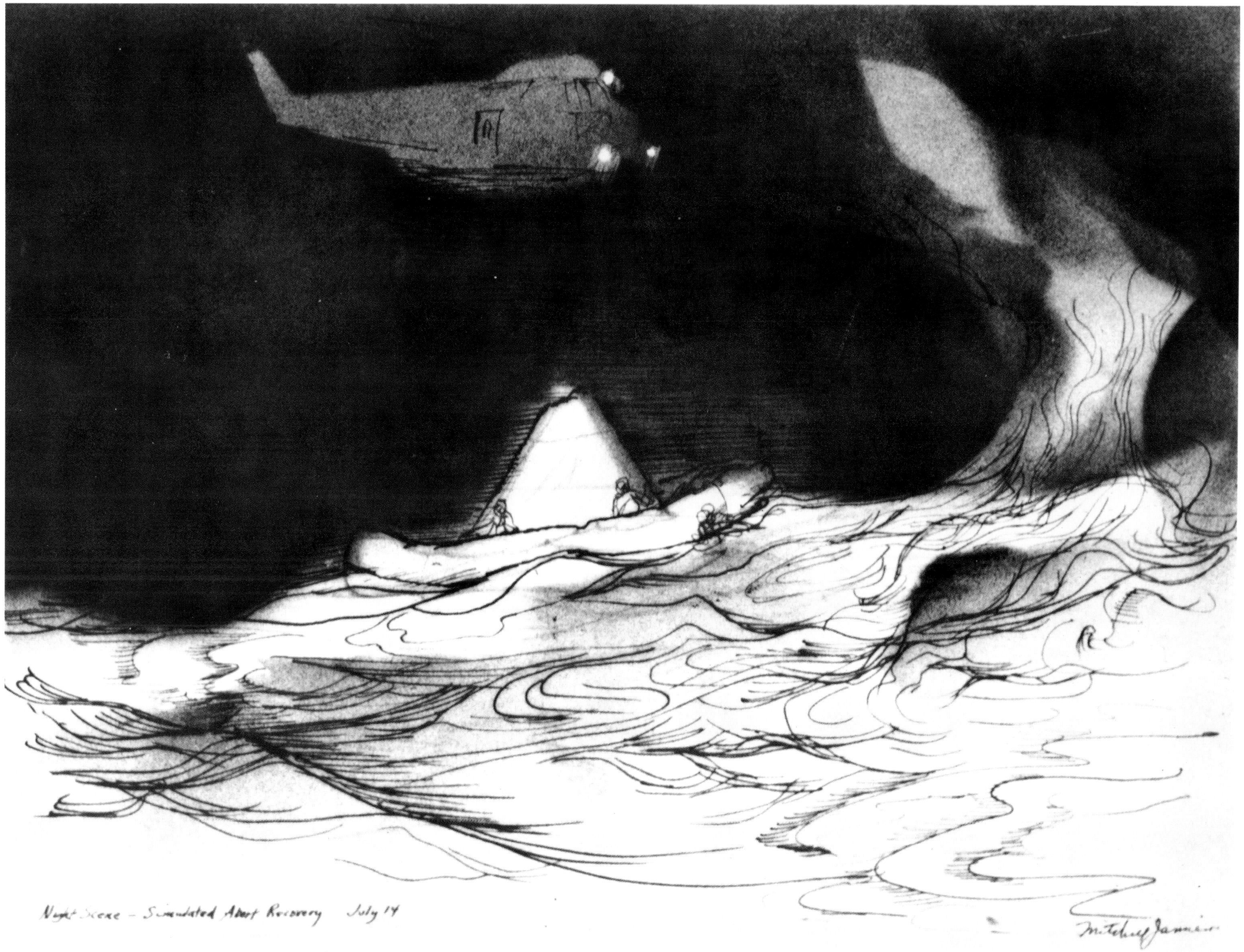

238. *Mitchell Jamieson*. PREDAWN PICKUP. Ink on paper, 12×18″

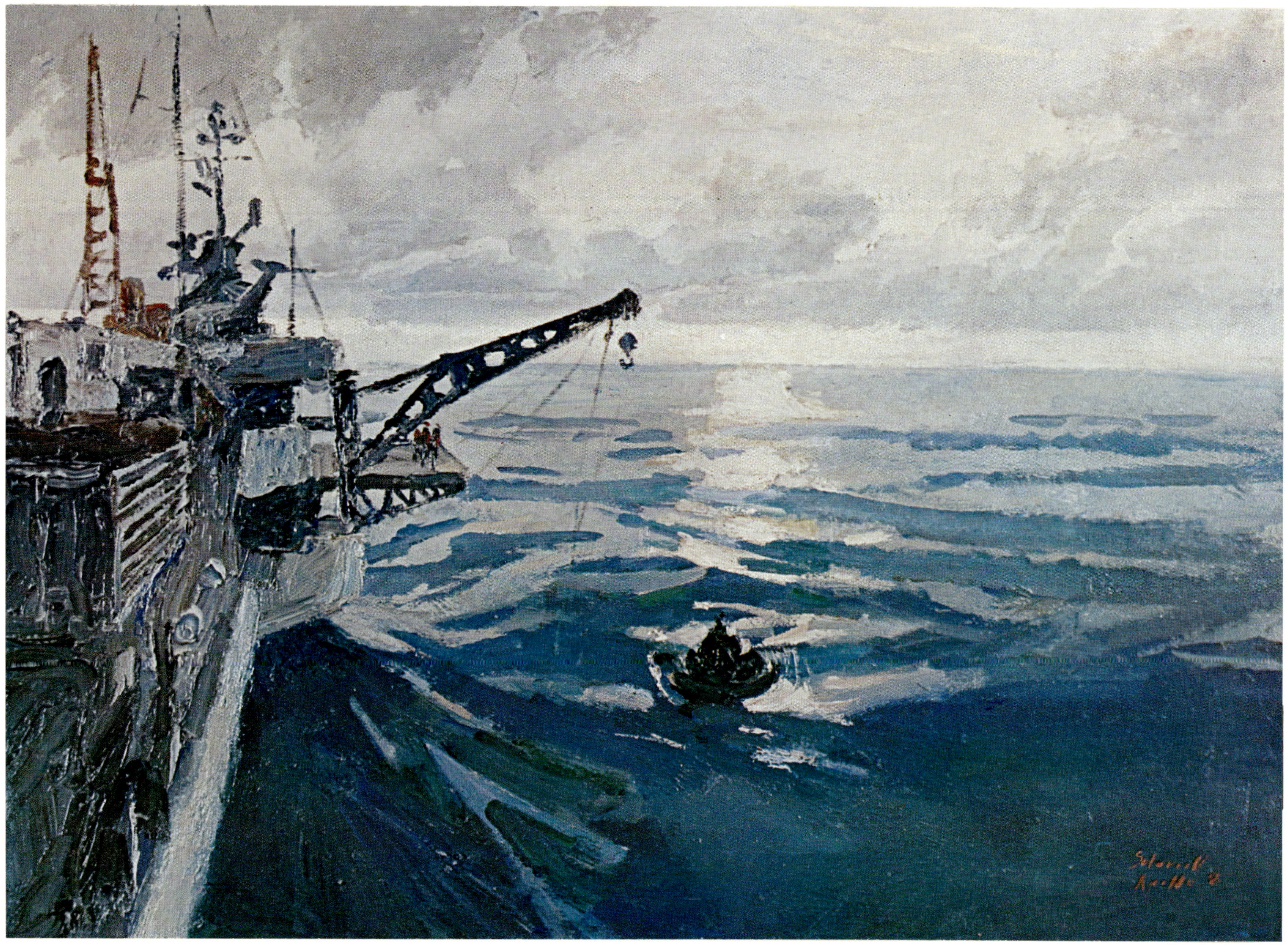

239.
Nicholas Solovioff
APOLLO 8 RETURNED
Oil on canvas, 14×20″

After the human cargo has been picked up by helicopters, the spacecraft becomes a scientific research problem, nursed by frogmen and supported by flotation rafts. The oily swells of the Pacific reflect the rising sun as the carrier maneuvers into position to hoist the Apollo 8 on board. Solovioff, well aware of the historic moment, has not forgotten to admire the beauty of the water reflecting the rosy-fingered dawn.

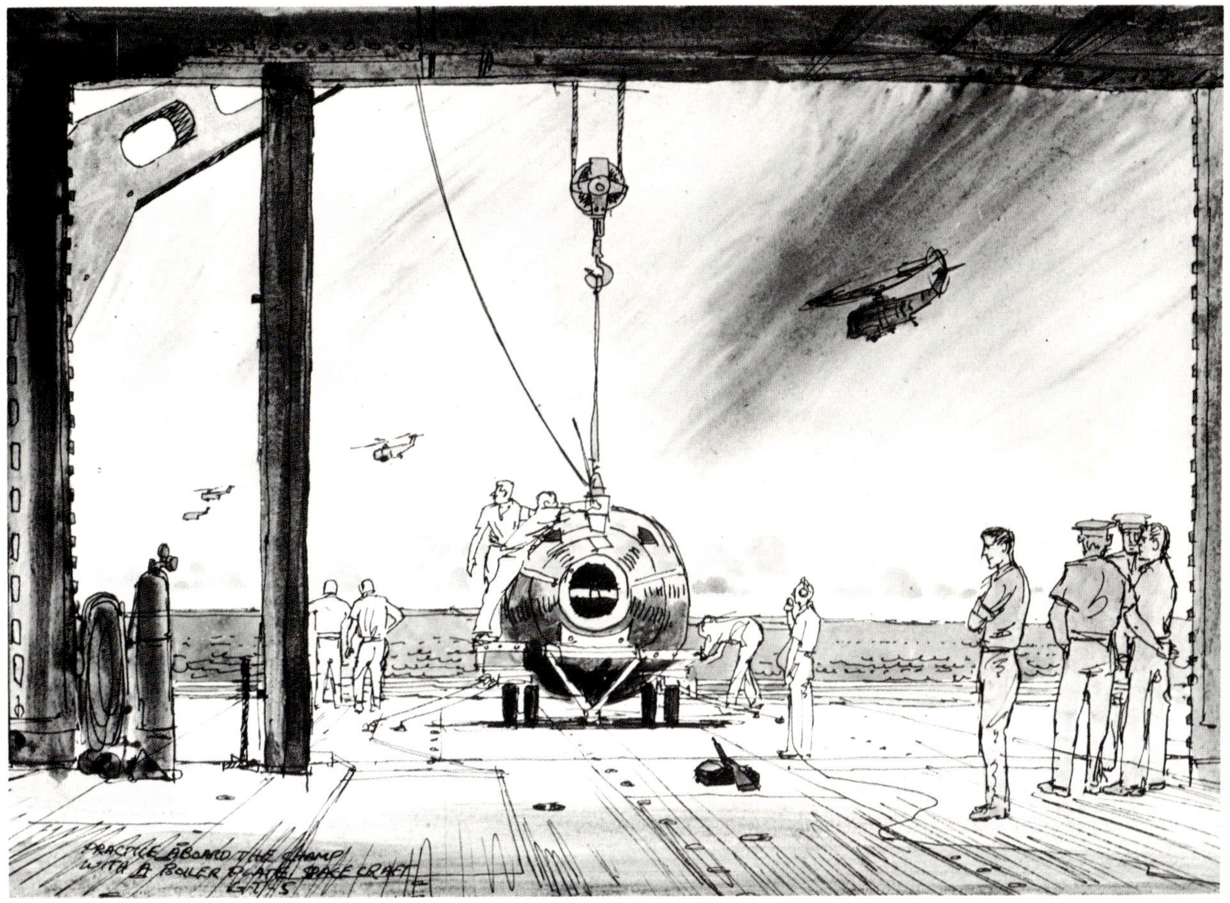

240.
Robert McCall. SPACECRAFT ON BOARD
Pen and wash on paper, 12×18″

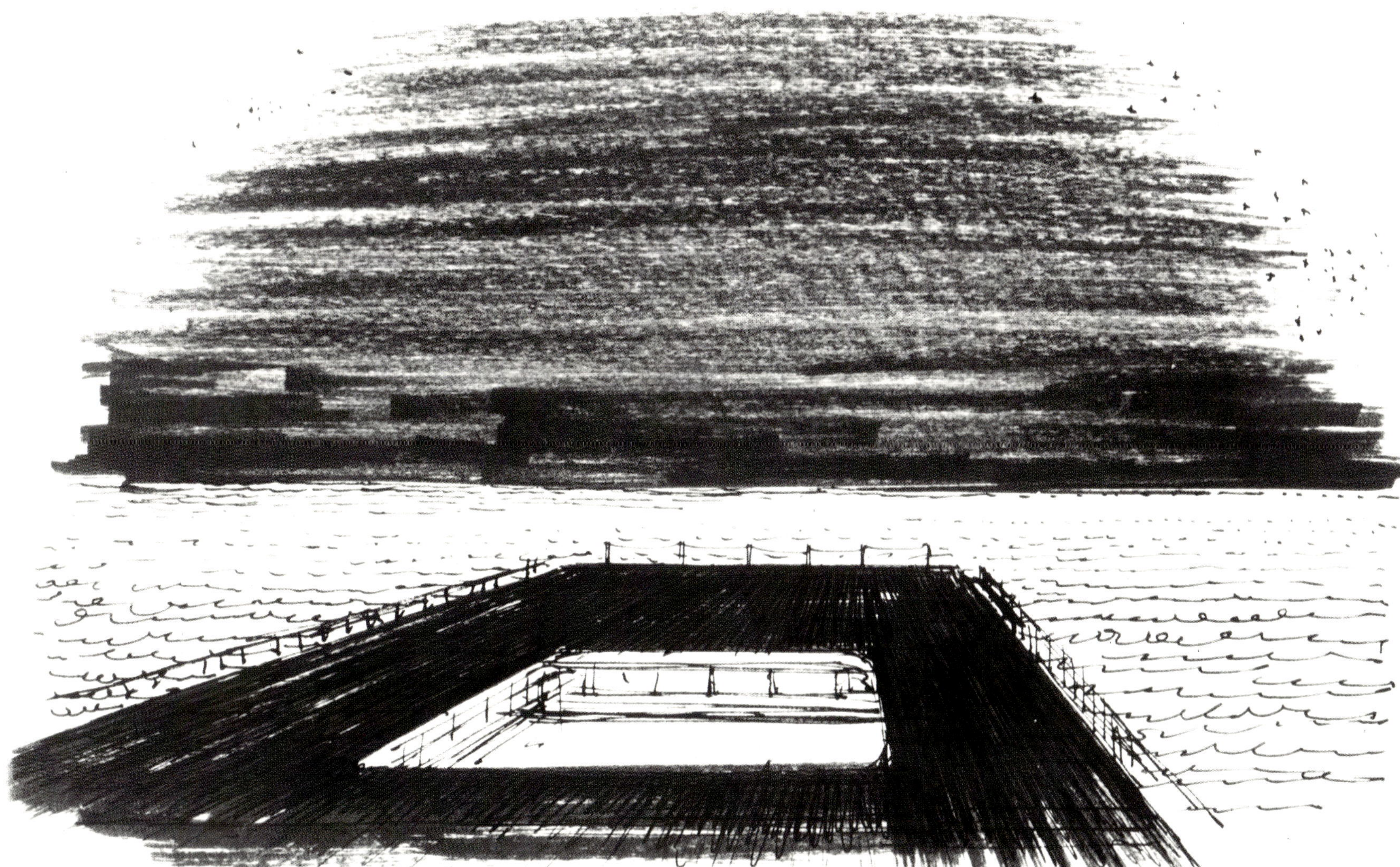

241.
Robert McCall
MOONLIT PACIFIC
Pen and ink on paper, 11×14″

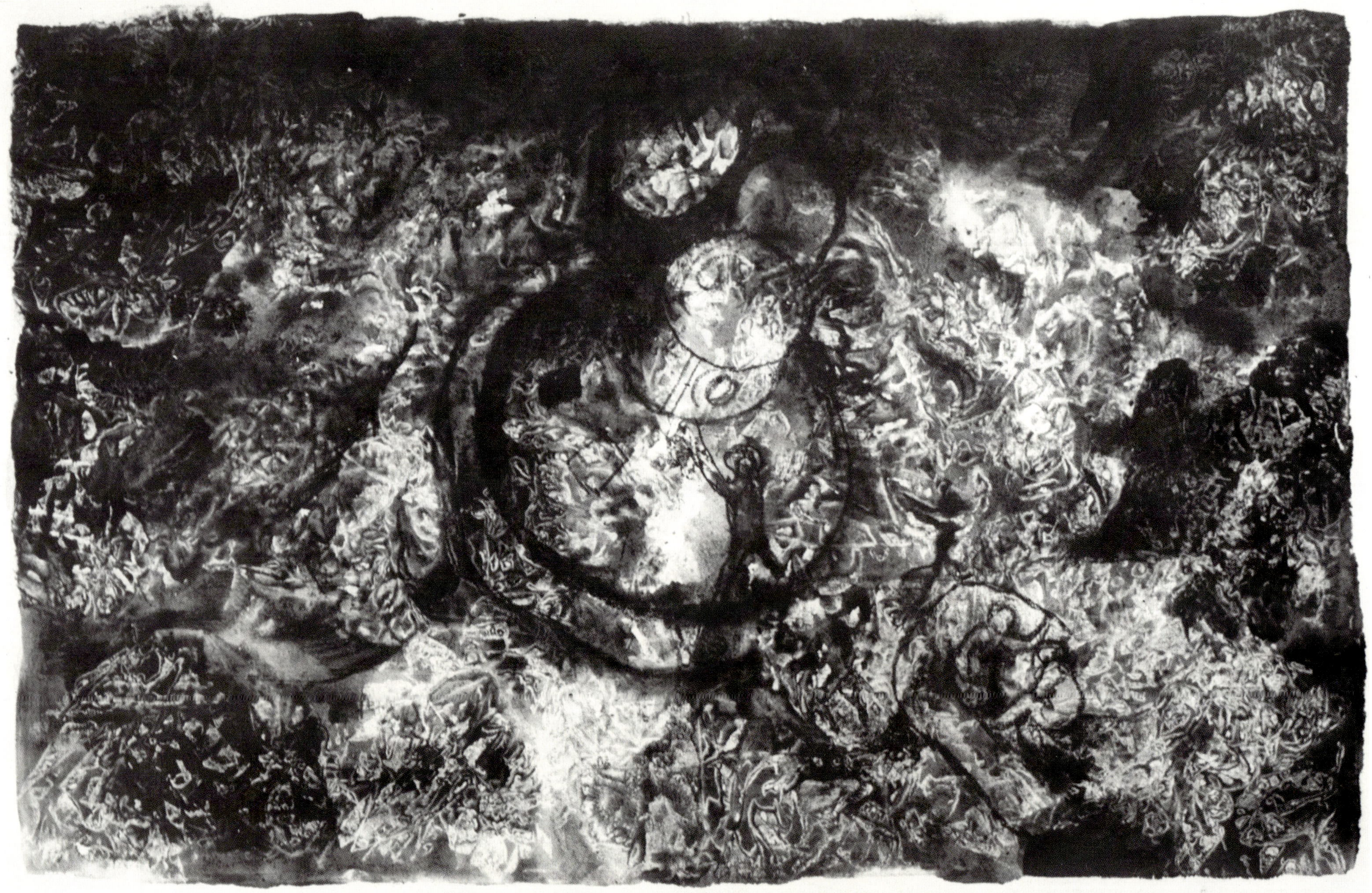

242. *Mitchell Jamieson.* FROGMEN. Ink on paper, 11×19″

243.
Leonard Dermott
APOLLO 9 COMING ABOARD
Acrylic on canvas, 30×36″

The Apollo program has stressed perfect training, techniques, and execution. Sometimes, however, there are slight mishaps, two of which Dermott has noted with quiet humor in this picture.

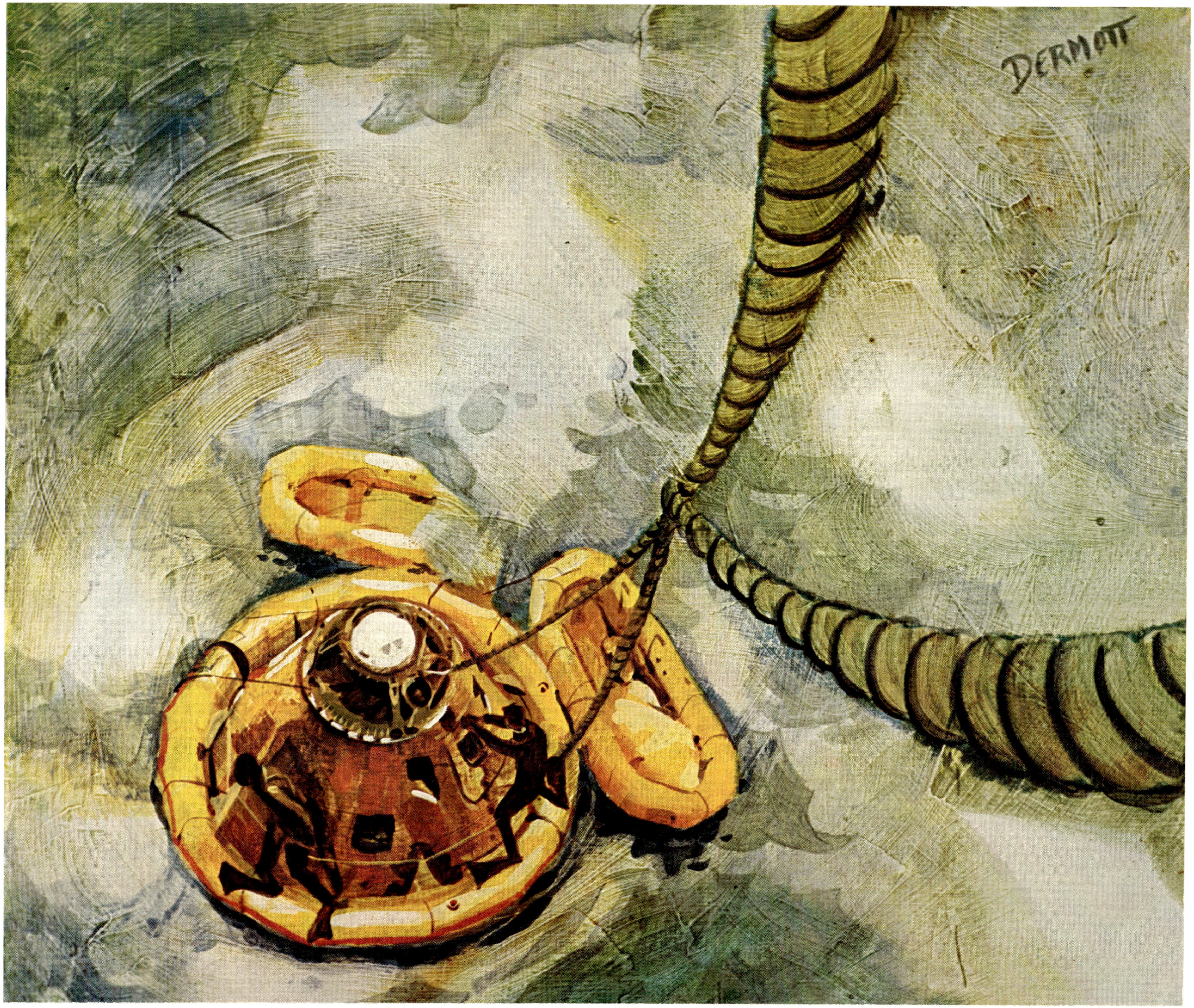
DERMOTT

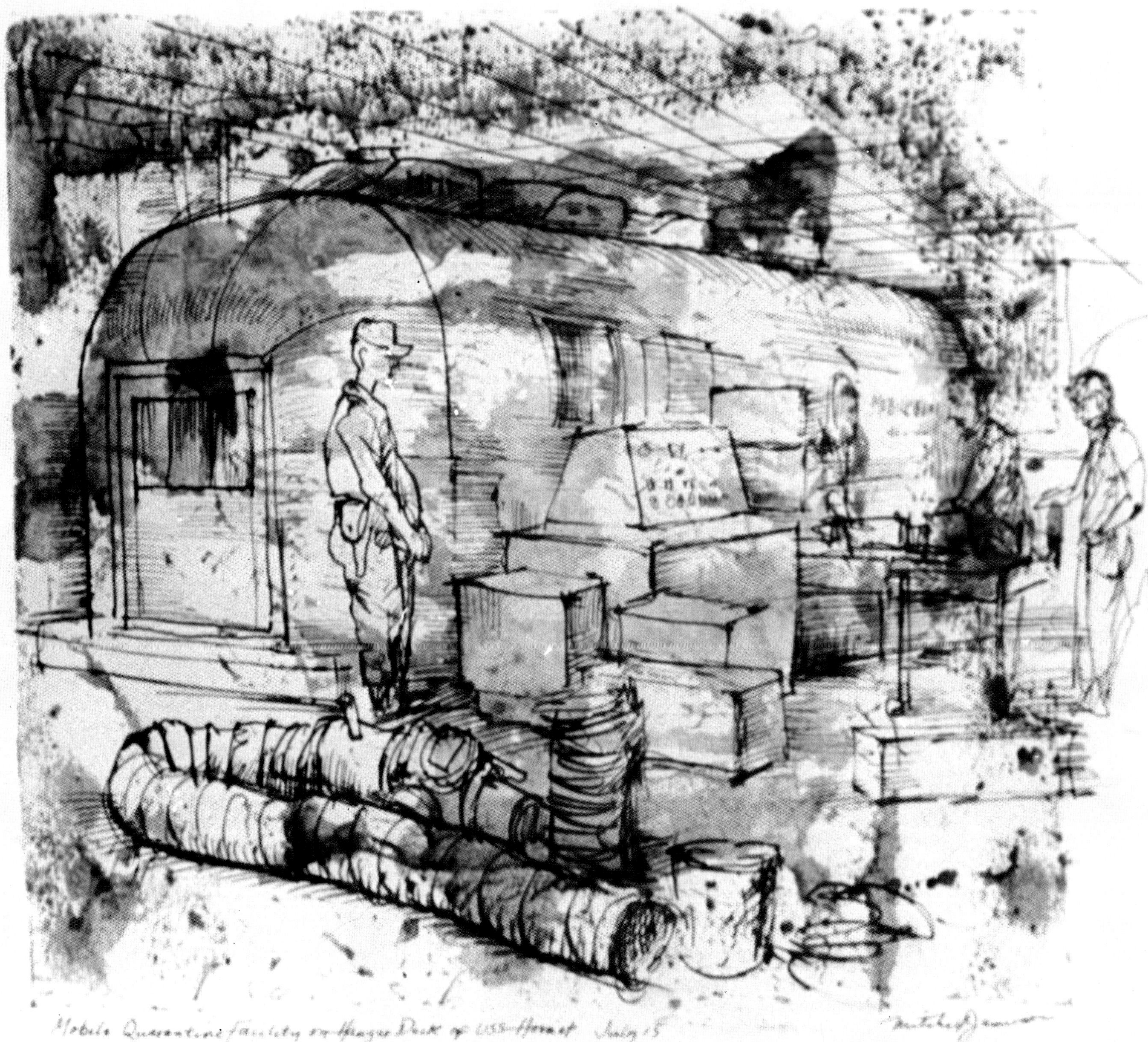

245.
Mitchell Jamieson
EQUATOR CEREMONIES
Ink on paper, 9½ × 14″

The carrier that picked up Apollo 11 had to cross the equator, and this meant that ancient rites, in stark contrast to the Space Age mission, were held, with newsmen the victims.

244.
Mitchell Jamieson
QUARANTINE FACILITY
Ink on paper, 9½ × 12″

Our children's children will probably be amused when they read that men of the 1960s thought there might be life on the moon. And yet history would never have pardoned us if one germ had gotten loose.

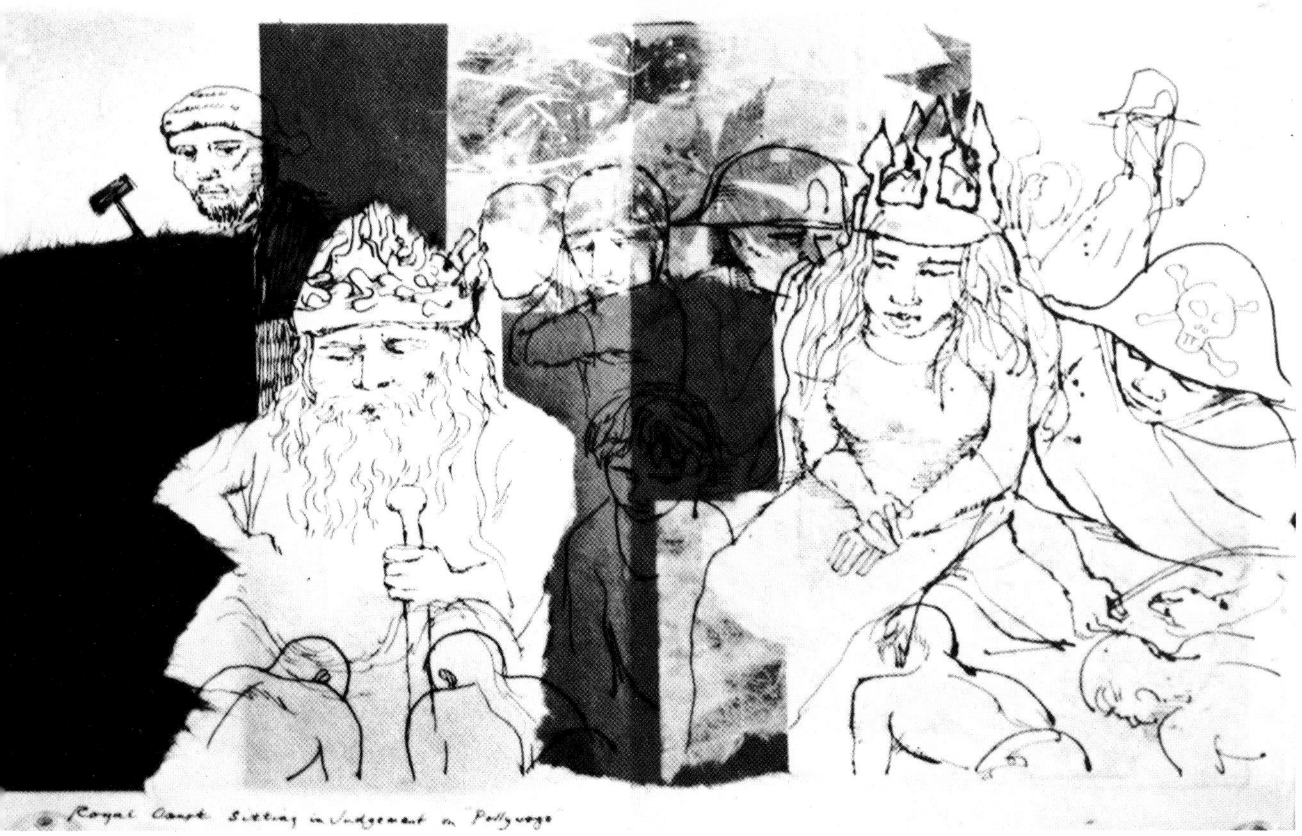

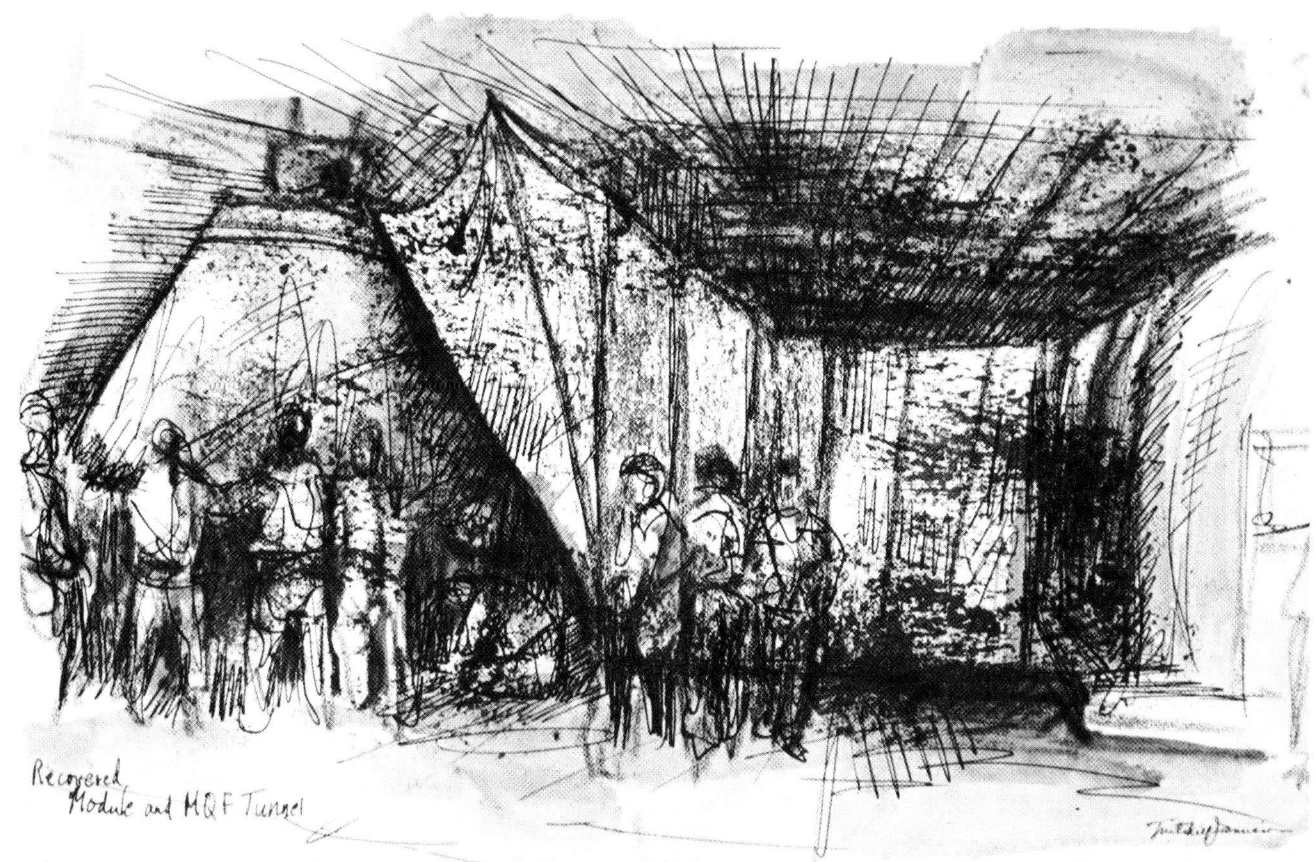

246.
Mitchell Jamieson
GERM-PROOF PASSAGE
Ink on paper, 9½ × 14″

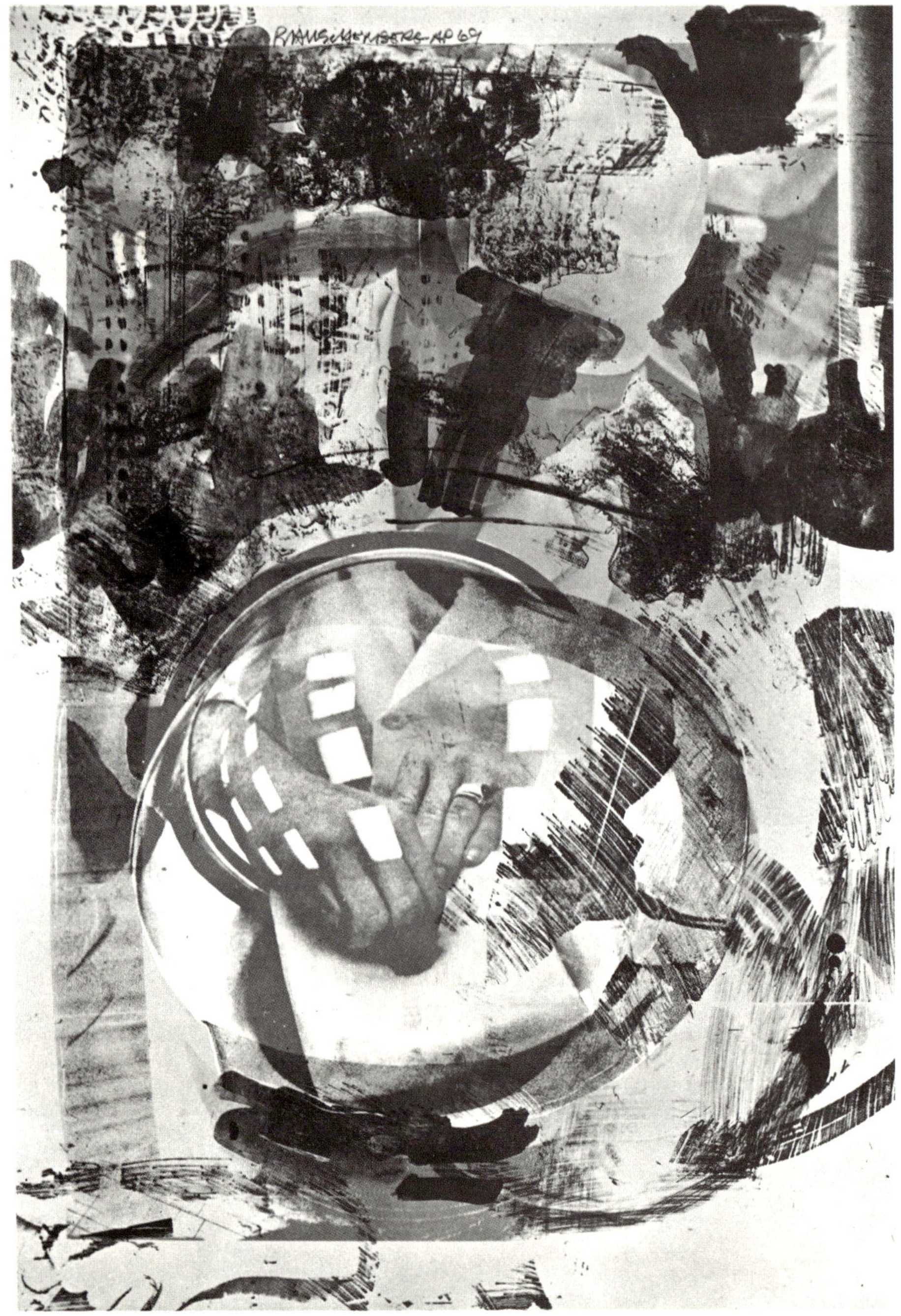

247. *Robert Rauschenberg.* SACK
Color lithograph on paper, 40×28″
From the Stoned Moon series reproduced courtesy of
Gemini G.E.L., Los Angeles, Calif.

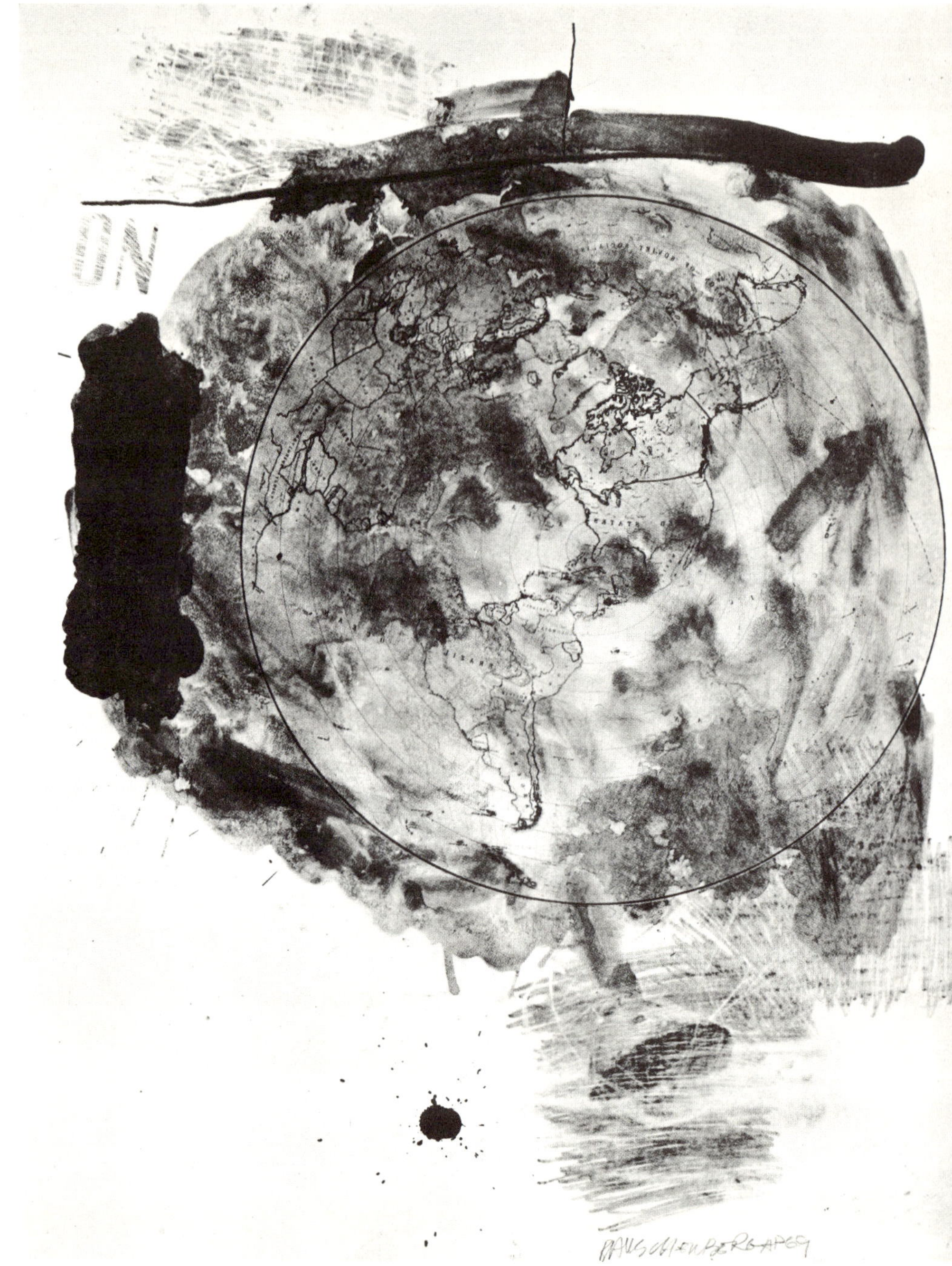

248. *Robert Rauschenberg.* MEDALLION
Color lithograph on paper, 32×25½″
From the Stoned Moon series reproduced courtesy of
Gemini G.E.L., Los Angeles, Calif.

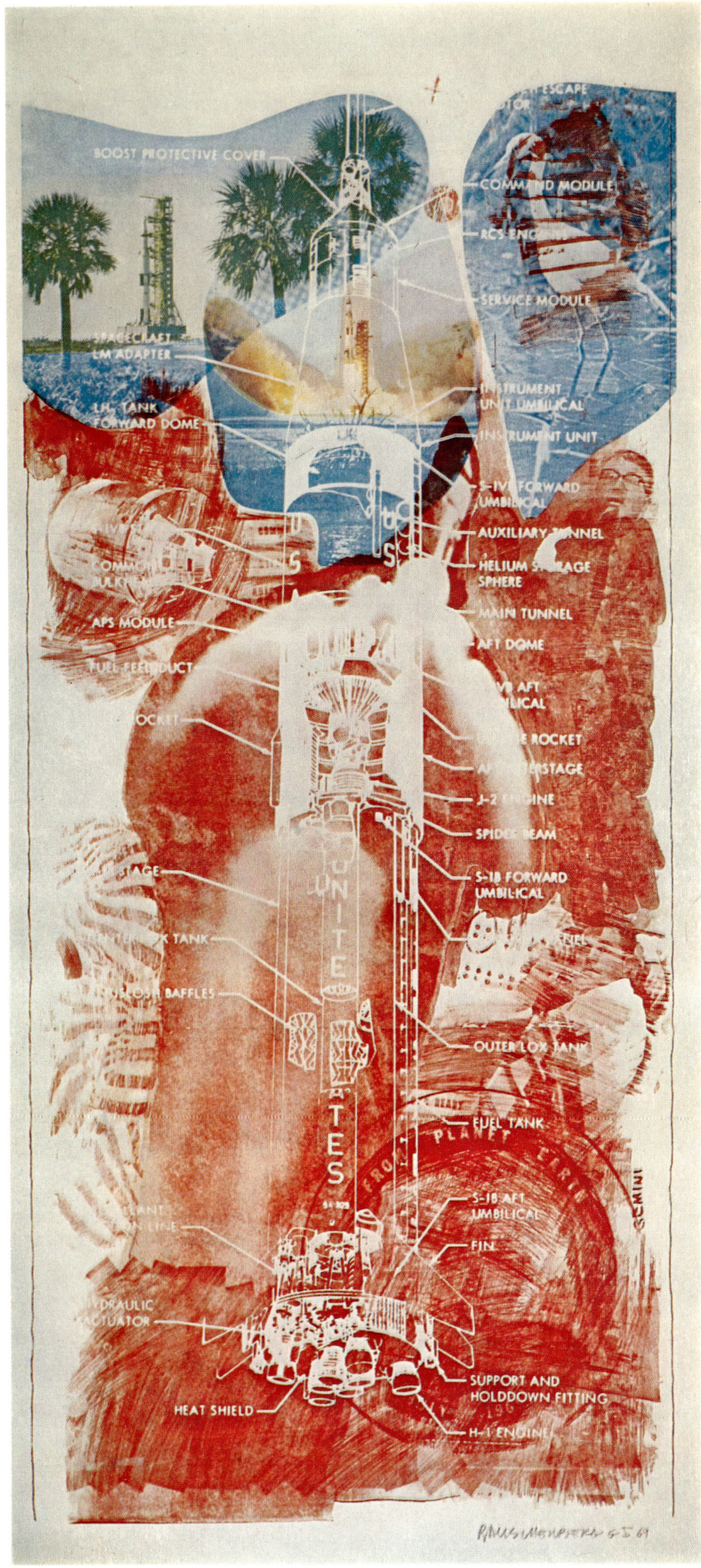

249.
Robert Rauschenberg. SKY GARDEN
Color lithograph on paper, 89×42″
From the Stoned Moon series
reproduced courtesy of Gemini G.E.L.,
Los Angeles, Calif.

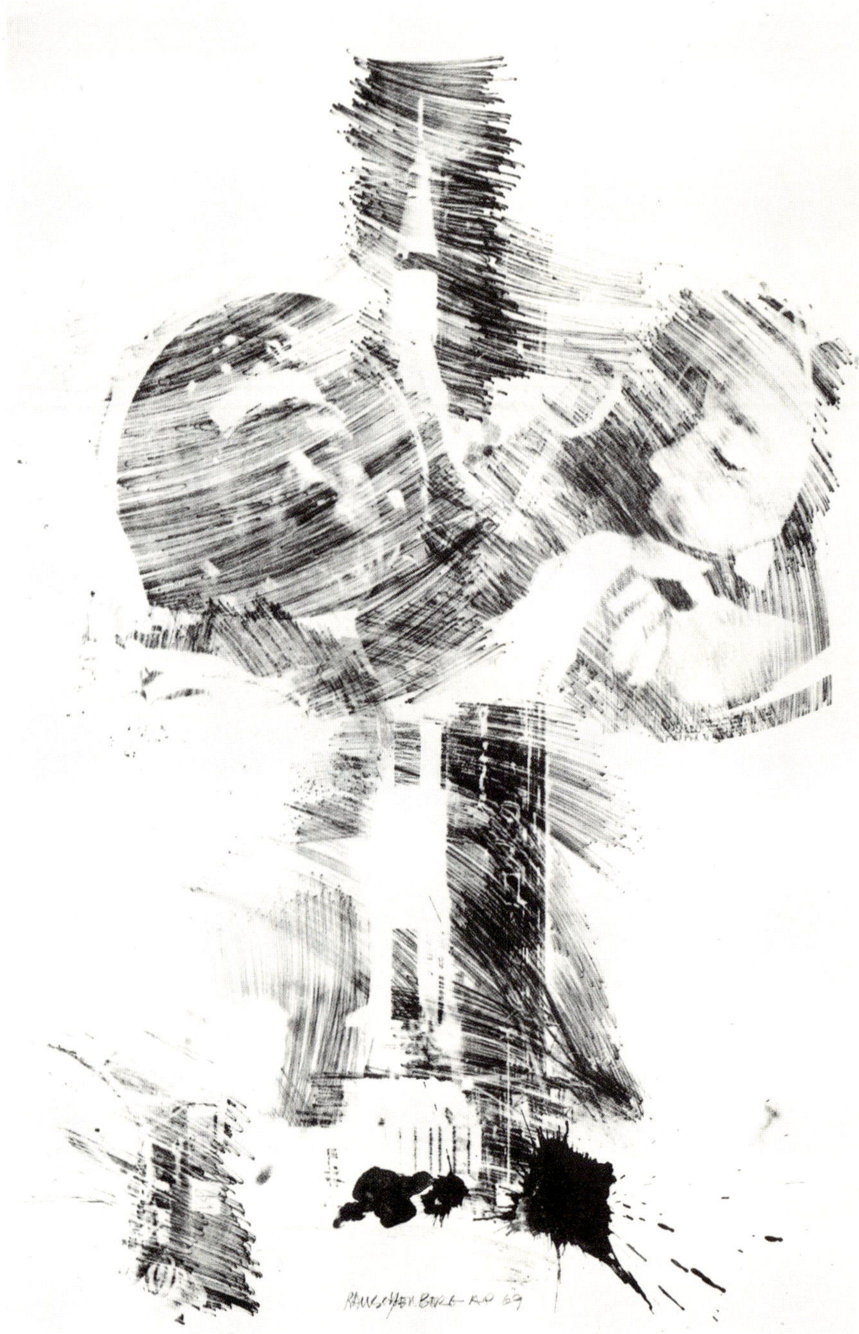

250. *Robert Rauschenberg*. MOON ROSE
Color lithograph on paper, 51×35″
From the Stoned Moon series reproduced courtesy of
Gemini G.E.L., Los Angeles, Calif.

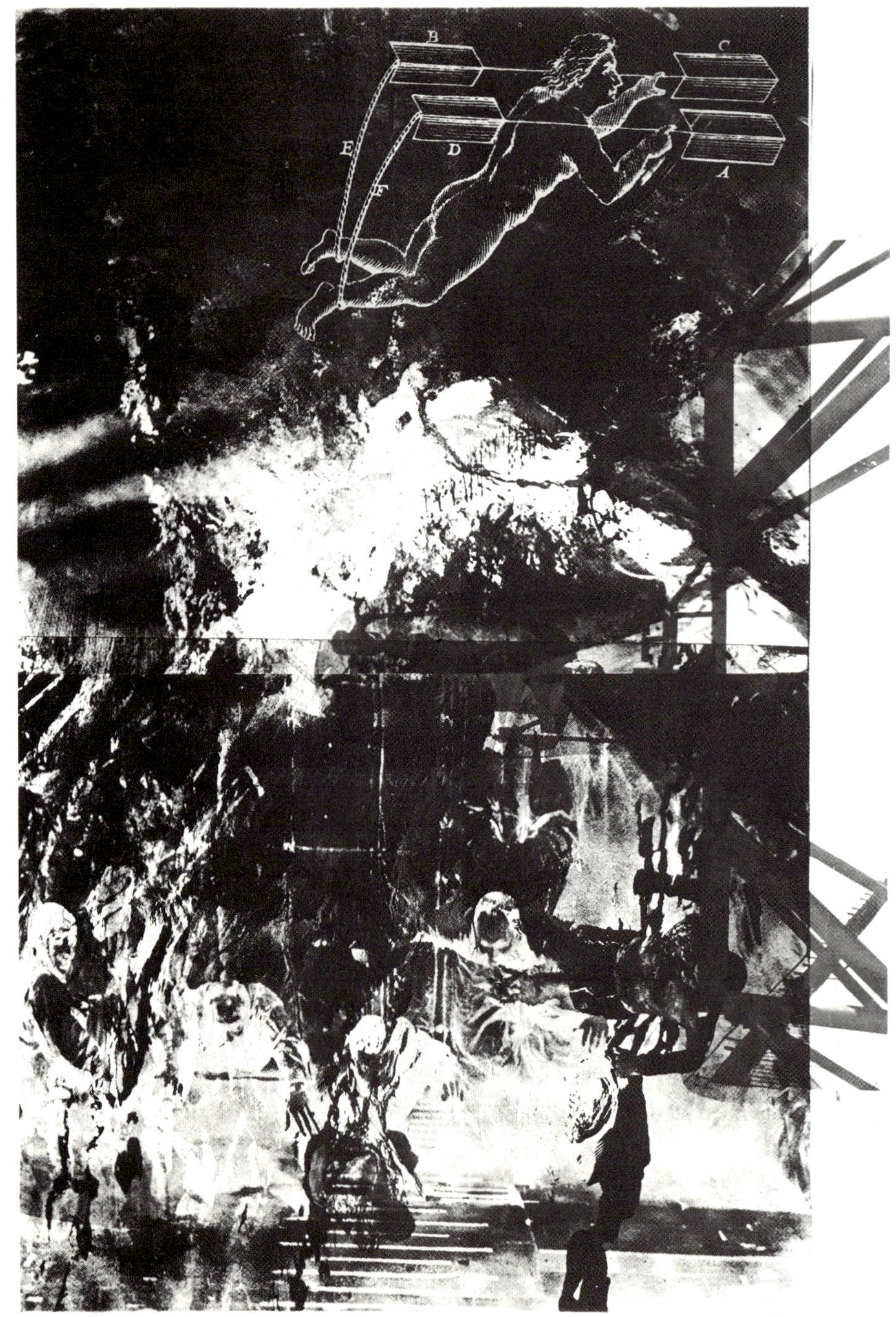

251. *Robert Rauschenberg*. BAIT
Color lithograph on paper, 35¼ ×26¼″
From the Stoned Moon series reproduced courtesy of
Gemini G.E.L., Los Angeles, Calif.

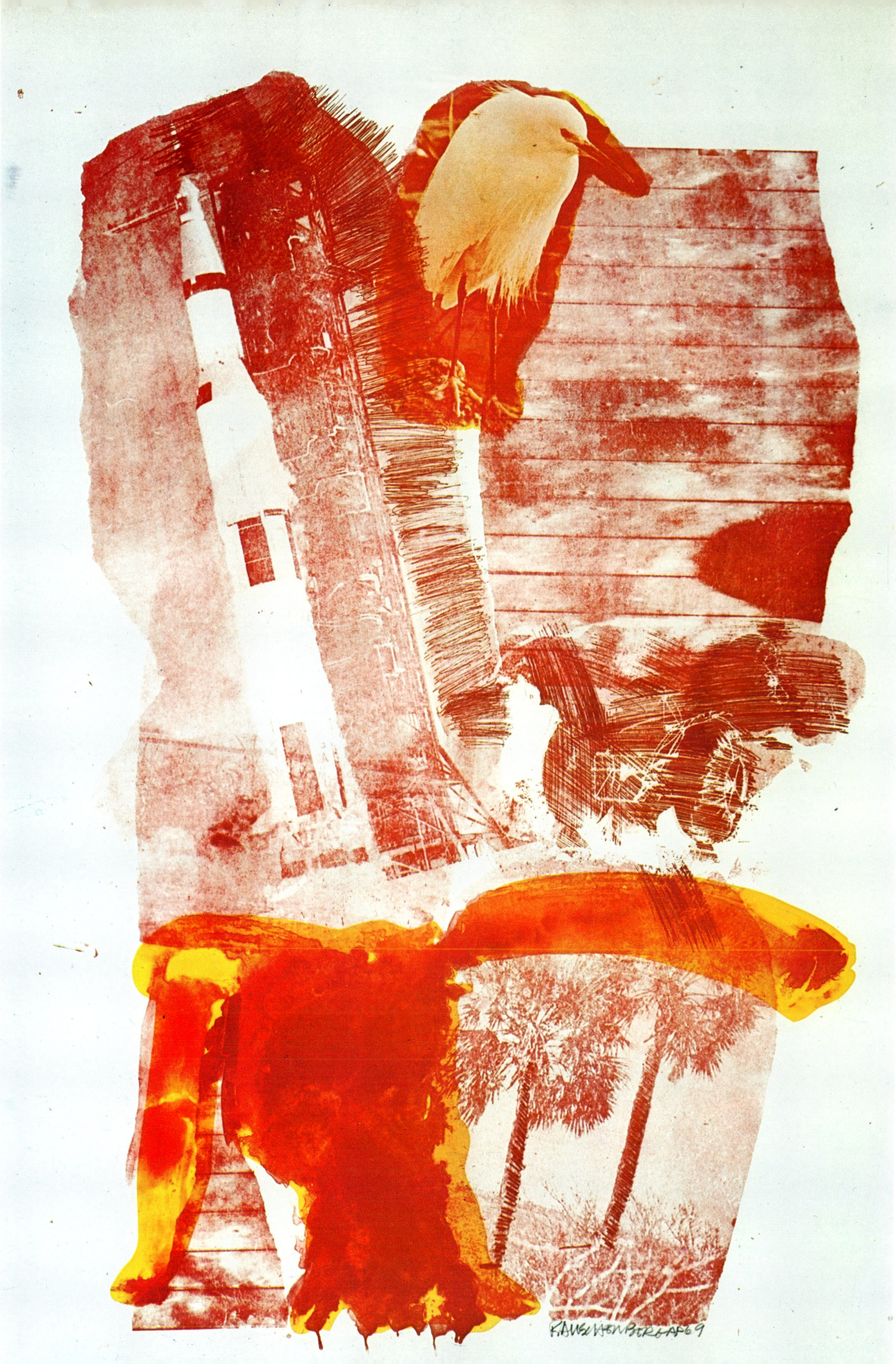

252.
Robert Rauschenberg. HYBRID
Color lithograph on paper, 54½ × 36″
From the Stoned Moon series
reproduced courtesy of Gemini G.E.L.,
Los Angeles, Calif.

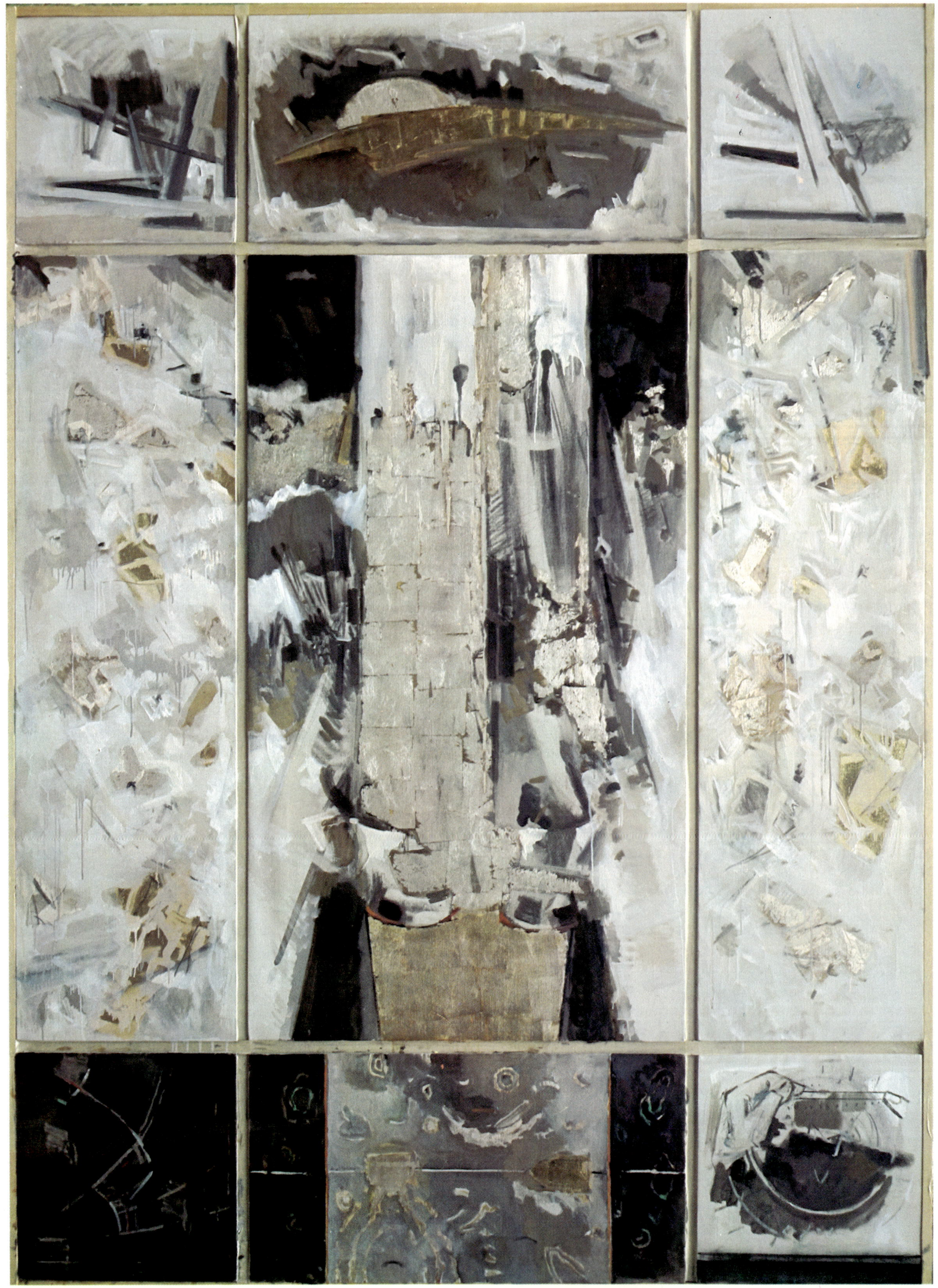

253.
Lamar Dodd
MOMENTS OF DECISION
Oil on canvas, 8×6″

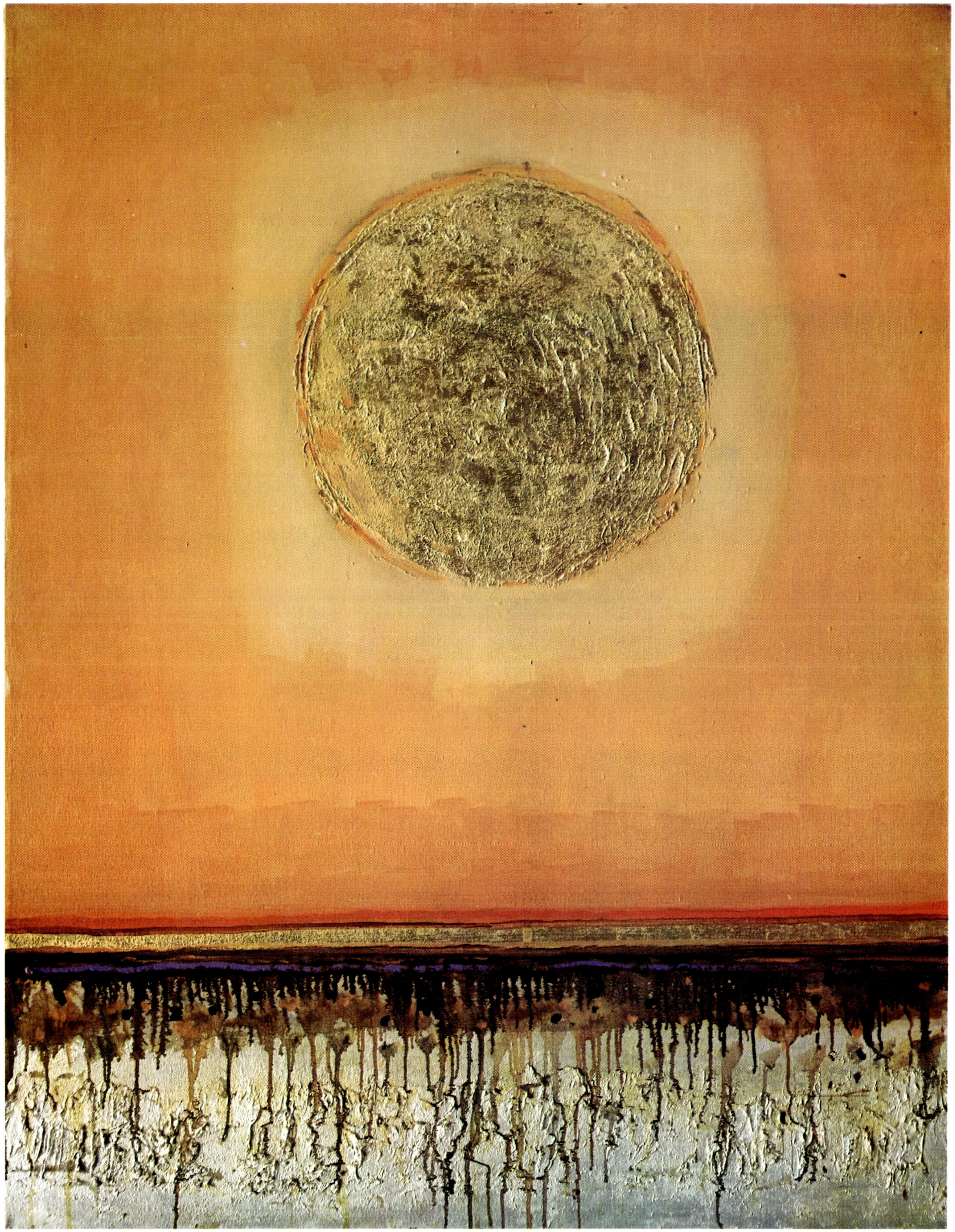

254.
Lamar Dodd
FUTURE TARGET
Oil on canvas, 6×4′

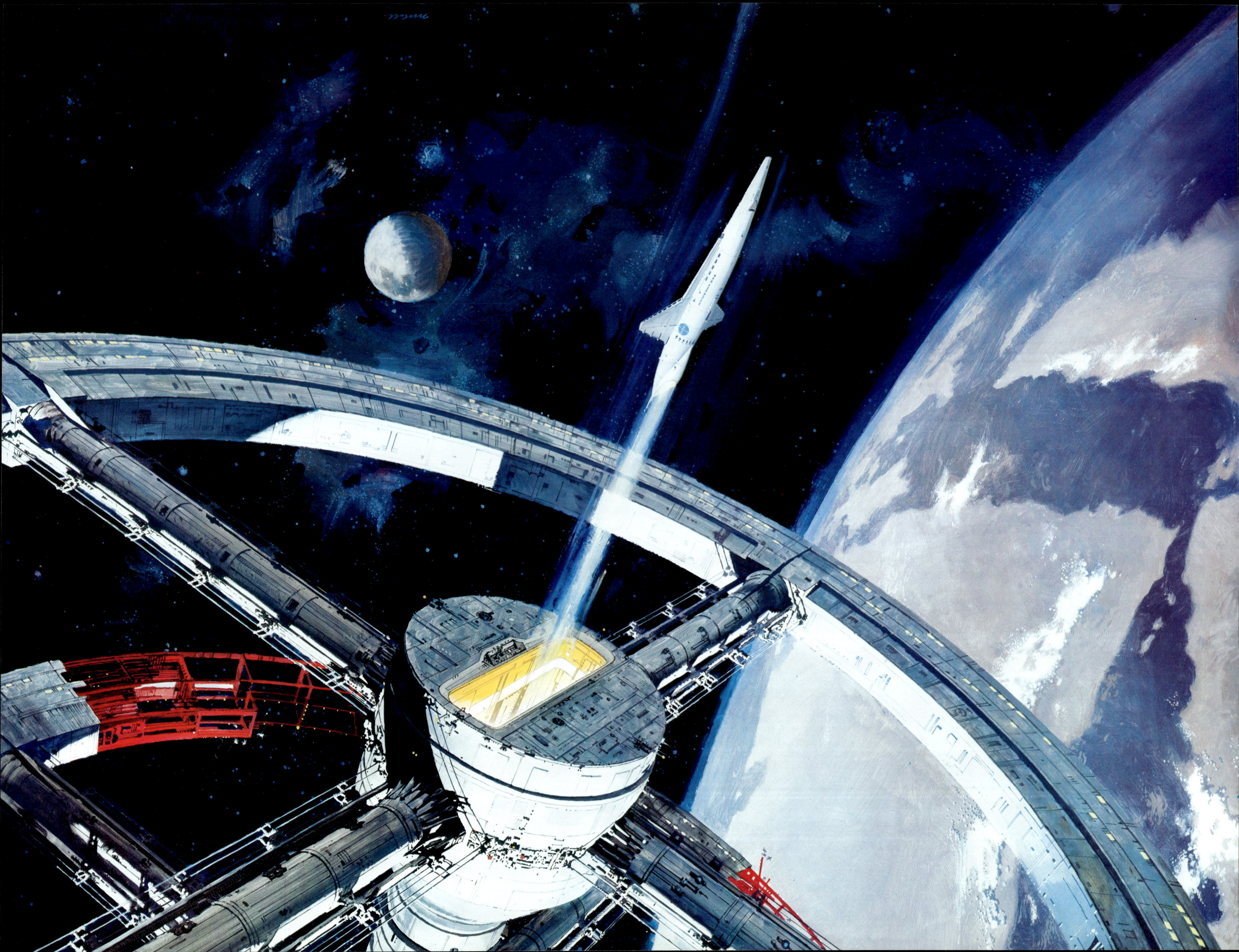

255.
Robert McCall
SPACE STATION #1
Mixed media on canvas, 53×40½″
Commissioned by Metro Goldwyn-Mayer
for the film *2001: A Space Odyssey*.
Reproduced courtesy of the Air and Space Museum,
Smithsonian Institution

The next phase of space exploration is still undecided, but probably, in the not too distant future, a space station, slowly revolving to provide man-made gravity, will be built to orbit the earth. It is hoped that one of its first visitors will be an artist.

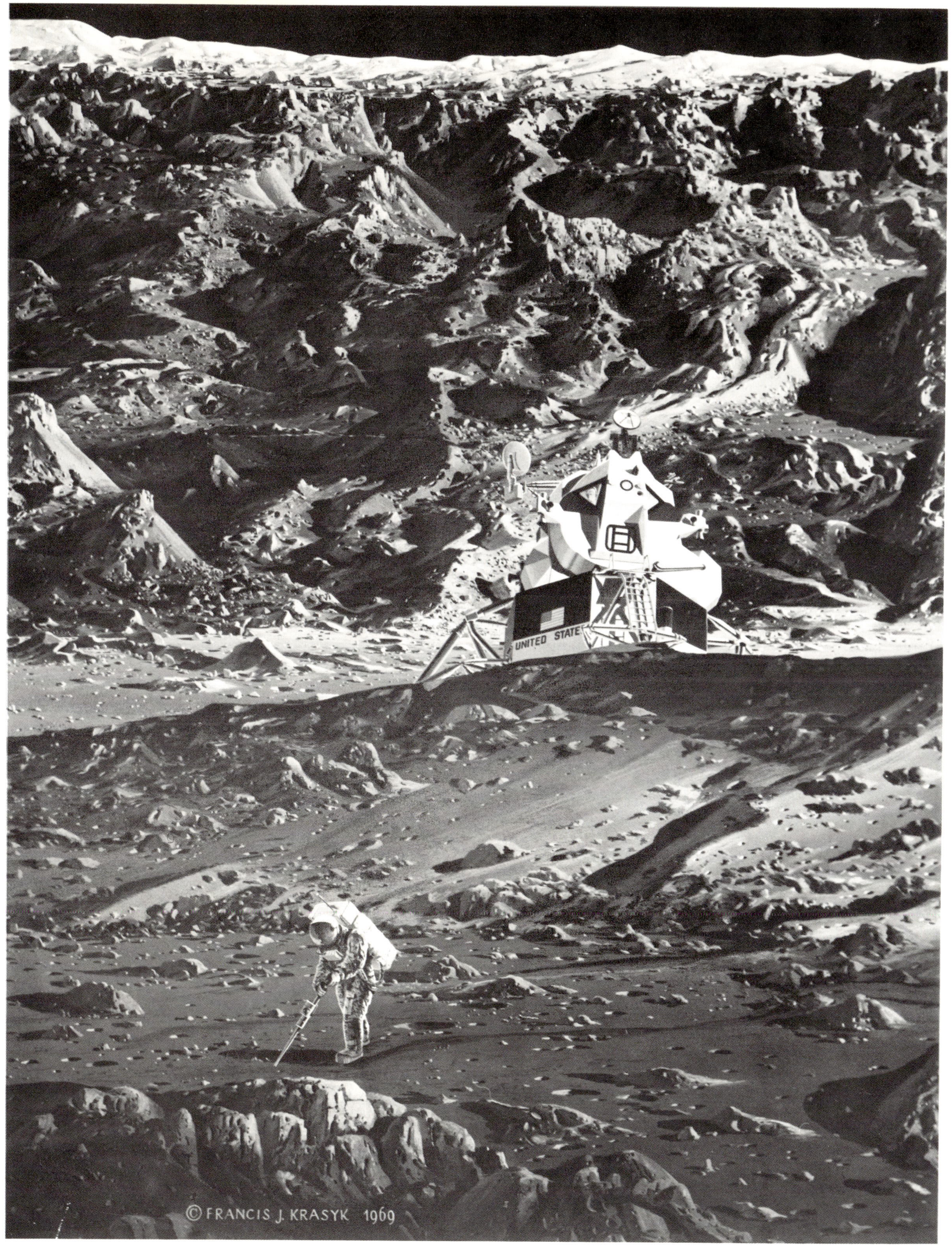

256.
Francis J. Krasyk
MAN IN COPERNICUS
Gouache on panel, 27×19″
Reproduced courtesy of the artist

Before man is through exploring the moon he may land inside the crater of Copernicus. This may be a dangerous mission bacause of the high walls and rough floor, but the scientific rewards might be incalculable. Who knows what the astronauts might find—gems of unimaginable beauty, traces of long extinct life?

257.
John Willenbecher
ALLEGORICAL SYSTEM: NIGHT
Construction, 19½ × 57½ × 4½″
Collection John G. Powers, Englewood Cliffs, N.J.
Reproduced courtesy of Richard Feigen Gallery, New York, N.Y.

It has been not only what the eyes may see in space that has drawn and held the imaginations of artists, but the laws and forces that keep heavenly bodies in their majestic orbits.

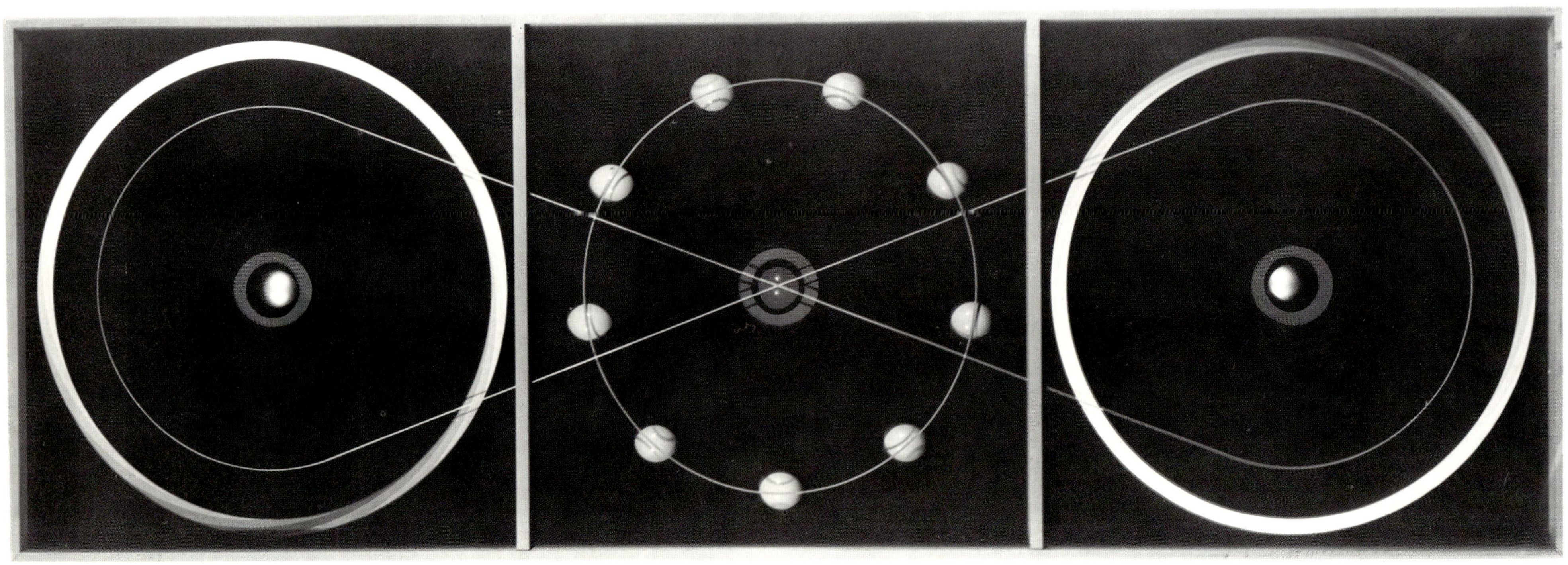

258.
Chesley Bonestell
A NEW WORLD
Mixed media on panel, 24 × 15″
Reproduced courtesy of Viking Press,
the artist, and the Air and Space Museum,
Smithsonian Institution

Beyond the moon lie man's next goals. During the 1930s a painterly prophet, astronomer, and artist of great ability thrilled magazine buyers with his imaginary views of outer space. Today, his forecasts are proving surprisingly accurate. The old master of space art is unquestionably Chesley Bonestell.

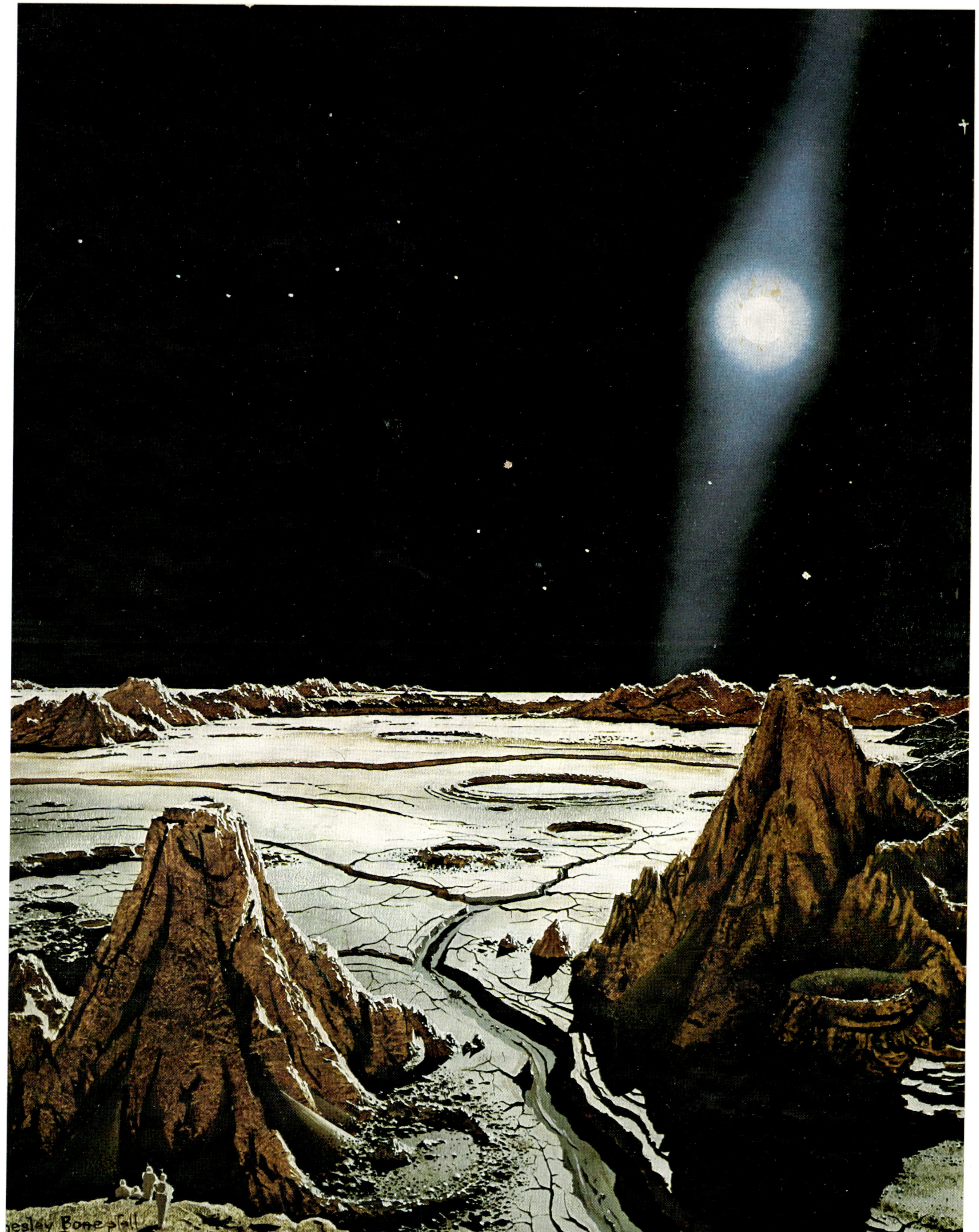

ACKNOWLEDGMENTS

My first debt is to the artists whose work is reproduced in this book. Many have given far more in time and talent than we had a right to expect. Some have become space addicts, and, at their own expense, have returned again and again to Cape Kennedy. I wish to thank these artists in particular for permission to reproduce in this volume work done on their own time and initiative. My second debt is to James Webb, Administrator of the National Aeronautics and Space Administration from 1961 to 1968, who first suggested the role that artists might play in the space program. I wish also to express my gratitude to his successor, Dr. Thomas O. Paine, who continued the support of the art program.

I am indebted to Mr. David Finley, Director Emeritus of the National Gallery of Art, and to the late Linton Wilson of the Fine Arts Commission, who recommended me for the role of consultant. My thanks are due to John Walker, Director Emeritus of the National Gallery, who agreed to let me work for brief periods on this assignment, and to his successor, J. Carter Brown, who has continued this policy and has himself visited Cape Kennedy with the artists on several occasions.

I am indebted to Julian Scheer, Assistant Administrator for Public Affairs for NASA, who from the beginning has supported the art program with understanding, help, and the necessary authorizations.

James Dean, my counterpart at NASA, Deputy Director, Media Development Division, more than anyone else has had the responsibility for making the program work. His constant and personal supervision of details, sincere interest, and never-failing willingness to cooperate, combined with a talent for tactful handling of difficult situations, prevented the art project from withering or lapsing into a routine operation.

A special word of thanks is due to Phil Weber, who with tact, patience, and knowledge has been the guide, confidant, and mentor of most of the artists working at Cape Kennedy.

INDEX OF ARTISTS

Numbers refer to plates.

Colorplates are marked with an asterisk.